T0073144

CHEMISTRY

CHEMISTRY
Our Past, Present, and Future

edited by

Choon Ho Do

Attila E. Pavlath

PAN STANFORD PUBLISHING

Published by

Pan Stanford Publishing Pte. Ltd.
Penthouse Level, Suntec Tower 3
8 Temasek Boulevard
Singapore 038988

Email: editorial@panstanford.com
Web: www.panstanford.com

British Library Cataloguing-in-Publication Data
A catalogue record for this book is available from the British Library.

Chemistry: Our Past, Present, and Future

ISBN 978-981-4774-08-6 (Hardcover)
ISBN 978-1-315-22932-4 (eBook)

Printed in the USA

Contents

PART III: CHEMISTRY AND ACTIVITIES

12. Problems and Solutions: Activities of Chemists and Educators for the Public

Attila E. Pavlath

13. What Can Chemistry Do for the Future?

Choon Ho Do

Preface

It is hard to imagine what our life would be without the numerous developments chemistry has provided in the past and is still providing for our everyday life. Just review our ordinary daily activities for a single day, from the moment we wake up to the time when we fall asleep. What would we do without toiletries, food, clothing, footwear, window glass, electric light, TV sets, telephones, etc.? All of these and much more are created by the wonders of chemistry. The list is very long and in their absence of chemical developments, we would be in the Stone Age.

Unfortunately, society is largely ignorant about the importance of chemistry. The discoveries of chemistry should be evident in various pieces of equipment frequently used every day; in materials that make our lives easier, more comfortable, and more pleasant; medications that ease pain and prevent and cure diseases, just to mention a few. Why is the general public unaware of all this? Are all of these advantages just taken for granted assuming that they just happen to be there? Apparently most people do not think about this. Why is it so difficult for the public to realize what life in our world would be like had chemistry not brought about those numerous discoveries? One does not have to be a scientist to be logical about this if these benefits are placed in the spotlight. Unfortunately, human nature is not perfect. It is a well-known fact that a catastrophe is more likely to capture the public's attention than a happy event. It is evident that the contributions of chemistry have to be publicized in plain language without any confusing chemical terms. That made us to realize the importance of this book as a means to enlighten anyone who picks it up.

Prof. Attila Pavlath, former president of the American Chemical Society and co-editor to this publication, was very much aware of the importance of public understanding of chemistry. During his presidency in 2001, his team launched a project titled "Technology Milestones from the Chemist's View" to disseminate the importance and application of chemistry in various fields to the public. The idea for the present book came from a paper by co-editor Prof. Choon H. Do, former President of the Korean Chemical Society, titled "Public Understanding of Chemistry and Communication Programs" as presented at the 14th Asian Chemical Congress held in Bangkok, Thailand, in September 2011. Consequently, the publisher kindly supported the development of a book including this concept. We realized it is our job and responsibility to change the public's perception of chemistry generally referred to as the Public Image of Chemistry. Hopefully, this book will enlighten the public and will support new chemical developments for a better future. The speed of the progressing of chemistry in connection with biology, physics, and engineering and its effects becomes faster and greater than ever before and will lead to better life for future generation.

The purpose of this book is to help its readers understand and appreciate chemistry, the discipline that builds the world, drives our lives, and keeps us healthy and safe. The book also aims to help the public realize that chemical knowledge is essential for the development of society and for the betterment of their lives now and in the future, too. The book also aims to help even a non-scientific person "think" or "consider" questions like "how, " "why," and "what" one can do for a better life—through chemistry.

We designed this book to lead readers into a broad range of chemistry topics in three parts divided into 14 chapters: In the first part, the book describes the relationships among chemistry and the Universe, our body, health, and our history. In the second part, consisting of eight chapters, the book describes the contributions of chemistry to the enrichment of our present life, which were made possible through the effects of chemistry in the fields of agriculture, food, energy, medicine, health, transportation, and communication.. In the third part, the book describes activities for the image of chemistry, for our future, and for chemistry in Africa, where ancient developments need to be further clarified.

We tried to make this book a tool for readers to understand in clear, non-technical language the benefits of chemistry with the hope that the reader's perception of chemistry would be based on correct information and not on media hype. Also, we have made an attempt to highlight the results of the discoveries rather than how chemists accomplished them. For example, we talk about the Haber–Bosch synthesis as the tool of making possible artificial fertilizers that enabled us to feed the world, not necessarily describing in detail the experimental conditions for doing so.

We were very fortunate to be able to gather excellent chapter contributors for this book and we thank them for their contributions: Prof. Sunney Chan, Dr. Erika Godor, Dr. Dorottya Godor, Dr. Temechegn Engida, Prof. Young Ha Kim, Prof. Mary Virginia Orna, Prof. Veronika Németh, Prof. Livia Simon Sarkadi, Prof. James Wei, and Dr. Andrew Yeh.

We thank the Publishers and Directors, Mr. Stanford Chong and Ms. Jenny Rompas, for giving us the opportunity to produce this book and Mr. Arvind Kanswal for his team's professional editing.

We acknowledge the permission of using many images and tables.

Choon Ho Do
Attila E. Pavlath

Introduction

Choon Ho Do[a] and Attila E. Pavlath[b]

[a]President of the Korean Chemical Society, 2010
[b]President of the American Chemical Society, 2001

Human beings are the results of a variety of chemical molecules and living by energies produced through chemical reactions of nutrients and oxygen in air and the use of materials, clothes, houses, cars, shoes, medicines, computers, TV, books, lightings, etc., made of chemical substances. We cannot be born, live, and survive without chemical reactions. The whole world is made of chemicals. The interactions of these chemicals are chemical reactions, and the study of the properties of these chemicals and reactions is chemistry.

Nevertheless, people do not think about these facts and are afraid of chemicals. They do not understand chemistry and chemicals correctly: They like pure water, clean air, medicines, perfumes and cosmetics, vitamins, and food supplements, but they are suspicious about or ignore fungicides, herbicides, artificial fertilizers, toxic substances, contaminants, etc. The materials we like and/or dislike are all chemicals.

Chemistry, as everything else in our life, is not perfect. There is always a risk, which may cause some problem. It may be an unexpected explosion in a chemical reaction or the release of a toxic material into our atmosphere. No action in our life has zero risk. We face many possible dangers from the day we are born. All medications, regardless of how beneficial they are, carry a risk clearly stated by the manufacturer by giving a long list of generally low but still more than zero probability of undesirable side effects. We accept those risks since we want the cure and we judge that the possible benefits outweigh the possible harmful side effects. To give an example, we pay little attention to the fact that in the United States every day more than 80 persons (30,000 persons per year) die in automobile accidents. Nevertheless, everyone gets confidently into the car regularly balancing the benefits against the possible dangers.

Then why do many people become suddenly concerned about and even hostile toward chemistry when some shortcomings of chemistry are reported once in a while? It is logical to assume that because the rare accidents related to chemicals are frequently reported in large headlines and the numerous benefits of chemistry generally are ignored or at best relegated to a note on the fifteenth page, they skew our mind about the ratio of good and bad. Frequently, people have become more and more convinced that chemistry represents a danger in our life. Therefore, we need to understand chemistry and chemicals and the consequence of their interactions.

Why is it important? You might assume, especially if you are not involved in any chemistry-related activities, that it is of no concern to you and it makes no difference in your life. However, this is not true! If the public does not understand the benefit of chemistry, they will not support its practice. The end result is less chemical research and development, less number of new products, and less improvement on existing ones we use in our life. Chemical scientists understand the governing principles and properties of chemicals, and chemical reactions occur among chemicals. They know HOW it is done. However, the public must realize WHAT chemistry has done and what the consequences are if chemistry is downplayed and also realize the reasons why we need continuous developments in chemistry.

Why do we need fresh air? Why do we need clean water to drink? Why do we need healthy food? Why do we need gasoline to drive a car? What does gasoline do for the car? What will happen if we breathe bad gas? What will happen to our body if we drink polluted water? What gives color pictures on TV when it is switched on? These questions are related to chemicals and chemical reactions. Whether you believe it or not, whether you like it or not, our life is sustained by chemical reactions and chemicals. Why? It is because our body and our Universe are made up of chemicals. Living organisms and inorganic materials are chemicals. We cannot avoid chemicals and we cannot live without chemicals. Every day we need amenities essential to our modern life, e.g., food, cloth, cars, housing, TV, smart phones. To produce these items, we have to produce more materials by better processes and in an economical way. All these are related to chemistry. We need chemicals and therefore it is necessary to understand chemistry for not only our existence sustenance but also improvement of our future.

The goal of this book is to show not just the important benefits of chemistry, but also reasons why chemistry is needed in the present time and in the future to solve problems ahead of us, whether you like it or not. Where and how to provide food for citizens to escape from hunger? How to get necessary energies for our living? How to develop medicines to cure diseases and healthy life? How to supply clean water to drink? What to do to prevent and reduce crimes? How to detect smuggling? How to detect fake items?

Sciences and technologies progress rapidly. So does society. You, as a human, whether you are a decision maker or an ordinary person, have to decide many things: not only scientific matters but also ethics for you and your family and society. That is why the public needs to understand chemistry to make a better choice or decision because most things and events are related to chemistry.

We hope that after reading this book, you will appreciate the benefits of chemistry. We also hope you will keep a proper balance between the benefits and problems in a rational and not sensationalist way. We have tried to make the book easily readable with joy and thoughtfulness. We hope this book will give you the tool to present the benefits of chemistry to your friends and neighbors in a simple non-scientific language and to help change their negative opinion about chemistry, too.

The book is composed of three parts consisting of 14 chapters. Following is a brief description of the parts and the chapters. Readers may start from any chapter because each chapter describes a different subject and is independent of each other.

Part I. Chemistry Inherited from the Universe. Images and perception on chemistry and comprehensive description are given for the need of understanding chemistry for our better future both personally and globally. This part is composed of the following three chapters.

Chapter 1. Chemistry in the Universe, Our Body, and Our Life. This chapter describes the chronological point of views of relationship between chemistry and the Universe, development of modern chemistry from alchemy, periodic table, chemical bonds, life, and biology. It also describes the relationship between biology and chemistry and the importance of the understanding of molecular biology for our future.

Chapter 2. Chemistry in Human History. Chemistry and the development of human history are discussed from philosophical and practical points of view. The chapter describes the ways in which humans took advantage of chemistry for their survival in their use of natural products, foods, materials, clothes, shelter, medicines, and communication.

Chapter 3. Did Chemistry Change the World? YES, chemistry did change our life from the world of cavemen to the modern world in terms of the development of materials. This chapter describes the developments related to wood, metals including copper, iron, aluminum, titanium, tungsten, lithium; bronze, plastics, leather, textile, glass, and ceramics.

Part II. Contributions of Chemistry. This part covers the every range of our life to realize how "chemistry made our present life possible" discussing the numerous achievements and effects of chemistry. The achievements of chemistry for human beings in the past are the basis of our present life and will be the driving force and lead to a better life in the future. The part is composed of the following eight chapters, from Chapter 4 to Chapter 11.

Chapter 4. Agriculture: Without chemistry helping the improvement of agricultural production by fertilizers and pesticides, supply of food, and benefits of new technologies, we would not have enough food for the ever-increasing population of the world. Chemistry also enhanced the flavor, appearance, and nutritional value of the food. It has also helped with livestock production, protection, and veterinary medical care.

Chapter 5. Food: Supply and Health: It is not enough to produce food. Chemistry's health is essential for its preservation in healthy, safe, and affordable conditions. It continuously meets the challenges of agricultural productivity, water, healthy food, food safety, process efficiency, and supply chain waste.

Chapter 6. Energy: Energy is needed everywhere: in residence, commerce, and industry. Chemistry plays numerous roles in providing energy for everything: the extraction and refining of fuels from nature, the processes in releasing energy, and the management of safety and the environment.

Chapter 7. Medication: Diagnosis: In addition to developing medicines, chemistry has helped doctors in diagnoses through clinical laboratories and various diagnostic instruments. History provides an insight into the development of the clinical application of chemistry and the development of X-ray, computer tomography, application of isotopes, positron emission tomography, and magnetic resonance imaging.

Chapter 8. Medication: Curing: Chemistry always had a major role in our health even before the development of novel pharmaceuticals, creation of new medical equipment, and refinement of diagnostic procedures for modern medicine. The development of pain management, inflammation management, psychotherapeutic agents, contraceptives, and a drug for diabetes, insulin, represents just a few examples.

Chapter 9. Regenerative Medicine: Repairing Body: Chemistry not only provides medicine to cure illness but also makes possible the development of artificial organs and biomaterials. This chapter discusses numerous examples such as artificial kidney, heart, and eye-lens. Regenerative medicine and tissue engineering are described, too. Cell therapy, bio-hybrid organs, bio-artificial liver, and artificial pancreas are a few examples of regenerative medicine.

Chapter 10. Transportation: To move from one place to another whether on foot or by some machine is based on chemistry, which contributed to transportation technology and left its stamp on the culture of human activities. The development of materials for vehicles, frames and wheels, fuels and engines, navigation system, including compass and GPS, building of roads using asphalt and concrete are the crucial factors.

Chapter 11. Communication and Entertainment: While communication might be thought as the result of physics, chemistry made the progress of electronic communication devices and storage devices possible. This chapter describes the role of chemistry in the development of cellular phones, electronics, fiberglass, paper, ink, recording sounds and pictures, copy machines, computers, and entertainment business, including movie films.

Part III. Chemistry and Activities. This part is composed of three chapters, from Chapter 12 to Chapter 14. It includes activities to improve the image of chemistry, roles of chemistry for the future, and activities in Africa.

Chapter 12. Problems and Solutions: Activities of Chemists and Educators for the Public: What is needed to understand chemistry and how can the public image of chemistry be improved? This chapter gives suggestions so the reader can help overcome sensationalist negative media publicity. Science teachers, chemists, and chemical engineers are urged to participate in the improvement of the image of chemistry and to create interest in science through science education.

Chapter 13. What Can Chemistry Do for the Future: While no one can accurately predict the future, this chapter discusses the possibilities in various areas such as artificial photosynthesis, new energy resources, materials, climate control, human life and health, human and Nature, change of desert, use of ocean and sea, and space travel. These topics are described in a systemic manner and not as individual chemical terms.

Chapter 14. Chemistry in Africa: Progress and Application: Chemistry is essential for progress in underdeveloped countries. This chapter lists methods and problems in chemical education. It discusses how various natural materials were used for health and agricultural purposes. Information is given on what type of research is being done to investigate needed materials for the special circumstances in those countries.

PART I
CHEMISTRY INHERITED FROM THE UNIVERSE

Chapter 1

Chemistry in the Universe, Our Body, and Our Life

Sunney I. Chan and Andrew P. Yeh

Division of Chemistry and Chemical Engineering, California Institute of Technology, 1200 East California Blvd., Pasadena, California 91125, USA

sunneychan@yahoo.com

Chemistry is all about atoms and the interactions between these atoms to form molecules that make up the world around us: the air we breathe; the water we drink; the food we consume; the clothing we wear; the sand we walk on; the minerals from which we derive our copper, aluminum, gold, platinum, and silver; the plastics we fabricate to make plexiglass sheets, adhesives, and containers; the mortar and bricks we use to build our homes; and so on. So it is meaningful to ask where all these atoms come from.

1.1 The Very Beginning of the Universe

According to cosmologists, the universe began with the "Big Bang" approximately 13.8 billion years ago. This is considered to be the age of the universe. After the initial expansion, the universe eventually cooled sufficiently to allow the formation of subatomic

Chemistry: Our Past, Present, and Future
Edited by Choon Ho Do and Attila E. Pavlath
Copyright © 2017 Pan Stanford Publishing Pte. Ltd.
ISBN 978-981-4774-08-6 (Hardcover), 978-1-315-22932-4 (eBook)
www.panstanford.com

particles: quarks, electron, proton, neutron, etc., and, later on, atoms of simple elements. Massive clouds of these primordial elements later coalesced via gravity to form stars and galaxies.[1]

The low-mass elements, hydrogen and helium, were produced by primordial nucleo-synthesis (i.e., the combination of protons and neutrons to form elemental nuclei) within the first several minutes of the Big Bang. The nuclei of the heavier elements were created later in stars by stellar nucleo-synthesis. With subsequent cooling of the universe, stable hydrogen and helium atoms began to form when the average energies of the nuclei and electrons became sufficiently low. This took hundreds of thousands of years. After about a billion years, giant clouds of the cold atomic hydrogen and helium gas began contracting under the influence of their mutual gravitational forces. As the clouds condensed to higher densities, they warmed, and when the temperature of the hydrogen gas reached a few million degrees, nuclear reactions started in the cores of these proto-stars. Eventually, the more massive chemical elements were formed in the cores of the very massive stars.

The ordinary matter in our universe is made up of 94 naturally occurring chemical elements. The details of how these various elements culminated within the planet of which we are a part remain an area of intellectual inquiry and still occupy the attention of researchers in geochemistry.

Approximately 73% of the mass of the visible universe is in the form of hydrogen, the atom with the lowest mass. Helium makes up about 25% of the mass. The remaining 2% are made up of the more massive elements. This might seem like a miniscule amount, but hardly less significant, as most of the atoms on planet earth and in our bodies are a part of this small portion of the matter of the universe.

1.2   Alchemy and Alchemists

The composition of matter has always been a fascination among humans. We live in a world of molecules, so chemistry has to be among one of our "closet friends." "Earth, water, air, and fire" were the first elements (Fig. 1.1).[2] The earliest chemists, often referred to as alchemists, were thought to be endowed with the power to transform matter from one form or another, including the "creation of the fabled philosopher's stone; the ability to transmute base metals into the noble metals (gold or silver); and the development of an elixir of life that would confer youth and longevity."[3]

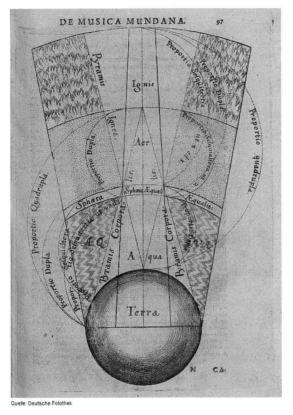

Quelle: Deutsche Fotothek

Figure 1.1 The classical elements in Babylonia: Segment of the macrocosm showing the elemental spheres of *Terra* (earth), *Aqua* (water), *Aer* (air), and *Ignis* (fire). Robert Fludd. 1617. Reproduced from Ref. 2.

Alchemy played a significant role in the development of early science. As a proto-science, it contributed a set of beliefs and theories that ultimately led to the development of modern chemistry and medicine. However, it was not until the 18th century before the foundations of modern chemistry were laid.

1.3 The Very Beginning of the Beginning of Modern Chemistry

Modern chemistry began with the discovery of nitrogen and oxygen, the two principal components of air. Daniel Rutherford, a Scottish chemist and physician, discovered nitrogen in 1772.[4] It was a fraction of air that was unable to explode. Joseph Priestley, a US chemist of British descent, discovered oxygen.[5] In a series of experiments culminating in 1774, Priestley found that air was not an elementary substance, but a mixture of gases. Among them was the colorless and highly reactive gas he called "phlogiston," and the air was "dephlogisticated" upon combustion. The great French chemist Antoine Lavoisier would soon give the name "oxygen" to this highly reactive component of air. Priestley's discovery answered "age-old questions of why and how things burn."

Nitrogen and oxygen are the main components of air, comprising 78% and 21% of the atmosphere, respectively. As we shall discuss later, nitrogen and oxygen are vital ingredients for living things and in manufacturing. The oxygen in the air we breathe is crucial to cellular respiration. Aerobic organisms such as humans cannot sustain life without oxygen. In contrast, nitrogen gas is very inert. However, nitrogen is required for the biosynthesis of the basic building blocks of plants, animals, and other life forms. Before it can be used, the nitrogen gas needs to be fixed and converted to more reactive forms. In nature, there exist nitrogen-fixing bacteria that convert nitrogen gas into ammonia. Nitrogen fixation is essential for agriculture and the manufacture of fertilizer.

Also there would be no life, or life forms as we know them, without water. For this reason, NASA's exploration for life on Mars began with the search for water. Water is the transparent fluid that forms the world's streams, lakes, and oceans, and while it covers about 70% of the earth's surface, only ~2.5% of this is freshwater. We also observe it as rain and snow, and it exists as fog, dew, and cloud as well. Water is the major constituent of the fluids of living things. In a normal adult, about 60% of the body weight comes from water.

The heating or boiling of liquid water with fire to generate steam fascinated early alchemists. However, the chemical composition of water as a composite or combination of two hydrogen atoms and one oxygen atom was not apparent until the early 19th century.

1.4 Dalton's Atomic Theory: Emergence of Modern Chemistry

John Dalton's atomic theory spawns the beginning of modern chemistry. The theory states that "matter consists of indivisible particles called atoms and that atoms of a given element are all identical and can neither be created nor destroyed. Compounds are formed by combination of atoms in simple ratios to give compound atoms (molecules)."[6] This theory is the basis of modern chemistry. Democritus first suggested the existence of the atom, but it took almost two millennia before the atom was placed on a solid footing, as the fundamental chemical building block by John Dalton (1766–1844). Atoms are no longer invisible and can be "seen" by electron microscopy and atomic force microscopy, two powerful imaging technologies, which were developed during the late 20th century, capable of seeing atoms one atom at a time (Fig. 1.2).

Subsequently, A. Lavoisier showed that water was made up of a combination of one oxygen atom and two hydrogen atoms.[9] Henry Cavendish had discovered hydrogen as a flammable and explosive gas in 1766. Lavoisier studied the reaction between oxygen and hydrogen and observed an explosion with the release of a colorless gas that rapidly condensed against the sides of the container to produce a colorless liquid. The liquid produced was water. He subsequently determined the composition of water by studying the reaction of Fe filings with water to form hydrogen and comparing the amount of hydrogen produced in this experiment with the mixture of oxygen and hydrogen released upon decomposing the same amount of water vapor over hot iron. The modern view of the structure of the water molecule is shown in Fig. 1.3.[10]

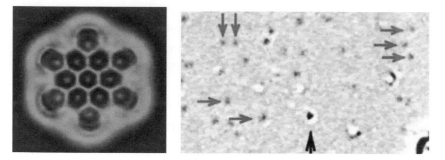

Figure 1.2   Left: A nanographene molecule exhibiting carbon atoms imaged by atom force microscopy (from [7]. Reprinted with permission from AAAS). Right: Single hydrogen atoms (highlighted by bold red arrows) seen on the surface of graphene, a sheet of carbon just one atom thick, imaged by transmission electron microscope (reprinted by permission from Macmillan Publishers Ltd: *Nature* [8], copyright (2008)).

Water molecule

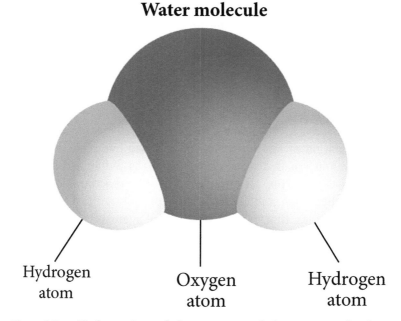

Hydrogen atom   Oxygen atom   Hydrogen atom

Figure 1.3   Modern view of the structure of the water molecule as represented by a space-filling model formed from two hydrogen atoms and one oxygen atom (or two parts of hydrogen and one part of oxygen) (taken from Ref. 10).

1.5   Periodic Table of Elements

The periodic table is a tabular arrangement of chemical elements organized on the basis of the number of protons in the nucleus, the electron configurations, and recurring chemical properties. The Russian chemist Dmitri Mendeleev developed the table in 1869 to illustrate the periodic trends in the properties of the known elements at the time.[11] The Mendeleev periodic table was a milestone in the systematization of the chemical properties of the elements, and it accelerated the pace of development of quantum theory to atomic physics and the understanding of chemical bonding. Mendeleev's periodic table has since been expanded and refined with the discovery or synthesis of further new elements and the development of new theoretical models to explain chemical behavior (Fig. 1.4).[12]

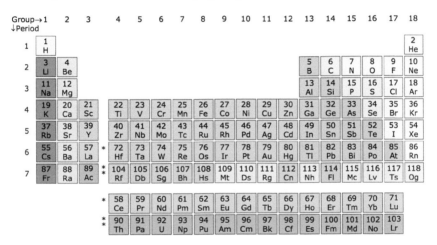

Figure 1.4   Modern periodic table of elements (taken from Ref. 12).

1.6   Quantum Chemistry and the Nature of the Chemical Bond

The beginning of the 20th century was an exciting period in modern physics with the development of quantum mechanics by Albert Einstein, Max Planck, Louis de Broglie, Neils Bohr, Erwin Schrödinger, Max Born, Pascual Jordan, David Hilbert, Paul Dirac, John von Neumann, Werner Heisenberg, Wolfgang Pauli,

Satyendra Nath Bose, and others.[13] The application of the new physics to the electronic structure of the atom quickly validated many of the predictions of the Mendeleev periodic table, and it did not take long before the new physics was applied to obtain an in-depth understanding of chemical bonding by Oyvind Burrau, Walter Heitler and Fritz London, Charles Coulson, Linus Pauling, Sir John Lennard-Jones, H. H. James and A. S. Coolidge, and others.[14] There was now an intellectual framework to study how atoms combine to form molecules, how molecules rearrange their atoms to form new ones, the nature of the chemical bond, and how the distribution of the electrons in molecules influences molecular structure and the interaction of molecules with light, and later, chemical reactivity. Modern chemistry was born, and what followed was a century of rapid and remarkable development in molecular science, not only in the conception of fundamental principles, but also the applications of these concepts to materials, biology, and other sciences. The role played by modern chemistry in the development of biology has been particularly significant.

Biology is the science of living cells. Each cell consists of a collection of molecules, both large and small, inorganic as well as organic molecules, self-assembled and structurally organized in three-dimensional space to provide a machinery that will sustain the flow of information, energy, and chemical signals within this space, as well as interactions of the various components with their surroundings, to create a quasi-steady state condition that varies with time. We call this condition "life." To sustain this "life," the cell must have the capability to grow and multiply, undergo development or differentiation into different states, and last, but not least, the ability to self-replicate or regenerate itself, according to a "blueprint" embedded within itself. Interestingly, to accomplish all this, the cell plays by the rules of chemistry. Nature has had 3.5 billion years to perfect the applications of these chemical principles in biology since the creation of primordial life.

1.7 The Emergence of Life

Life as we know it would not have happened without water. However, the exact origin of our planet's water, which covers about 70% of the earth's surface, is still a mystery to scientists. There exist several hypotheses as to how water may have accumulated on the earth's surface over the past 4.6 billion years in sufficient quantity to form oceans. The prevailing consensus is that the water was delivered to the primordial earth via violent collisions with asteroids from the outer solar system shortly after the formation of the planet.[15] Astronomers have recently found the first evidence of ice on an asteroid.[16] Hundreds of thousands of asteroids are known to be orbiting in the asteroid belt between the inner and outer solar system.

The hydrogen in water is paramount, but volatile elements, such as nitrogen and carbon, are also crucial to the origins of life. Researchers examining hydrogen and nitrogen isotope ratios in meteorites have now made a giant stride toward clarifying this issue. Isotopes of a given element are atoms containing the same number of protons but different number of neutrons in the nuclei. For example, deuterium is a heavier form of the element hydrogen with one neutron compared to normal hydrogen, which has no neutrons. Similarly, there is an isotope of the element nitrogen that has one more neutron (called nitrogen-15). A research team led by C. M. O'D. Alexander at the Carnegie Institution of Washington, United States, has recently compared the hydrogen isotope ratios in chondritic meteorites, ancient fragments of asteroids, with those that have been measured in comets.[17] A meteorite is a solid piece of debris from asteroids or comets that originates from outer space and survived its impact with the earth's surface (Fig. 1.5).[18] These researchers showed that the proportion of deuterium to hydrogen was lower in the meteorites than in comets. Comets are born in the outer solar system, where the extreme cold results in frozen water having a higher proportion of deuterium than does ice formed in the less cold regions. Alexander et al. also analyzed the nitrogen isotope ratios of the chondritic meteorites. The hydrogen and nitrogen isotopic compositions observed for these chondrites match closely with what we see on earth, suggesting that asteroids, the parent bodies of these chondritic meteorites, were the dominant source of volatile hydrogen, carbon, and nitrogen.

The next question is how life on earth began. This is still an unsettled issue. The most popular theory is abiogenesis, which hypothesizes that "the natural process of life arises from nonliving matter, such as simple organic compounds."[19] In 1953, Stanley Miller and Harold Urey demonstrated that most amino acids, basic chemicals of life, could be synthesized in conditions designed and intended to simulate the atmosphere and oceans of the early planet earth.[20] But how did the first organisms develop from the primordial soup? Electric sparks can generate amino acids and sugars from an atmosphere loaded with water, methane, ammonia, and hydrogen, as was shown in the celebrated Miller–Urey experiment,[21] suggesting that lightning might have helped create the key building blocks of life on earth in its early days. Over millions of years, larger and more complex molecules could form. The first molecules of life might have met on clay, according to an idea put forth by Alexander Graham Cairns-Smith, an organic chemist working at the University of Glasgow in Scotland.[22] Cairns-Smith suggests that surfaces of mineral crystals in clay might have concentrated these organic compounds together and arranged them into organized patterns. Later, organic molecules took over this job and organized themselves.

Figure 1.5   The Hoba meteorite in Namibia, the largest known impact meteorite (2.7 m long and 60 tons in weight) (taken from Ref. 18).

There is another theory suggesting that life may have begun at deep submarine hydrothermal vents at the bottom of the seas and oceans.[23,24] These deep-sea vents spew out an abundance of key hydrogen-rich molecules, which the rocky alcoves at the ocean floor could then have concentrated and provided the mineral catalysts for critical reactions. These vents are rich in chemical and thermal energy and sustain vibrant ecosystems, even to this day.

On the other hand, complex organic molecules have been found both in interstellar space and in the solar system. Thus, it is possible that microscopic life might have existed throughout the universe and distributed to the planet by meteoroids, asteroids, and other small bodies in the solar system, providing the starting material for the development of life on earth. Perhaps then, life did not begin on earth at all but was brought here from elsewhere in space.[25]

In any case, it seems that life on earth began more than 3 billion years ago, evolving from the most basic of microbes into a dazzling array of complexity over time. The earliest undisputed evidence of life on earth comes from microbial mat fossils 3.48 billion years old, found in the sandstones of the western part of Australia.[26] A microbial mat is a multi-layered sheet of microorganisms (mainly bacteria and archaea) that grow at the interfaces between different types of material (predominantly moist surfaces).

Self-replication is an important element of life. So scientists are naturally interested in uncovering how self-replicating molecules or their components came into existence. The generally accepted view is that current life on earth descended from an RNA world, although life derived from RNA might not have been the first life to exist. Some scientists, who focus on understanding how catalysis in chemical systems in the early earth might have provided the precursor molecules necessary for self-replication, theorize that "metabolism" came first and speculate that the biochemistry of life may have begun shortly after the Big Bang, during an "inhabitable" period when the age of the universe was only 10–17 million years old.

1.8 Modern Biology Is All Chemistry

Nowadays, we know that DNA needs proteins in order to form and, vice versa, that proteins require DNA to form, so how could these have formed without each other? The answer may be RNA, which can store information like DNA, serve as an enzyme like proteins, and help create both DNA and proteins. Later, DNA and proteins succeeded this "RNA world," because together they are more efficient in facilitating the gamut of chemistry needed to sustain activities of the cell and to propagate life. RNA still exists and performs several functions in organisms, including acting as an on-off switch for some genes. The question still remains as to how RNA got here in the first place.

In any case, we now know from the work of James Watson and Francis Crick that the main role of DNA is to store information on how other molecules should be made and arranged.[27] Genetic sequences in DNA are essentially instructions on how amino acids should be arranged in proteins. Along the way, nature discovered "self-replication" by exploiting the "laws or principles" of chemistry to control the specificity of recognition between molecules and the kinetic control of chemical reactions between biological molecules, without which life as we know it cannot be sustained.

In fact, nature even "discovered" some chemistry of its own. For example, the first use of transition elements such as iron, manganese, nickel, cobalt, copper, zinc, and molybdenum to promote complex chemistry (e.g., production of molecular hydrogen from water, conversion of methane into methanol, and fixing of nitrogen to form ammonia) appeared during the early days of the planets when the atmosphere was highly reducing, billions of years before all this chemistry was "re-discovered" by chemists during the past century. The modern field of organometallic chemistry was founded during the 1950s, but it turned out nature already knew how to combine the transition metals with sulfur, carbon monoxide, and cyanide to form organometallic compounds in different ways to carry out difficult and complex chemical reactions in the reducing environment of early earth. So we can always learn "new" chemistry from nature.

In order to sustain life as we know it, it turns out water is an absolute requirement. Water is an unusual solvent, as it is a liquid milieu that can accommodate many compounds, including inorganic salts, amino acids, and nucleic acids, and other building blocks of molecules essential for the subsistence of life, allowing them to come together so that they can recognize one another for chemical reactions, namely to rearrange and exchange atoms to form new molecules. Frequently, individual water molecules are even involved in the chemistry. Yet, the water molecules in liquid water form strong hydrogen bonds with one another to create a unique three-dimensional network (Fig. 1.6).[28] To accommodate simple organic molecules, such as those that might be the precursors required to make the molecules essential for life, the hydrogen-bonded structure might need to be disrupted. To begin with, for these simple organic molecules to be "soluble" in the water solution, typically, they contain certain polar groups that can disrupt the local water structure to form hydrogen bonds with individual water molecules. However, because of the structure of water, the carbon-containing parts of these simple organic molecules tend to be solvated by a layer, or layers, of hydrogen-bonded water molecules. Moreover, in order to minimize the surface area in contact with the water structure, these organic molecules prefer to associate with one another to form intermolecular complexes that are in turn solvated by layers of water molecules. In other words, in an aqueous environment, there exists a force in the solution that brings organic molecules together so that the probability of chemistry between these molecules is materially enhanced. Perhaps, this was the beginning of some of the early primitive molecules that were important in initiating life during the early days of the earth. When the chemistry between these carbon-containing organic molecules takes place on the surface of minerals in clays, which provides basic and acidic groups to accelerate the chemistry, the probability of chemistry is even further enhanced.

This hydrophobic force (a force resulting from "phobia" of water), as this driving force has come to be known in modern chemistry, operates at all levels of the chemistry of biological systems. For example, the secondary and three-dimensional structures of DNA, RNA, and protein molecules are unstable if these molecules are taken out of water (Fig. 1.7, left). In DNA and RNA, the double helix will not be formed without water, as there will be insufficient driving force for each strand of the double helix to form the secondary structure(s) to allow recognition of a complementary strand. To store the genetic

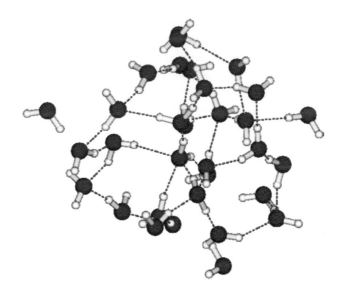

Figure 1.6   Local structure of liquid water: The hydrogen-bonded network (reproduced with permission from T. S. Pennanen, J. Vaara, P. Lantto, A. J. Sillanpää, K. Laasonen, J. Jokisaari, Nuclear Magnetic Shielding and Quadrupole Coupling Tensors in Liquid Water: A Combined Molecular Dynamics Simulation and Quantum Chemical Study, *J. Am. Chem. Soc.*, 2004, 126(35), 11093–11102, DOI: 10.1021/ja048049i).

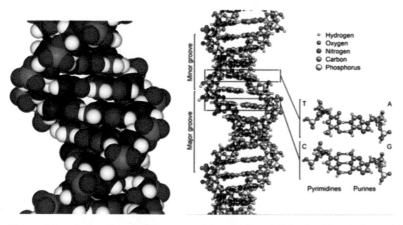

Figure 1.7   Left: Space-filling model of the DNA double helix of Watson and Crick. Right: Two complementary strands of nucleic acid molecules form a double-helical structure held together by A-T and G-C base pairs (taken from Ref. 29).

information and to form a new complementary strand to pass the information to a second generation, nature has devised an ingenious scheme to use molecular structure to control molecular recognition. The secondary structures of the DNA and RNA are stabilized by *vertical* stacking forces between adjacent nucleic acid bases A, T (or U), G, and C; and the double helix is formed by complementary base-pairing between the A and T (or U) and between the G and C via the formation of *horizontal* hydrogen bonds across the two strands, where A and G denote the purine bases adenine and guanine; and T, U, and C, the pyrimidine bases thymine, uracil, and cytosine, respectively (Fig. 1.7, right).[29] Similarly, proteins will not form their three-dimensional structures without the hydrophobic force, without which there would be no enzymes. With most water-soluble proteins, the side chains of hydrophobic amino acids are sequestered to form a hydrophobic interior core with the polar amino acids facing outside and the three-dimensional structure is stabilized by hydrogen-bonding between amino acids using their peptide backbone and side chains (Fig. 1.8).[30] Enzymes are the most important catalysts in nature. Without enzymes, biological chemistry will not take place and there will be no life.

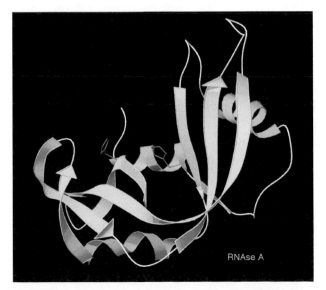

RNAse A

Figure 1.8   X-ray structure of bovine pancreatic ribonuclease A, an enzyme that cuts RNA (taken from Ref. 30).

1.9   Molecular Building Blocks of a Cell

The biological cell is made up of many different kinds of molecules, including inorganic salts, small molecules such as amino acids (the building blocks of proteins), nucleosides and nucleotides (the building blocks of the genetic materials DNA and RNA), lipids and phospholipids of the biological membrane, sugars, carbohydrates, other small metabolites. A schematic of an animal cell is depicted in Fig. 1.9 (top), which highlights some of the molecules as well as the way they are organized to give different cellular components, such as the nucleus, where the chromosomes are packaged; the mitochondrion, the powerhouse of the cell where ATP is synthesized; the lysosome, which breaks down all kinds of cellular debris, including proteins and nucleic acids; the endoplasmic reticulum, which participates in protein synthesis and protein sorting and transport, etc. The phospholipid and other lipids, such as cholesterol, form the cell membrane that encases the cytoplasm of the cell. The hydrocarbon-like tails of the phospholipids are extremely hydrophobic, and these molecules spontaneously assemble to form a bilayer leaflet, in which many membrane proteins reside (Fig. 1.9, bottom). The polar head-groups of the phospholipids interact with the surrounding water, without which the cell membrane would break up.

The DNA and RNA prefer to organize themselves into Watson–Crick double helices stabilized by the vertical stacking forces between adjacent nucleic bases as well as the horizontal hydrogen-bonding interactions across the two polymeric strands, as noted earlier. Variants of the Watson–Crick structures are also found, but some features of the stacking forces and hydrogen-bonding interactions are always preserved. Proteins usually fold into three-dimensional structures in the cytoplasm as well as in the membranes due to interplay between the same kinds of forces, although we now know that many proteins are intrinsically disordered. These three-dimensional structures bestow on these macromolecules their distinct shapes and surfaces to facilitate their interactions with one another, and with other molecules, small and large. For example, proteins bind to DNA and RNA, and proteins bind to other proteins to tune their functions. Without these interactions, there would be no flow of information from the DNA, no signaling between molecules, and no response to input of molecular and chemical signals from outside the cell, the result of all of which would be no chemistry and no life.

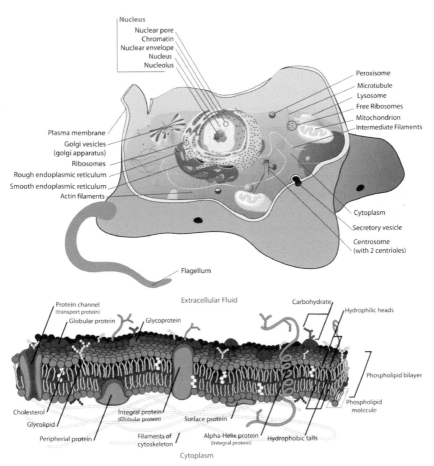

Figure 1.9   Top: Schematic of an animal cell (taken from Ref. 31). Bottom: Cross section of a typical membrane (taken from Ref. 32).

1.10   Central Dogma of Molecular Biology

James Watson and Francis Crick discovered the structure of the DNA double helix.[27] This was the beginning of modern biology, often referred to as molecular biology. From here, Crick formulated the central dogma of modern biology, in which he hypothesized how the flow of genetic information takes place within a biological system (Fig. 1.10a).[33,34] In the simplest of terms, the central dogma of molecular biology is "DNA makes RNA, which in turn makes proteins." Crick's hypothesis does not preclude the reverse flow of information from RNA to DNA; rather, it only rules out the flow from protein to RNA or DNA. In other words, an RNA transcript made originally from DNA can also be reverse-transcribed back into DNA, but the translation of the RNA transcript into protein is irreversible (i.e., RNA cannot be made from protein). Thus, nature has designed a versatile and elegant machinery for self-replication as well as for conversion of the genetic information stored in the genome into RNA and proteins for biological function, and it accomplishes all these processes with high fidelity.

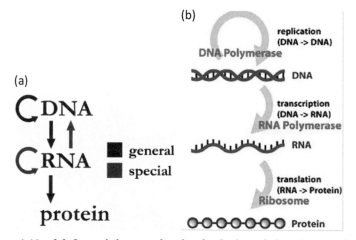

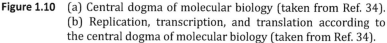

Figure 1.10 (a) Central dogma of molecular biology (taken from Ref. 34). (b) Replication, transcription, and translation according to the central dogma of molecular biology (taken from Ref. 34).

We now know the following: (i) The DNA is the genetic material that dictates the blueprint of life. The sequence of the DNA contains all the information that determines the makeup of our cells, our tissues, our organs, our mental disposition and social behavior, genetic disorders or syndromes, and even our predisposition to diseases such as cancers and neurodegeneration. Our own genome, for example, is packaged into 23 pairs of chromosomes in the nucleus of the cell (Fig. 1.11).[35] Each chromosome contains a multitude of genes encoding for a specific protein or specifying a designated cellular function. (ii) The cell contains the machinery to copy a DNA molecule to create a duplicate copy, namely, to replicate itself in a process called "replication." (iii) It also contains the machinery to transfer the sequence of each gene into RNA by a process called "transcription." (iv) This RNA message could be translated into a polypeptide of amino acids to form a protein in a process called "translation." A protein is typically a sequence of amino acids linked together by peptide bonds, with the order of these amino acids being programmed by the gene sequence. The translation is carried out one amino acid at a time according to the genetic code in which the sequence of three contiguous nucleic acid bases in the RNA transcript is translated into a specific amino acid (Fig. 1.10b).[34] Once synthesized, the protein could undergo further chemical modifications, in a process that has come to be known as post-translational modification. Each chemical modification of the protein introduces a message into the protein molecule, which might be used as a signal for recognition of a partner protein for transfer of some information or allow the protein to be targeted to a specific location of the cell. For example, attachment of a complex carbohydrate or a linkage of simple sugars to the protein after translation is used to target the protein for transport to the cell membrane, where it becomes a determinant for recognition by other cells. Within the cell, there is a trafficking machinery to transport the proteins to the location of the cell designated for their specific functions. All these processes involve bringing macromolecules together to promote the outcome with high fidelity according to the laws of chemistry.

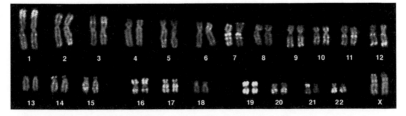

Figure 1.11 The 23 pairs of human chromosomes (taken from Ref. 35).

1.11   Applications of Molecular Biology

One of the most powerful applications of molecular biology emerged when scientists learned how to incorporate individual genes into plasmids of bacteria, e.g., *Escherichia coli*, and allow the replication of the microorganism to mass produce the protein encoded by the gene (gene cloning) (Fig. 1.12).[36] The gene DNA sequence may be created by chemical synthesis, or obtained from plants, bacteria, or human. In this manner, it is possible to prepare large quantities of the protein of interest. This strategy can also be used to obtain variant proteins in which one or more amino acids have been substituted or replaced by others. This recombinant DNA technology spawned a new area of applied molecular biology and biotechnology that uses microorganisms to perform specific industrial processes.

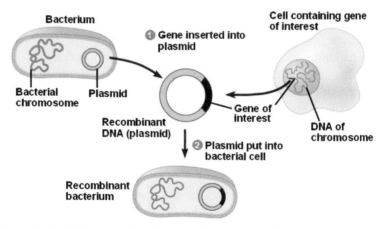

Figure 1.12   DNA molecular cloning (taken from Ref. 37).

Before DNA molecular cloning was possible, molecular biologists had to learn how to splice genes. Gene splicing is the technology of preparing recombinant DNA in vitro (i.e., outside the cell in solution, by cutting up DNA molecules and joining fragments together). The DNA molecules could come from more than one organism (i.e., originate from any species). For example, plant DNA may be joined to bacterial DNA, or human DNA may be joined with fungal DNA. The DNA sequence incorporated may also be prepared by chemical synthesis. DNA sequences may be removed from an organism using enzymes called nucleases and they can be spliced together with the use of DNA ligases.

With the technology of gene splicing in hand, it is now possible to manipulate the genome of an organism using biotechnology. Genes can be excised (i.e., "knocked out"), inserted (i.e., "knocked in"), and mutated with precision. This has been a powerful method to study the relationship of the gene with phenotype. The use of this genetic engineering technology to generate genetically modified organisms (GMO) has also led to genetically modified bacteria, mice, zebrafish, orchids, and agricultural crops. It is conceivable that genetic engineering will someday be applied to humans to correct genetic disorders or the predisposition of an individual to certain medical syndromes (e.g., via gene therapy).

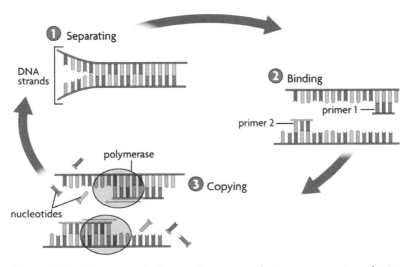

Figure 1.13 The use of the polymerase chain reaction to obtain recombinant DNA (taken from Ref. 38).

One of the revolutionary methods developed in molecular biology, a method by which large quantities of DNA can be obtained, is by the *polymerase chain reaction* (PCR).[38] With this technology, PCR can be used to amplify a single copy or a few copies of a piece of DNA by several orders of magnitude, generating thousands to millions of copies of a particular DNA sequence (Fig. 1.13). Developed in 1983 by Kary Mullis,[39] PCR is now a common and often an indispensable technique used in medical and biological research laboratories for a variety of applications. As described, this technology differs from molecular cloning in that PCR replicates DNA in the test tube, free of living cells, while molecular cloning involves replication of the DNA within a living cell. However, PCR is now often part of the molecular cloning process, e.g., it is used to generate large quantities of a DNA sequence that is to be cloned into a vector of a bacterium.

1.12   Genomics, Proteomics, and Bioinformatics

During the past two decades, scientists have developed the technologies to sequence the entire genome of an organism, namely, the sequence of all the genes packaged into each chromosome. When the sequencing of the human genome was finally announced in 2001, there was considerable excitement because we could now begin to understand the molecular basis of life, diseases, and human behavior. Thus, we have embarked on a new era of human biology: molecular medicine. We can now compare the genomes of individuals to understand their differences in phenotypes, their different abilities to combat infectious diseases, their predispositions to develop certain types of cancers, their different sensitivities to various medications used to control various metabolic disorders, and even their social behaviors.

Aside from the human, scientists have now worked out the genomes of many other biological species, including different fungi, bacteria, plants, and animals. Life science is life science, and the methods and technologies developed by molecular biologists are applicable to the study of any species of life. This is hardly surprising, as the methods and technologies employed are all based on the science of chemistry and the applications of chemical principles. Basically, the same strategies and methods are used to obtain proteins of interest from microbes in large quantities, develop plants that are more resistant to drought, and understand the molecular basis of resistance of microorganisms to certain antibiotics.

With the genomes of different species of life (including fungi, bacteria, plants, and animals) known, it is now possible to compare the evolution of various life species. Unknown gene sequences can be searched and compared across different species to glean what their functions and protein structural folds might be. Also, the likelihood of existence of a protein in a given organism can be assessed from the genome data bank. Data-mining methods have been developed for this research, creating a new field called bioinformatics.

In the recent years, powerful methods have also emerged for high throughput sequencing of proteins based on mass spectrometry of proteins and peptides. This is a bottom-up approach to deriving sequence information about proteins and peptides of interest. An interesting application of this technology is the identification of biomarkers from patients with various disorders, especially cancers. The hypothesis is that the onset of or predisposition to certain cancers in an individual might be uncovered by the detection of these biomarkers for early diagnosis and for the development of drugs and therapeutics to treat the disease.

Although modern research in the life sciences is now done according to the new paradigm of genomics, proteomics, and bioinformatics, chemistry continues to play a major role in the advancement of the field. Chemical synthesis, the development of new methodologies, and the study of reaction mechanisms and structure–activity relationships are expected to foster future developments in the life sciences in the years to come.

1.13   Enzymes as Perfect Catalysts

Chemical reactions are often catalyzed by a spectator molecule, called a catalyst, to allow the chemical transformations of interests to proceed at a significantly faster rate. This onlooker participates intimately in the mechanism of the reaction, but it is regenerated during the chemical transformation and is not, thus, consumed. Enzymes are the catalysts of a cell. They are usually protein molecules folded into a three-dimensional structure to embrace a so-called active site that can recognize a substrate and transform it into a desired product needed by the cell (Fig. 1.14).[40] In the absence of enzymes, chemical transformations would only happen very slowly, if at all. Thus, enzymes are extremely important in the function of a cell. In humans, some diseases are related to a deficiency in a specific enzyme.

Typically, in a cell, with the assistance of enzymes, chemical transformations occur efficiently and with high fidelity and under controlled conditions determined by the concentration of specific enzymes. Chemical reactions in a cell are very efficient. In this manner, the chemistry of life proceeds in an orderly fashion. Other macromolecules can also behave as catalysts. It is generally accepted that the first enzymes are RNA molecules, which can mediate self-replication. The three-dimensional structures of many enzymes are now known to atomic resolution. Biochemists have also learned to engineer them to tune their chemical reactivity and also to make them more stable so that they can operate under harsh conditions such as higher temperatures or in organic solvents.

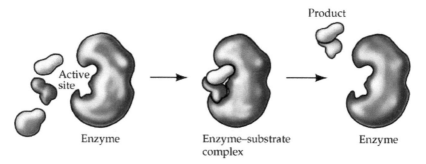

Figure 1.14 Lock and key mechanism of an enzyme. Reproduced from Ref. 41. By permission of Oxford University Press, USA.

1.14 Can We Learn from Biology to Develop More Efficient Chemical Catalysts for Chemical Manufacturing?

Chemical catalysis is a time-honored field. Many chemical reactions are mediated by catalysts. For example, a reaction might be catalyzed by a proton (H^+) in solution, base (OH^-), or even a small inorganic cation like Li^+. Chemical catalysts can be more complex molecules with complicated structural scaffolds, and the chemical reaction can be promoted on the surface of a solid or at an interface. Nevertheless, many difficult chemical transformations can benefit from a suitable catalyst so that the chemistry can take place in water. These include the controlled oxidation of methane into methanol and the fixation of N_2 to form ammonia under ambient conditions. These chemical transformations occur in microbes with high efficiency. For example, the enzyme methane monooxygenase converts methane into methanol efficiently at room temperature. The enzyme nitrogenase converts N_2 into ammonia. At the moment, these chemical transformations are carried out industrially under high temperatures and pressures and the processes are not particularly atom or energy efficient. Recently, we have learned from the structure of the active site of the particulate methane monooxygenase how to develop a biomimetic catalyst that can mediate the efficient conversion of methane into methanol under ambient conditions (Fig. 1.15).[42] This is but one example of how we can learn from nature to develop catalysts to perform these processes in solution. We believe this is a new field of rationally designed catalysts. For example, in recent years, chemists have learned to develop catalysts for asymmetric synthesis, an area of particular significance for the synthesis of drugs and pharmaceuticals to intervene with health disorders. Being able to synthesize the proper stereo-isomer is critical in the development of such drugs.

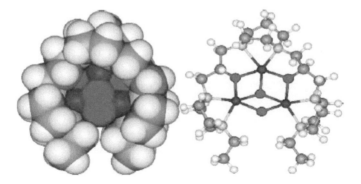

Figure 1.15 A tricopper–dioxygen complex capable of mediating efficient conversion of methane into methanol under ambient conditions, mimicking the catalytic site of the particulate methane monooxygenase in methanotrophic bacteria when the tricopper cluster is activated by dioxygen. Atoms: copper (•); oxygen (•); nitrogen (•); carbon (•); and hydrogen (•). Reproduced with permission from Ref. 42, copyright 2013 John Wiley & Sons.

1.15   Greener Chemistry

In the years to come, chemists need to synthesize their molecules under greener conditions. We believe catalysts, especially catalysts that can function in water, are needed to usher in this new era of chemistry. In this work, we think that we can learn from nature. In recent years, we have already learned from methanotrophic bacteria to develop a catalyst for the conversion of methane into methanol. Such a catalyst can be used eventually to liquefy our abundant supply of methane stored on planet earth. In the foreseeable future, this catalyst can be used to mitigate the methane emissions from permafrost in the receding glaciers, the methane gas produced by methanogenic bacteria in garbage dumps, and the methane produced in rice paddies to lessen the emission of this green house into the atmosphere and stratosphere.

1.16   Future of Chemistry

Chemistry is a central science. The laws or principles of chemistry apply to everything we do. In fact, history has shown that progress in the development of all scientific disciplines becomes accelerated when the methods of the research is "molecularized." Witness the progress that has been made in recent years in geological sciences, environmental sciences, and materials sciences, not to mention biological sciences that we have focused on in this chapter. Since there are no boundaries in molecular science, we have every reason to stay optimistic about the future of chemistry.

References

1. Big Bang. *Wikipedia, The Free Encyclopedia.* http://en.wikipedia.org/wiki/Big_Bang (accessed June 9, 2015).
2. Classical element. *Wikipedia, The Free Encyclopedia.* http://en.wikipedia.org/wiki/Classical_element (accessed June 9, 2015).
3. Alchemy. *Wikipedia, The Free Encyclopedia.* http://en.wikipedia.org/wiki/Alchemy (accessed June 9, 2015).
4. Who discovered nitrogen? http://discovery.yukozimo.com/who-discovered-nitrogen (accessed June 9, 2015).

5. American Chemical Society. International Historic Chemical Landmarks. Discovery of oxygen by Joseph Priestley. http://www.acs.org/content/acs/en/education/whatischemistry/landmarks/josephpriestley oxygen.html (accessed June 9, 2015).

6. Dalton's atomic theory. http://www.iun.edu/~cpanhd/C101webnotes /composition/Dalton.html (accessed June 9, 2015).

7. L. Gross, F. Mohn, N. Moll, B. Schuler, A. Criado, E. Guitián, D. Peña, A. Gourdon, G. Meyer, *Science* (2012), **337**, 1326–1329.

8. J. C. Meyer, C. O. Girit, M. F. Crommie, A. Zettl, *Nature* (2008), **454**, 319–322.

9. The discovery of the composition of water. http://www.chm.bris.ac.uk/webprojects2001/hossain/water.htm (accessed June 9, 2015).

10. The chemical world. http://wps.prenhall.com/wps/media/objects /476/488316/ch01.html (accessed June 9, 2015).

11. History of the periodic table. *Wikipedia, The Free Encyclopedia.* http://en.wikipedia.org/wiki/History_of_the_periodic_table (accessed June 9, 2015).

12. Periodic table. *Wikipedia, The Free Encyclopedia.* http://en.wikipedia. org/wiki/Periodic_table (accessed June 9, 2015).

13. Quantum mechanics. *Wikipedia, The Free Encyclopedia.* http:// en.wikipedia.org/wiki/Quantum_mechanics (accessed June 9, 2015).

14. Chemical bond. *Wikipedia, The Free Encyclopedia.* http://en.wikipedia. org/wiki/Chemical_bond (accessed June 9, 2015).

15. Where did earth's water come from? http://www.livescience.com/ 33391-where-did-water-come-from.html (accessed June 9, 2015).

16. H. Campins, K. Hargrove, N. Pinilla-Alonso, E.S. Howell, M.S. Kelley, J. Licandro, T. Mothé-Diniz, Y. Fernández, J. Ziffer, *Nature* (2010), **464**, 1320–1321; A.S. Rivkin, J.P. Emery, *Nature* (2010), **464**, 1322–1323.

17. C. M. O'D. Alexander, R. Bowden, M. L. Fogel, K. T. Howard, C. D. K. Herd, L. R. Nittler, *Science* (2012), **337**, 721–723.

18. Meteorites. *Wikipedia, The Free Encyclopedia.* http://en.wikipedia. org/wiki/Meteorite (accessed June 9, 2015).

19. Abiogenesis. *Wikipedia, The Free Encyclopedia.* http://en.wikipedia. org/wiki/Abiogenesis (accessed June 9, 2015).

20. S. L. Miller, H. C. Urey, *Science* (1959), **130**, 245–251.

21. Miller–Urey experiment. *Wikipedia, The Free Encyclopedia.* http:// en.wikipedia.org/wiki/Miller%E2%80%93Urey_experiment (accessed June 9, 2015).

22. Graham Cairns-Smith. *Wikipedia, The Free Encyclopedia.* http://en.wikipedia.org/wiki/Graham_Cairns-Smith (accessed June 9, 2015).

23. 7 theories on the origin of life: Deep sea vents. http://www.livescience.com/13363-7-theories-origin-life.html (accessed June 9, 2015).

24. Hydrothermal vent. *Wikipedia, The Free Encyclopedia.* http://en.wikipedia.org/wiki/Hydrothermal_vent (accessed June 9, 2015).

25. Panspermia. *Wikipedia, The Free Encyclopedia.* http://en.wikipedia.org/wiki/Panspermia (accessed June 9, 2015).

26. N. Noffke, D. Christian, D. Wacey, R. M. Hazen, *Astrobiology* (2013), **13** (12), 1103–1124.

27. J. D. Watson, F. H. Crick, *Nature* (1953), **171**, 737–738.

28. Local structure of liquid water through NMR parameters. http://www.oulu.fi/physics/node/664 (accessed June 9, 2015).

29. DNA. *Wikipedia, The Free Encyclopedia.* http://en.wikipedia.org/wiki/DNA (accessed June 9, 2015).

30. Kinemage. *Wikipedia, The Free Encyclopedia.* http://en.wikipedia.org/wiki/Kinemage (accessed June 9, 2015).

31. Animal cell structure. https://commons.wikimedia.org/wiki/File:Animal_cell_structure_en.svg.

32. Cell membrane. *Wikipedia, The Free Encyclopedia.* http://en.wikipedia.org/wiki/Cell_membrane (accessed June 9, 2015).

33. F. Crick, *Nature* (1970), **227**, 561–563.

34. Central dogma of molecular biology. *Wikipedia, The Free Encyclopedia.* http://en.wikipedia.org/wiki/Central_dogma_of_molecular_biology (accessed June 9, 2015).

35. Autosome. *Wikipedia, The Free Encyclopedia.* http://en.wikipedia.org/wiki/Autosome (accessed June 9, 2015).

36. Molecular cloning. *Wikipedia, The Free Encyclopedia.* http://en.wikipedia.org/wiki/Molecular_cloning (accessed June 9, 2015).

37. DNA molecular cloning, bio1151.nicerweb.com.

38. Polymerase chain reaction. *Wikipedia, The Free Encyclopedia.* http://en.wikipedia.org/wiki/Polymerase_chain_reaction (accessed June 9, 2015).

39. K. F. Mullis, F. Faloona, S. Scharf, R. Saiki, G. Horn, H. Erlich, *Cold Spring Harbor Symposium in Quantitative Biology* (1986), **51**, 263–273.

40. Enzyme. *Wikipedia, The Free Encyclopedia.* http://en.wikipedia.org/wiki/Enzyme (accessed June 9, 2015).

41. Cooper, G.M., *The Cell: A Molecular Approach*, 2nd ed. (2000) Sinauer Associates, Inc., Sunderland, MA, USA.

42. S. I. Chan, Y.-J. Lu, P. Nagababu, S. Maji, M.-C. Hung, M. M. Lee, I.-J. Hsu, P. D. Minh, J. C.-H. Lai, K. Y. Ng, S. Ramalingam, S. S.-F. Yu, M. K. Chan, *Angewandte Chemie International Edition* (2013), **52**, 3731–3735.

Chapter 2

Chemistry in Human History

Mary Virginia Orna

Department of Chemistry, College of New Rochelle,
New Rochelle, New York 10805, USA

maryvirginiaorna@gmail.com

2.1 The Value of History

2.1.1 Valuing the Past

Why do we value the past? Why is history important? We need history to look at the past in order to find out how we arrived at the present. And, of course, the trajectory of this past helps us to determine where we might go in the future and how to improve it. Modern science, technology, and medicine are intimately bound up in a giant knotted web of human activity all over the globe, and chemistry is a proud partner in this web. In fact, the history of chemistry is the history of everything: Chemistry has been the constant companion of civilization in its development. Modern life, as we know it, would not exist without the continual use of chemistry and chemical technology from the earliest times—whether their practitioners knew they were using them or not. Without chemistry, we would still be in the Stone Age.

Chemistry: Our Past, Present, and Future
Edited by Choon Ho Do and Attila E. Pavlath
Copyright © 2017 Pan Stanford Publishing Pte. Ltd.
ISBN 978-981-4774-08-6 (Hardcover), 978-1-315-22932-4 (eBook)
www.panstanford.com

2.1.2 Purpose of Human Activity

In this context, we have to ask why the people who went before us did what they did? What were their purposes in developing mathematical, medical, and astronomical expertise? They were not interested in these areas for themselves, but for the eminently practical purposes of survival, commerce, trade, healing, agriculture, war, and spiritual power. Many of chemistry's (or actually alchemy's) early techniques are now denied importance or validity and relegated to the realm of magic or superstition. However, we have to remember that over the millennia, all attempts to understand and/or control the world we live in were done gropingly. Mistakes were made, but in doing these mistakes, people learned and adjusted and, most importantly, kept oral traditions or written records about these attempts.

2.1.3 Building on Precedent

It follows that everyone has predecessors. Everyone builds upon the shoulders of those who have gone before. One example is the development of medical practice. Hippocrates (460–370 BCE) is often credited with the title "Father of Medicine," a title he received from his followers. They themselves contributed greatly to the profession by insisting on detailed case reports and developing a huge storehouse of practical experience that enabled them to predict the course of an illness, but these data did not help in understanding the reasons for disease and ill health.

2.1.4 Materials on Hand

To accomplish the functions of society, people used what they had available. For example, the Egyptians used papyrus, a sedge plant (*Cyperus papyrus*) that grows in the Nile delta, for making useful objects such as sandals and baskets, but there is evidence that as much as 6000 years ago, they also used it to make a writing material, a use we commemorate today with the word "paper." This was accomplished by cutting the reeds into strips, aligning them parallel to one another to form sheets, and soaking them in water to allow for slight decomposition to facilitate their adhesion to one another. And once you transform a native object into something else by way of what we now call chemical change, here you have chemistry! The ancient Babylonians did the Egyptians' one better. The material they had on hand was clay, lots of it, from which they could form clay tablets that turned out to be a durable material soft enough to receive impressions, while wet and hard enough to constitute a permanent record of those impressions when dry. One might say that in this single technique, we have the origins of mathematics and information technology.[1] It was a somewhat long step from raw clay to the creation of pottery, but when it was done, voilà! Chemical changes through vitrification of the clay mass.

2.1.5 Origins of Material Culture

So we see, after all, that human beings have been doing chemistry throughout the world for thousands of years before the science of chemistry was born. Material culture has its origins in large part from chemical technologies. Many ancient peoples developed a variety of chemical technologies before they developed writing. Some of these technologies left long-lasting traces: mining and metallurgy, pottery, clay artifacts and tablets, glass, and inorganic pigments. Some of them left less durable artifacts, such as textiles, organic dyes, paper, and various food-making processes (including baking and fermentation).

2.2 Uses of Natural Substances

2.2.1 Introductory Remarks

Obviously volumes have been written about the history of natural substances and their uses throughout the ages. This brief chapter can only touch upon a few of these substances and how they were transformed by chemistry to produce materials that transformed the world. Some examples are foodstuffs, clothing, metals, glass, and medicines.

2.2.2 Observation Results in Application

Even before we had more knowledge of how chemicals can help our life, chemicals contained in natural products were used as essential factors in our life. It was not known what actual chemical compounds or which actual chemical processes were involved. Nevertheless, observation of accidental effect led to application. In agriculture, animal waste was found to increase food production, but there was no knowledge that its nitrogen content was responsible for it. The preparation of various metals, such as copper or iron, from their ores was done without any metallurgical science. The developers of gunpowder did not have any understanding what chemical reaction was taking place. In the ancient civilizations of the Chinese, Indians, and North Africa, the medicine men used hundreds of herbs and plant materials for health care and for preventing diseases. The earliest known written document is a 4000-year-old Sumerian clay tablet that records remedies for various illnesses. The barks of certain trees were used as painkiller without knowing that it contained what we know today as aspirin's forerunner, salicylic acid. Throughout human history, natural products, including terrestrial plants, animal products, and marine organisms, and products of micro-organismal fermentation have been used in traditional medicines. The accidental conversion of grape juice to a pleasant drink, i.e., wine, is even recorded in the Bible during the Great Flood.

2.2.3 Improving on Nature through Knowledge

When chemists developed more investigative knowledge, such as analyzing natural products and determining the components responsible for their beneficial effects, the components were synthesized and made available in larger quantities and more economically to the population. In addition, chemists were able to improve on nature and develop more efficient components for various aspects of our everyday life, such as transportation, communication, medicines, and food.

2.3 The Basics of Survival

2.3.1 Inventiveness

Obviously, human activity through countless millennia was concentrated on survival. Hostile elements from weather and seismic conditions, predatory animals, other human and perhaps subhuman groups, illness, disease, accidents, lack of nutrition, food, clothing, shelter, and the basic necessities of life, all took their toll. It has been estimated that the average lifespan (discounting infant mortality) for a prehistoric human being was less than 30 years, and those who survived beyond that magic number did so minus teeth. And the magic number did not change very much even up to the time of the Roman Empire. So how do we account for the fact that puny little human beings continued in the ascendancy seemingly against all odds? Of course, the answer is brain power and, in particular, inventiveness with respect to the use of material objects so that virtually every material substance could contribute in some way to survival.

2.3.2 Control of the Environment

Survival depends, in large part, on how well an individual or a group can control its environment. Groups that lived in caves had a better chance than those who lived out on the open steppes. People who had fire managed better than those who did not. Those who had plentiful combustible material were more likely to survive than those who did not. In fact, archaeological evidence points to a key factor in the growth of a civilization: the hearth. The hearth was a source of warmth and protection from the elements. It was the place where food could be cooked or tools fashioned. Casting a look far into the future, we can see that the hearth was the birthplace of chemical thermodynamics, fuel chemistry, tribology, food chemistry, and so much more. And once people began to experiment with the mixing of various types of fuels with other substances, more possibilities opened up. One of these was gunpowder, one of the four iconic inventions attributed to Chinese ingenuity.[2]

2.3.3 How Gunpowder Changed the World

The invention of gunpowder is one of the classic examples of how a substance developed mainly out of curiosity and for entertainment purposes could morph first into an instrument of war and destruction, then into an instrument that made possible enormous building and construction projects, and finally into a means of rewarding world-class research leading to the betterment of the human condition. The history of gunpowder is a classic example of a substance that has both enormous positive and negative effects on society.

This explosive material, the first ever devised and the only one available for over a thousand years, redrew the map of the world, changed the nature of warfare, and set in motion events and theories that would change our view of the nature of reality. Gunpowder, of course, was not called by this name until after the invention of guns. Nor was it called black powder until after other explosives were devised that lacked the lustrous appearance of the free carbon it contained. But regardless of what we call it, this is a mixture of two elements (carbon and sulfur) and a compound (preferably potassium nitrate) capable of producing a big bang under favorable conditions: intimate mixing, proper grain size, and little or no moisture adhering to the grains. This mixture, in varying proportions, was invented in China before 1000 CE and due to its startling new relationship to fire, it found uses there for both driving off evil spirits and in pyrotechnic entertainment.

Gunpowder, which represented one of the greatest advances in the manipulation of natural materials, remained a deep mystery for many centuries, largely because chemistry, which would give a theoretical and practical basis to the action of gunpowder, was still a nascent body of knowledge with a long way to go. The first known recipe for gunpowder comes from a 1249 CE work of alchemist-Franciscan monk Roger Bacon, though it is thought that his information came from the Far East by way of Arabic manuscripts. The explosive Bacon described was composed of six parts of saltpeter to five parts of charcoal and five parts of sulfur. Later versions used more saltpeter, eventually stabilizing at a ratio of 15:3:2, which is close to the stoichiometric ratio of 15:2.7:2.4.

Gunpowder works by the rapid burning of sulfur, which requires the lowest ignition temperature, and then of carbon at higher temperatures, in oxygen released by the thermal disintegration of the potassium nitrate. Hence the designation "low" explosive that reacts by deflagration as opposed to "high" explosives that react by detonation (and not to be invented until the 19th century). The resulting hot gases (nitrogen and carbon dioxide) expand so rapidly that they push anything they come across—including cannon balls and castle walls—out of the way.

And thus the lines of development of the Chinese entertainment industry and the European military arts converged. Almost overnight, military history became the history of the use of this explosive new mixture in new and more violent ways. Perhaps no one invention has revolutionized the conduct of warfare as has gunpowder. War moved from hand-to-hand combat to killing at a distance; castle walls, once resistant to the most persistent siege, were the easy prey of cannonballs. And those who prevailed were those who grasped the import of this powerful new weapon and did not scruple to use it to the utmost.

So, is gunpowder all bad? Is it used only to destroy? Actually, this magical and incendiary substance also became the focus for scientific investigation, and its manufacture and use gave rise to technological advances of a multidisciplinary nature. For good or for ill, gunpowder has influenced the development of engineering, metallurgy, physics, chemistry, trade, industry, societal structure, and economics. Perhaps the most important, and yet the least obvious, impact was its influence on the theory and nature of reality. Study of the reaction of gunpowder led to the realization in 1665 CE that fire is not an element, thus overthrowing the millennia-old "four element" hypothesis of Aristotle (384–322 BCE). Thoughts about the driving force of gunpowder eventually led Christiaan Huygens (1629–1695) to the idea of the internal combustion engine, and on the basis of its action, Robert Hooke (1635–1703) laid the foundation for the first coherent theory of combustion, an idea that Antoine Laurent Lavoisier (1743–1794) would later, and more famously, exploit in the overthrow of the phlogiston theory.

So, as with most other things in life, gunpowder shows us two faces: It was a momentous and irreversible milestone in the course of history that became a focus of scientific investigation and, at the same time, fueled the slave trade. It is the energy behind those "leaden messengers that ride upon the violent speed of fire,"[3] and at the same time, it is that magic substance that gives delight to millions of spectators at pyrotechnic displays. With the advent of the high explosives, such as nitroglycerine, and of nuclear explosives that can produce much bigger bangs, perhaps it is poetic justice that the major use of gunpowder today is that for which it developed in China a millennium ago: the source of oohs! and ahs! on major celebratory events.[4]

But wait! What about gunpowder's first cousin, dynamite? Invented by Alfred Nobel (1833–1896) in 1866, dynamite was the first high explosive that could be safely managed and controlled. Its use led to major building projects that would otherwise have been prohibitive, such as the Panama Canal and the Seward Highway. Dynamite made Nobel a very wealthy man, but its destructive use in so many quarters of the world horrified him so much that he established a legacy meant to counter such use: the Nobel Prizes. According to Nobel's will, these prizes were meant to be bestowed on the individuals whose work conferred "the greatest benefit to mankind." One glance at the major advances in medicine, physiology, physics, and chemistry that have been recognized by the Prize makes one realize how the human will directed toward good can counterbalance perverted and evil purposes.[5]

2.4 Control of the Necessities of Life: Food

2.4.1 In the Beginning

We know that our ancient forebears were, by necessity, hunters and gatherers. They were not free to determine their diet; rather, their diet was determined by environmental factors and the fact that humans derive from a line of higher primates able to use a wide range of plant and animal foods.[6] This versatility with respect to diet has been verified by the archaeological record where methods such as stable isotope ratio analysis can determine the presumed proportion of vegetable to animal content in selected sites and over a vast range of time. Admittedly, these data are sketchy and also must take into account deterioration of organic materials over periods ranging from hundreds to thousands of years.[7] However, the evidence seems to point to a daily diet basically composed of plant foods with an occasional intake of animal matter as the opportunity arose. This is hardly the picture painted by the proponents of today's "caveman" or "paleo" diet, which places an emphasis on meat and seafood.

Figure 2.1   One invention leads to another. Reproduced with permission from Thermochemistry (module), p. 64. *SourceBook*, a section of *ChemSource*. Mary Virginia Orna, editor and copyright holder. With the permission of the editor/copyright holder.

2.4.2   Agriculture

That food has played a major role in the evolution of civilization has never been disputed. A community can continue to exist as long as there is enough food; a community will flourish only if there is a surplus of food to support non-food producers, practitioners of technologies such as artisans, metal workers, weavers, bakers, potters, and those who record everything: the scribes. Such growth was only possible with the advent of agriculture, made possible in turn by the invention of the plough.[8] And once there was a surplus of food, means to preserve and store it had to be developed. All of this technology depended on chemistry, although chemistry as a science was still a long way off in the future. Another concern as human communities grew larger was lack of sanitation; often, raw sewage was dumped into the community's source of drinking water, giving rise to illness and disease. A chemical process to overcome this difficulty was the development of the art of brewing; sugar- and starch-containing grains and fruits fermented in the presence of (wild) yeast, and the resulting alcohol content of the brew rendered it fit to drink by killing off the microorganisms contained therein.

Of course, the pleasant effects of such fermented drinks was not lost on the practitioners either. Let us examine just one of these technologies, food preservation, based mainly, in ancient times, on the use of salt.

2.4.3   Salt

The Wadi Natrun is a valley in Egypt containing several lakes, many of which shrink considerably in the summer, leaving behind evaporates rich in natron. The deposits are a mixture of sodium salts, primarily carbonate. The chemical symbol for sodium, Na, derives from this word, *natrun* in Arabic, *natron* in Greek, *natrium* in Latin. Natron was an important commodity in Ancient Egypt, finding use as a cleaning agent for body, teeth, and gums. Because it can absorb water from organic cellular material, it was used extensively for mummification and this same desiccating property made it useful for the preservation of meat and fish. The more common sodium chloride became the desiccant of choice worldwide and was much sought after in the ancient world. In addition, salt is essential to the diet of herbivores and, therefore, essential to farmers who raise domestic animals. When meat or fish is immersed in salt, the salt gradually removes the water from the organic material by the process of osmosis, thus killing the bacteria liable to cause the foodstuff to rot. Food can be preserved for long periods of time when salted, and so salted food can be transported long distances without fear of spoilage. Before the advent of refrigeration and other food preservation methods, salt was the only available choice.

Salt deposits can be found in many parts of the world either in solid form such as salt domes that can be mined, as brine wells that can be pumped out, or as shallow sea beds that can evaporate to produce a solid. The salt is seldom pure; it consists of several sodium compounds as well as small fractions of potassium, calcium, and magnesium compounds. Salt mining is a major industry because presently salt is one of the most important raw materials in manufacturing plastics, paper, soap, synthetic rubber, and many other products. These latter day uses of salt are a far cry from dietary supplement and food preservative, and could only come about through our understanding of the nature of salt through chemistry.[9]

"OUR PLAN IS TO EXTRACT SULPHATES, BROMIDES,
COPPER, SILVER AND GOLD FROM SEA WATER. ALL
WE'VE MANAGED TO GET SO FAR, HOWEVER, IS SALT."

Figure 2.2   Everything is relative, even the value of a chemical commodity. Reproduced with permission from The Chemistry of Seawater (module), p. 40. *SourceBook*, a section of *Chem Source*. Mary Virginia Orna, editor and copyright holder. With the permission of the editor/copyright holder.

2.5   Control of the Necessities of Life: Clothing

2.5.1   Introduction

The manufacture of silk may not involve chemistry as directly and obviously as the manufacture of porcelain or bronze, but chemistry comes into the process at many stages from raising mulberries and silkworms to printing and dyeing—not to mention the chemistry carried out by the silkworms in making the fibers, chemistry that humans have attempted to mimic with only limited success. Silk is obviously a coveted textile fiber that people will take a lot of trouble to produce—and why? Due to its molecular pleated-sheet structure, clothing made of silk provides coolness in summer and warmth in winter. It became such a major Chinese export in the Middle Ages that the road by which it arrived in Europe was called the Silk Road. Other natural fibers used in clothing are wool, flax, cotton, cashmere, camel hair, hemp, and almost a dozen others. Each one has played a role in history and finds uses today despite the development of synthetic fibers. Animal hides were also used as clothing from the earliest times and, in fact, were probably the easiest, and hence earliest, form of clothing for prehistoric humans. Chemistry entered into their use when the hides began to be treated with oak bark to create a more durable substance; the ancient tanners ("*tannum*" is Latin for oak bark) did not know that they were chemically transforming the protein chains in the hides by cross-linking with the polyphenols in the oak bark to produce durable, flexible leather.[10]

2.5.2   Nature of Polymers

All fibers are polymeric materials, that is, their molecules are made up of repeating subunits that form long strands that confer properties on the aggregate that are quite different from those of the individual units. For example, wool consists of long chains of amino acids, small molecules linked together by virtue of what is called a peptide linkage. Each amino acid has its own chemical properties that may render it acidic, basic, or neutral, or hydrophobic or hydrophilic. These individual properties determine the overall behavior of the wool fibers with respect to receptivity to dyes, behavior in water, flexibility, and so forth. The polymer subunits in plant fibers are carbohydrates (sugars) as opposed to amino acids, and therefore, plant fibers have very different properties from animal fibers. For countless of years, the materials that made up human clothing were confined to either animal hides or to the natural fibers available in a given area. It was only in the 20th century that advances in polymer science gave rise to synthetic fibers such as nylon, rayon, polyesters, or Kevlar. Understanding the properties of natural fibers and developing the technology to produce synthetic fibers belong to the world of chemistry.

2.5.3   Coloring Fibers

It would be a drab, drab world and the fashion centers of Paris and New York would not exist if chemists had not succeeded in coloring the fibers that go into clothing manufacture. While many techniques existed and were well-documented, the versatility of dyestuffs was limited by availability and in some instances to the difficulty of obtaining the colorant. The only natural blue organic colorants in antiquity, often used as dyes, were indigo and woad, both containing the identical coloring matter, indigo, or indigotin. It was probably the oldest coloring matter known. The chemical identity of this colorant was, of course, unknown until the advent of modern chemistry, and both materials were thought to be distinctly different from one another. Indigo is, to this day, one of the few naturally occurring dyes in wide use, principally to color blue jeans. Carmine is a red coloring material that has been used from ancient times in both hemispheres. In the Old World, the source of the red color was the kermes female scale insect (*Kermes vermilio*), a parasite that lived chiefly on the kermes or scarlet oak (*Quercus coccifera*). It has been known at least since the days of Moses; the color is mentioned in Exodus 26:1 in the directions for making the curtains for the Temple. It was used widely by color-makers in ancient Greece and Rome, and since it was abundant on the Iberian Peninsula, the Spaniards paid half their tribute to Rome with kermes grains. In the New World, the source of the color was the wingless female scale insect *Dactylopius coccus*, indigenous to Mexico, Central and South America, and found chiefly on two host plants, the prickly pear and torch thistle cacti. It was in use as a textile dye from at least the old Peruvian Paracas culture that dates back to 700 BCE. Carmine cochineal coming from the New World was a very important Spanish import to the point where cochineal-laden galleons were the prime targets of pirates on the high seas. It achieved fame by its use as the colorant for the British redcoats and, ironically, for the Star Spangled Banner.

2.5.4 Emergence of the Dye Industry

The use of dyestuffs in local and cottage industries was, with some few exceptions, based on what materials were available locally. Coloring fabrics was time-consuming and expensive, to say nothing of the fact that the colors produced had a very limited range. Large-scale use of major natural dyestuffs was made possible by widespread cultivation of the pertinent plant species such as the madder and indigo plants for the dyestuffs themselves, and cactus plants as hosts for the scale insect, *Dactylopius cacti*, that fed upon them. As the appetite for these natural products grew, European entrepreneurs attempted to either improve upon the natural or to synthesize dyes in the laboratory. One of the more successful innovators, William Henry Perkin (1838–1907), began his research largely in the dark and with a very different purpose in mind. Perkin's mentor at college had a fertile research mind that suggested a problem that the precocious Perkin could dig into: do humankind a favor by finding a synthetic pathway to quinine, a much sought-after drug for the relief of malaria and some parasitic ailments. Hence, in 1856, when he began to try to synthesize quinine as it were by brute force—treatment of his starting materials with strong acids and powerful oxidizing agents—he was just plain lucky that they rearranged themselves into something useful, a magnificent purple coloring material, and he was very astute in recognizing that fact. It was not long before this color, initially called "mauve," became the most desirable color in the fashion houses of London and Paris and quickly spawned a long list of other dramatic synthetic colors over the following decades.[11]

Figure 2.3   What is the difference between a color and a hue? Reproduced with permission from The Chemistry of Color (module), p. 54. *SourceBook*, a section of *ChemSource*. Mary Virginia Orna, editor and copyright holder. With the permission of the editor/copyright holder.

By the end of the century, in 1897, Adolf von Baeyer (1835–1917), professor of chemistry first at Strasbourg and later at Munich, succeeded in finding a commercially viable pathway to synthetic indigo. The demand for natural indigo withered dramatically, and in fact, by about 90% within a period of only 15 years. This dealt a lethal blow to Great Britain's Indian monopoly, rendering a lucrative colonial possession into a dependency in a very short period of time. In fact, by 1914 only 150,000 acres of Indian soil were given over to indigo cultivation as compared with about 1,700,000 acres at the time of synthetic indigo's appearance in 1897. The economy in the United States was unaffected by this chemical development because, although indigo had been a viable cash crop in the Carolinas and Louisiana in the 18th century, cotton replaced it in the 19th century as commercially more successful.

By 1900, the dye industry had matured, at least in Germany, in such a way that its influence was to say the least, historic. It did not spring to life unexpectedly and in a vacuum, but it evolved with the help of and to meet the needs of other industries such as heavy chemicals, illuminating gas, and textiles. It also owed its existence to the early pioneers in organic chemical research such as Justus von Liebig (1803–73) and his students at Giessen. It grew rapidly because of the capitalistic system with intense competition among dye companies, investment of venture capital 19th century style, and the freedom to embark on risky ventures with great rewards if they met with success. The industry was embedded in an economic system that, in enticing greater consumption, led to frantic research, which led in turn to an increased pace of organic chemical discovery and to innovations particularly in process.

Here are some of the areas that were completely transformed by the dye industry:

2.5.4.1   Fashion

Prior to aniline dyes, color austerity was the order of the day: Only the rich could afford the rainbow of colors available in the natural dyes. But the ordinary person's desire for the riot of color, now inexpensive and available to all, coupled with a clarity, variety, and stability not found in the natural dyes, greatly stimulated dye research.

2.5.4.2   The educational structure of industrial society

The needs of the dye industry stimulated educational reforms that were already under way at the university level so that at the end of the 19th century, Germany had the finest system of scientific and engineering education worldwide, and this system influenced higher learning in all parts of the world.

2.5.4.3 The social structure

By employing large numbers of academically trained chemists in its plants and by growing a white-collar managerial proletariat, the dye industry displaced the independent tradesman and raised the educational level of the masses; this carried over into other disciplines as well.

2.5.4.4 Political action

The dye industry organized and supported lobbies, exerted strong influence on legislation, and helped establish a model patent system.

2.5.4.5 Industrial research

The dye industry originated and developed the industrial research laboratory and the research team, an organizational structure that prevails to this day.

2.5.4.6 Power

Through its hegemony in the dye industry, Germany became the foremost industrial power in Europe, leading to its almost becoming the foremost power in Europe. World War I was in actual fact a battle of technologies that promoted growth in other branches of the chemical industry, while the dye industry itself declined in importance.

Not all of these developments were unequivocally positive. Through the dye industry, Germany became a giant war machine, manufacturing explosives, poison gases, photographic film, drugs, natural product substitutes, and a whole host of other products born of its technology. The industrial juggernaut also led to massive pollution of the air and of water sources; it led to a scarring of the earth; above all, it led to an unprecedented increase in chronic diseases, to deaths by heavy metal poisoning, to birth defects, and so much more. The rivers that ran red with industrial effluent also ran red with blood. We might say that the developments in the European dye industry of the late 19th century steered the course of 20th century history right up to and beyond World War II.[12]

2.6   Control of the Necessities of Life: Shelter

2.6.1   Introduction

The unprecedented mushrooming of synthetic dyes and pigments, the theoretical developments that gave rise to structural characterization of molecules, and the ability to relate color and constitution of chemical substances were a great boost to the subdiscipline we call organic chemistry. Growth in the areas of structural materials, particularly wood, stone, metal, and glass, paralleled these developments. Structural clay, wood, stone, and metals were, of course, used for other purposes long before they became incorporated into dwellings, mainly as tools and weapons. For example, there is ample archaeological evidence that flint was an important commodity from Neolithic times, when flint knapping was developed into a fine art.[13]

2.6.2   Clay, Stone, and Wood

Structural clay is soft clay material that has been formed into bricks or blocks and then fired. The firing process is a chemical process that transforms the clay, typically a hydroxy aluminosilicate such as kaolinite, through a series of solid phase transitions into a vitrified cristobalite mass.[14] Its availability, ease of working and transformation, and stability rendered it an ideal building material from prehistoric times up to the present time. **Stone**, equally available as a building material, provides us evidence of this use without the need to excavate. One has only to look at the famous Egyptian step-pyramid of King Djoser, the oldest extant monumental structure made of stone, dating from 2800 BCE, or at the more famous structure at Stonehenge, dating from 2600–2000 BCE. How does chemistry interface with stone? Not necessarily by changing it, but by determining its nature and structure, and thereby seeking for ways to preserve it or improve upon it. Chemists now know that stone comes in three varieties: silicious, consisting mainly of silica, SiO_2 (quartz is an example); argillageous, consisting mainly of alumina, Al_2O_3 (clay and slate are examples); and calcareous, mainly composed of lime, calcium carbonate, $CaCO_3$ (limestone and marble are examples). We now know from the nature of marble that it needs to be protected from acid rain. Artificial or engineered stone is now commonplace, especially in areas where there is a scarcity of stone, such as China. Modern cast stone is used as a structural material because of the ease of molding it into whatever size and shape is necessary for the building. Though **wood** has been a structural material from time immemorial, both for dwellings and for boats, it also had a number of other uses because it is combustible. It has always served as a fuel, but since charcoal was so necessary for the iron and glass industries of the early modern era, there was a timber crisis in Europe by the end of the 16th century. Natural wood's use as a building material has always had its limitations. Supply never kept up with demand because it takes much less time to chop down a tree than to grow another one. Tree size also limited the measure of some buildings. For example, the California Mission churches are long, but very narrow because their width was determined by the available timber that made up the roof beams. Natural wood differs in its chemical composition from tree species to tree species, but generally it consists of three major compounds: cellulose, hemicellulose, and lignin. When wood is used to make high-quality paper pulp, the lignin, a complex polymer of aromatic alcohols, must be removed because it is responsible for the yellowing of paper. Considering that the vast majority of printed matter in libraries around the world is made of paper made from wood pulp, an intimate understanding of wood chemistry, hence paper chemistry, is necessary for paper conservation. Chemistry in our own day has taken a hand in the manufacture of engineered wood.

Figure 2.4   Step-Pyramid of King Djoser, Saqqara, Egypt. Photograph by Dennis G. Jarvis, under license by Creative Commons (http://commons.wikimedia.org/wiki/File:Egypt-12B-021_-_Step_Pyramid_of_Djoser.jpg).

2.6.3 Metals

For thousands of years, human beings were dependent on wood, stone, and clay. However, metals quickly took their place for many useful items, particularly tools and weapons, as smelting techniques developed. The only metals that occur in the free state are gold and silver, and occasionally copper. All three seem to have been discovered between 9000 and 7000 BCE in the form of nuggets or lumps that, unlike stone, did not chip, but deformed and changed shape like clay objects. All other metals are concealed in their ores, often as oxides, and the relationship between a metal and its oxide is not apparent by mere visual observation. Gradually ancient workers found out how to smelt copper by mixing copper ores with wood or charcoal in a firepit.[15] If the fire were hot enough, the carbon in the wood or charcoal would combine with the oxygen in the copper ore to produce carbon dioxide or carbon monoxide, releasing elemental copper in the process. The order of production of free metals over the centuries follows the order of the magnitude of the free energy of formation of the metal oxides, which becomes less negative as the temperature increases. At the same time, the free energy of combustion of carbon to either carbon dioxide or carbon monoxide is inversely proportional to the temperature, that is, as the temperature increases, carbon becomes a better reducing agent. The temperature at which the curve for metal formation and carbon combustion meet is the temperature at which one can obtain the free metal. These curves have been summarized in a so-called Ellingham Diagram (plot of the standard free energy of formation of the metal oxide versus temperature), which allows one to estimate the temperature necessary for the successful smelting of a metal.[16] A glance at the diagram indicates that the free energy of the formation of metal oxides follows the electrochemical series. It is no wonder that elemental gold, silver, and copper appeared first, the latter around 9000 BCE in Western and Central Asia. Next came tin and lead, neither of which had much impact on civilization. However, since copper and tin constitute the alloy bronze, a harder more durable metal than either of its constituents, the bronze age dawned at around 3500 BCE with the production of many useful artifacts such as saws, nails, and weapons. Next in the procession of free metals came iron, liberated in blast furnaces at temperatures exceeding 2000°C, thus initiating the industrial revolution, but produced at much lower temperatures as so-called "bloomery" iron and lending its name to the Iron Age that dates to about 1200 BCE. Iron alloyed with about 1% carbon is called steel, and it is this alloy that could eventually be produced cheaply enough to become the structural material of choice for buildings and for bridges or for any structure that demanded great strength. None of these developments would have been possible without chemistry, initially for understanding the nature and properties of the metal, how best to produce it, and then how best to modify it for useful purposes.

The iconic symbol of the industrial revolution was the Ironbridge over the River Severn in England erected by Abraham Darby III in 1779. Though other metals are used for structures, such as aluminum and titanium, chiefly as alloys, iron as steel remains the kingpin of the construction industry.

Figure 2.5 Ironbridge at Ironbridge Gorge, Shropshire, UK. This was the first bridge made of iron ever to be erected (http://commons.wikimedia.org/wiki/File:The_Iron_Bridge_%282772205816%29.jpg).

2.6.4 Glass

Our last key structural material is glass, a substance that many people contend actually changed the world. Think for a moment about the many objects made of glass that we take for granted. Eyeglasses have extended the intellectual life of millions by approximately 40 years. It would be unthinkable to lack glass windows in domestic and public buildings unless you could substitute a more modern synthetic material. But humans did not always have glass at their beck and call.

Glass origins are shrouded in myth and mystery. Pliny the Elder reports that it was discovered on the shores of the Eastern Mediterranean by some sailors carrying a load of niter who used blocks of said material to build a cooking fire on the sand, thus providing the flux to high-melting silica that would produce the molten transparent semi-liquid we call glass. Another tradition says that glass was accidentally produced by heating sand with kelp, a seaweed whose ash contains large amounts of sodium carbonate, and voilà! Whatever the origins, one must look with skepticism on some of these easy ways of producing a substance that requires temperatures far higher than ordinary cooking or camp fires to produce.

Apart from the myths, archaeological finds indicate that glass appeared somewhere in Mesopotamia and the technique of producing it spread throughout the Mediterranean region. Trial and error gradually dictated the necessity of mixing silica (SiO_2) with a fluxing agent that would allow it to melt at lower temperatures, normally soda (Na_2CO_3), and a stabilizing agent, lime ($CaCO_3$), without which the glass would simply dissolve in water.[17] Soda-lime glass seems to be among the earliest glasses known, and the appearance of glass itself can be dated to the second millennium BCE.

The Romans took up the trade of glass-making sometime around 300 BCE. Their first techniques involved casting glass into molds, but gradually their methods became more refined, and these were spread throughout the Mediterranean world. As more and more glass artifacts were found, archaeologists thought that they were objects of trade finding their way into northern Europe by the normal trade routes. Subsequent chemical analysis of many of these objects indicated that the glass artifacts were actually of local origin. So not only the artifacts, but also the art of glass-making was a very portable commodity.[18]

Roman glass-making declined during the period of the barbarian invasions. Meanwhile, refugees from mainland cities of northeastern Italy managed to tame the salt marshes at the head of the Adriatic and develop, over several centuries, into a seafaring nation optimally positioned to dominate the trade between the East and the West. Thus Venice, in its ascendancy, had access to the high-quality materials from Egypt and the Levant that would allow it to develop unique glass materials that were sturdy, colorless, and extremely thin. At one point, Venice was a city in which, from end to end, glass-making furnaces were lighting up the sky.[19]

Glass technology took off from Venice so that today, wherever the raw material of silica is readily available, you have glassworks, and these glassworks specialize in glasses that can be hammered, sawed, bent, and shaped in any way, shape, or form desired by the architect to make an ideal structural building material. The use of glass in public buildings and office complexes has steadily increased over the past few decades, and the trend looks set to continue. Glass is an inexpensive material, which offers many superior properties in different applications. It is environmentally friendly and fully recyclable, an increasingly important consideration with the growing emphasis on lifespan thinking. There is a strong trend in modern architecture toward transparent structures, which allow natural light to enter buildings and on other hand open up the natural landscape to end-users inside. Glass creates an airiness, provides a sense of space that can only be achieved with larger glass sizes and in lighter support structures. All of this is possible today through the chemistry that can create such a versatile material.

2.7 Communication

Although Chapter 11 is dedicated to the role of chemistry in communication technology, I would like to offer a timeline here based on the one found in Wikipedia[20] but with several modifications. If one examines the modes of indirect communication developed over the centuries, we can break them down into (a) symbolic, (b) graphic, and (c) electronic modes. Hence, Tables 2.1–2.3 will trace the development of each of these modes and the chemistry employed in them.

Table 2.1   Symbolic communication timeline

Date	Mode of communication	Chemistry involved
Before 3500 BCE	Communication through paintings on walls in caves	Use of pigments; first archaeological evidence of a chemical process other than combustion
26–37 CE	Roman Emperor Tiberius governs the empire by signaling messages by metal mirrors from Capri to the mainland	Ore smelting to produce pure metals that could be burnished to a smoothness necessary to the task
1520 CE	Ships on Ferdinand Magellan's voyage signal each other by firing cannons	Chinese invention of gunpowder (see Section 2.3.3)
1792 CE	Claude Chappe establishes first long-distance semaphore telegraph line	Wooden rods at varying positions conveyed the information; wood had to be worked and protected chemically

Table 2.2   Graphic communication timeline

Date	Mode of communication	Chemistry involved
Before 3500 BCE	Communication through paintings on walls in caves	Use of pigments; first archaeological evidence of a chemical process other than combustion
Ca. 3500 BCE	Sumerians develop cuneiform writing and the Egyptians develop hieroglyphic writing	Cuneiform was inscribed on clay tablets (chemistry described in Section 2.6.2); Egyptians wrote on stone or papyrus (chemistry described in Section 2.1.4)
Ca. 1600 BCE	The Phoenicians develop an alphabet	Unknown writing surface, but necessary chemical intervention
105 CE	Ts'ai Lun invents paper in China	Pulped rags, hemp, plant fibers, and silk were manipulated polymers

(*Continued*)

Table 2.2 (*Continued*)

Date	Mode of communication	Chemistry involved
7th Century CE	Hindu-Malayan empires write legal documents on copper plate scrolls	Metal-working technology
751 CE	Paper is introduced to the Muslim world after the Battle of Talas	Paper-making technology
1305 CE	The Chinese develop wooden block movable-type printing	Woodworking technology
1450 CE	Johannes Gutenberg finishes a printing press with metal movable type	Metal-working and alloying technology
1844	Charles Fenerty produces paper from wood pulp	Polymer manipulation

Table 2.3 Electronic communication timeline

Date	Mode of communication	Chemistry involved
1831 CE	Joseph Henry proposes and builds an electric telegraph	In 1800, Alessandro Volta built the first battery, making electric current available through a chemical reaction. In 1831, Michael Faraday and Joseph Henry simultaneously discovered electromagnetic induction, the principle behind the electric telegraph.
1843 CE	Samuel Morse builds the first long-distance electric telegraph line	Electrochemistry and electromagnetic induction

The remainder of communication advances are documented in Chapter 3, remembering that if it were not for the advent of the voltaic pile (battery) and the development of advanced synthetic materials, both products of chemistry, we would not have the communication devices, such as computers and cellular telephones, that we enjoy today.

2.8   Measurement

2.8.1   Introduction

For practical reasons related to trade and commerce as well as survival, measurement has been very important from the dawn of pre-history. The fundamental units of measure from the beginning were recognized as length, mass, and time. The human body lent itself very nicely for measuring length: the foot is obvious; the lick is the distance between the tip of the thumb and the tip of the index finger; the hand, once five fingers in width, is now four, and used to measure the height of horses. For measurements of weight, the human body provides no such easy approximations as for length. But nature steps in. Grains of wheat are reasonably standard in size. Weight can be expressed with some degree of accuracy in terms of a number of grains—a measure still used by jewelers. As with measurements of length, a lump of metal could be kept in the temple as an official standard for a given number of grains. Copies of this could be cast and weighed in the balance for perfect accuracy. But it is easier to deceive a customer about weight, and metal can all too easily be removed to distort the scales. An inspectorate of weights and measures was from the start a practical necessity and has remained so in all societies. The preservation of the standard kilogram at Sèvres, France, is a continuation of the temple practice. Various methods of measuring time were developed over time (excuse the pun), a favored method being the sundial. Timekeeping assumed great importance in medieval monastic society, and for many centuries, the church was the master timekeeper. Much later, construction of meridian lines allowed cities to signal to the populace the precise time the sun crossed the meridian each day, and observation over long periods allowed for prediction of solar movement and eventually prediction of eclipses. Development of the technology of making clocks and watches democratized time, and today, one can purchase a perfectly good timepiece for the price of a sandwich. Where does chemistry fit in with measurement? A few examples will suffice.

2.8.2 Quantitative Experimentation

In the 17th century, Johann Baptista van Helmont (1579–1644) conducted a famous experiment that historians of science say was far ahead of its time. Van Helmont grew a willow tree in a clay vase with 200 pounds (91 kg) of soil. After 5 years, he dried the soil and found that its weight had decreased by only 2 ounces (0.06 kg): Water alone had, therefore, been sufficient to produce 160 pounds (73 kg) of wood, bark, and roots (plus fallen leaves that he did not weigh). Presumably, he claimed, there were minerals in the water he fed to the tree. Today, we know that plants form carbohydrates from atmospheric carbon dioxide, but their mineral content is derived from soil, not air. Lacking controls, it is difficult to presume the origin of the necessary mineral content for plant growth. However, what is remarkable about this experiment is the fact that he measured quantities both before and after he completed his work; that he measured at all is simply incredible because measurements in alchemy, which was the "science" of the day, were not considered important, nor was there any standard way to perform them. This relatively simple experiment foreshadowed the concept of conservation of matter, a fundamental tenet of all chemistry.

2.8.3 Conservation of Mass

In 1786, Immanuel Kant (1724–1804) wrote "Chemistry can become nothing more than a systematic art or an experimental doctrine, but never a true science, because its principles are merely empirical and...are incapable of the application of mathematics."[21] Despite Kant's rigid view, chemists have managed to formulate a number of general laws, the first of which was the Law of Conservation of Mass in 1774, generally attributed to the French scientist Antoine Laurent Lavoisier (1743–94). Although recent literature has called into question Lavoisier's originality, particularly with respect to his being credited with the formulation of the Law of Conservation of Mass,[22] one must admit that he accomplished his revolution by a consistent application of the principle of conservation of weight as a way of determining and confirming the results of chemical experiments. He was the first to make it a systematic instrument of experimental investigation and confirmation.[23] Justus von Liebig (1803–73) said of him "He discovered no new body, no new property, no natural phenomenon previously unknown. His immortal glory consists in this—he infused into the body of science a new spirit."[24] And it was only through the development of a system of chemistry with logical interrelationships and accurate qualitative and quantitative analyses that a considerable number of pure substances could be identified and incorporated into a consistent whole. In 1785, Lavoisier stressed that his published results were based on repeated weighing and measuring experiments that were the only criteria for admitting anything in physics and chemistry. It is well-known that he spent a fair amount of his fortune on the best scientific apparatus that money could buy, often designing the apparatus himself and then having it purpose-built by a specialist. Lavoisier's insistence on the need for precision instruments was not lost on those who followed him, although some of his pieces of apparatus were so unique and so complex that scientists of lesser means were forced to improvise and, therefore, sacrifice both accuracy and precision.

2.9 Chemistry's Role in the Development of Medicine

2.9.1 From the Medical School in Salerno to Universities

As the influence of Islam spread throughout the Mediterranean, Western medicine recovered ancient medical theories, while at the same time learning from the clinical and educational developments of Arab medicine. The first medical school in Europe was established at Salerno (9th century), originally as a monastic dispensary. It was an outstanding medical institution and a cultural model for medical training. Universities began as centers of secular education, and medical faculties developed from the 12th century onward at Montpellier, Bologna, and Paris. The great novelty of medical curricula was the teaching of anatomy.

2.9.2 The Body in Movement

Girolamo Mercuriale (1530–1606) emphasized the importance of exercise as a means of preventing or curing orthopedic illnesses.

In De arte gymnastica,[25] he identified three kinds of gymnastics (military, medical, athletic), which, if correctly practiced, could help cure physical defects. Girolamo Fabrizio d'Acquapendente (1537–1619) studied the motor system and the treatment of deformities of the limbs. He also built the first anatomical theater, in the shape of an amphitheater, in Padua in 1594. Anatomical dissections became an integral part of the course in anatomy and surgery. Via anatomical studies, D'Acquapendente discovered the valves of the veins and their positioning so as to obstruct the flow of venous blood toward the peripheral parts of the body.[26]

2.9.3 Healing the Body, Healing the Soul: Hospitals

From the 15th century onward, hospitals increasingly began to act as institutions for healthcare although they did not lose their original function as places of "hospitality" for the poor, foundlings, and foreigners in difficulty. Provisional emergency infirmaries known as lazar houses were set up in times of plague both to treat the sick and isolate them from the rest of the population in an attempt to stop the spread of the disease. In the 16th century, the spread of syphilis led to the creation of specialized hospitals known as "incurables." During the 17th century, hospitals increasingly became centers for medical research. Military medicine became a scientific discipline in the 18th century. Camps were equipped with first-aid units, and the patients were then transferred to hospitals set up in nearby towns. After being washed in boiled water, wounds from firearms and sharp weapons were stitched and bandaged. The development of new techniques and instruments made it possible to avoid amputation by applying tourniquets.

2.9.4 Anesthetics and Antiseptics

Modern dentistry began with dental surgery. The use of nitrous oxide as anesthetics by Horace Wells (1813–48) in 1844[27] and of ether by William T. G. Morton in 1846[28] in the painless extraction of tumors revolutionized dental surgery. New materials led to developments in prosthetics. X-rays, campaigns for dental hygiene, water fluoridation, and the discovery of the mechanism of tooth decay and treatment followed swiftly.

Subsequent to the demonstration by Ignaz Semmelweis (1818–63) of the effectiveness of washing of hands with chlorinated water before a gynecological examination so as to avoid infection, and to the studies of Louis Pasteur (1822–95) on the germs involved in putrefaction, Joseph Lister[29] (1828–1912) introduced chemical antisepsis by means of carbolic acid in the treatment of wounds in 1869. Asepsis based on sterilization by steam, pioneered by Ernst von Bergmann (1836–1907), led to a radical improvement in surgical practice.

2.9.5 Laboratory

The laboratory became central to clinical progress with Claude Bernard (1813–78). Experiments in vivo carried out in accordance with the principles of the exact sciences made it possible to study the physiopathological processes and metabolism of organs, tissues, and cells. The idea of homeostasis, the organism's capacity for self-regulation, was foreshadowed. Physiology became diversified into the independent sectors of biochemistry, endocrinology, and neurophysiology.

2.9.6 Pharmacology

Modern pharmacology was born with organic chemistry. The application of chemistry in physiological studies in the 19th century led to experimental pharmacology with the isolation of pure substances and the synthesis of organic substances and agents with medicinal effects. While the biotechnological production of drugs became possible with the advent of molecular biology, investigation of the genes' controlling individual reactions to drugs became the basis of pharmacogenomics and its idea of tailor-made medication.

Here are some steps along the way:

- Jokichi Takamine (1854–1922) crystallized the hormone adrenalin in 1901.
- Edward Calvin Kendall (1886–1972) isolated thyroxine in 1914.
- Frederick Grant Banting (1891–1941) and Charles Herbert Best (1899–1978) demonstrated the hypoglycemizing action of pancreatic extracts in 1921.
- Bertram Collip (1892–1965) succeeded in purifying insulin in 1922 after demonstrating its effectiveness on dogs with their pancreases removed.

- Gregory Goodwin Pincus (1903–67) demonstrated in 1950 that progesterone can be used to prevent ovulation in rabbits.
- Carl Djerassi (b. 1923) and George Rosenkrantz produced norethindrone, a synthetic progesterone usable as an oral contraceptive in 1951.

The juice of the opium poppy and its derivatives have been used to counter pain since ancient times. F. W. A. Setürner isolated morphine from opium between 1805 and 1817, and a period of the use of analgesics administered by the newly invented hypodermic syringe began. Albert Niemann (1834–61) isolated cocaine from coca leaves in 1860, and its anesthetic properties were described by Karl Koller (1857–1944) and Sigmund Freud (1856–1939) between 1880 and 1884. William Stewart Halsted (1852–1922) demonstrated in 1885 that cocaine interrupts the conduction of sensory nerves and thus introduced the surgical technique of block anesthesia.

Figure 2.6   Sticking to the rules may be inconvenient. Reproduced with permission from *Medicinal Chemistry* (module), p. 59. *SourceBook*, a section of *ChemSource*. Mary Virginia Orna, editor and copyright holder. With the permission of the editor/copyright holder.

Pierre Paul Broca (1824–80), Jean-Martin Charcot (1825–93), Camillo Golgi (1843–1926), and Sigmund Freud studied, over the course of nearly a century, cerebral anatomic structures, nerve synapses and chemical mediators, receptors and their molecular regulation with a view to discovering how thought and memory are developed and preserved. Giulio Bizzozero (1846–1901) described the nerve cells as static "perennial cells" that do not divide. Research into the physiology of the neuron brought to light "neuronal plasticity" stimulated by the nerve growth factor, and identified at the molecular level chemical mediators and receptors of messages and cellular "channels" that open and close to transport them.[30] The relations between brain and mind have themselves become objects of investigation. Meanwhile, increasingly advanced instruments such as computerized tomography (CT), positron emission tomography (PET), and nuclear magnetic resonance (NMR) "photograph" the brain as it thinks or suffers.

The revolution in the treatment of mental illnesses began in 1949 with the discovery of the antimanic effects of lithium salts by the Australian psychiatrist John Cade, although "taking the waters" rich in lithium salts was known to be effective against "madness" from ancient times. The effectiveness of chlorpromazine as an antipsychotic drug was demonstrated over the next few years. Imipramine, the first anti-depressant, was introduced in 1955. The anxiolytic effects of benzodiazepines were established between 1957 and 1960. Synthesized in the mid-1970s, fluoxetine is the active ingredient of Prozac, the prototype of the modern selective serotonin reuptake inhibitors (SSRI), marketed in the United States since 1987.

2.10 Conclusion

The practice of chemistry is ubiquitous in human societies, and it has been from the beginning of civilization. Therefore, the chemical–historical "tourist" can find material for study wherever human material culture has been well preserved. It is hoped that this comprehensive catalog of how chemistry has impacted human civilization from the dawn of history will give the ordinary citizen an appreciation of the chemical enterprise as expressed in the mission and vision of the American Chemical

Society: "Improving people's lives through the transforming power of chemistry."

References

1. Fara, P. *Science: A Four Thousand Year History*. Oxford University Press: New York, 2009.
2. "'Four great inventions' at Olympic opening warmly-welcomed," *People's Daily* Online, August 15, 2008, http://english.peopledaily.com.cn/90001/90776/6476950.html.
3. Shakespeare, W. *All's Well That Ends Well*, III.2.
4. Kelly, J. *Gunpowder: Alchemy, Bombards & Pyrotechnics: The History of the Explosive that Changed the World*. Basic Books: Cambridge, MA, 2004.
5. www.nobelprize.org
6. Milton, K. Hunter-gatherer diets: A different perspective. *Am. J. Clin. Nutr.*, 2000, 71(3), pp. 665–667.
7. Burton, J. H. "Trace Elements in Bone as Paleodietary Indicators." In Orna, M. V. (Ed.), *Archaeological Chemistry: Organic, Inorganic, and Biochemical Analysis*; *ACS Symposium Series 625*. American Chemical Society: Washington, D.C., 1996; pp. 327–333.
8. Burke, J. *Connections*. Little, Brown, and Company: Boston, 1978; p. 10; p. 63.
9. Kurlanski, M. *Salt: A World History*. Walker & Co.: New York, 2002; pp. 41–46; 291–317.
10. Lambert, J. B. *Traces of the Past*. Addison-Wesley: Reading, MA, 1997; pp. 147–152.
11. Garfield, S. *Mauve: How One Man Invented a Color that Changed the World*. W. W. Norton: New York, 2001.
12. Orna, M. V. *The Chemical History of Color*. Springer: New York and Heidelberg, 2013; pp. 88–91.
13. Lambert, J. B. *Ibid.*; pp. 9–11.
14. Orna, M. V. *Chemistry and Artists' Colors*, 3rd Ed. ChemSource, Inc.: New Rochelle, NY, 2013; pp. 373–383.
15. Salzberg, H. W. *From Caveman to Chemist*. American Chemical Society: Washington, D.C., 1991; p. 10.
16. Greenwood, N. N., and Earnshaw, A. *Chemistry of the Elements*. Pergamon Press: New York, 1984; p. 327.

17. Kolb, K. E., and Kolb, D. K. *Glass. Its Many Facets.* Enslow Publishers: Hillside, NJ, 1988; p. 11.

18. Lambert, J. B. *Ibid.,* Chapter 5.

19. Rasmussen, S. *How Glass Changed the World.* Springer: Heidelberg, 2012; p. 43.

20. http://en.wikipedia.org/wiki/Timeline_of_communication_technology

21. Kant, I. In *Sämmtliche Werke*; Rosenkranz, K.; Schubert, F. W., Eds.; Leopold Voss: Leipzig, Germany, 1840; vol. 5, p. 310.

22. Bensaude-Vincent, B., and Simon, J. *Chemistry—The Impure Science,* 2nd Ed. Imperial College Press: London, 2012; pp. 86–88.

23. Siegfried, R. Lavoisier and the conservation of weight principle. *Bull. Hist. Chem.,* 1989, 5, pp. 24–31.

24. As quoted by Jaffe, B. *Crucibles: The Story of Chemistry,* 4th revised edition. Dover Publications: New York, 1976, p. 72.

25. Mercuriale, G. *De arte gymnastica.* Venice, 1569 (Reprinted by ILTE Industria Libraria Tipografica Editrice: Turin, Italy, 1960).

26. D'Acquapendente, G. F. *De venarum ostiolis.* Laurentius Pasquati: Padua, 1603.

27. Wells, H. *A History of Discovery of the Application of Nitrous Oxide Gas, Ether, and Other Vapors to Surgical Operations.* J. Gaylord Wells: Hartford, CT, 1847 (available as a Google ebook).

28. Wolfe, R. J. *Tarnished Idol: William T. G. Morton and the Introduction of Surgical Anesthesia.* Norman Publishing: San Anselmo, CA, 2000.

29. http://www.historylearningsite.co.uk/joseph_lister.htm.

30. Gross, C. G. *Brain, Vision, Memory: Tales in the History of Neuroscience.* MIT Press: Cambridge, MA, 1999.

Chapter 3

Did Chemistry Change the World?

Attila E. Pavlath

President of the American Chemical Society, 2001

Attila@pavlath.org, AttilaPavlath@yahoo.com

3.1   Introduction

The question is really not whether chemistry changed the world or not, but how. When we compare the world of the cavemen, the conditions of their living, with those of the most advanced society of the modern world, there cannot be any question about the gigantic difference. However, even though you see and comprehend the changes, if you are not a chemist, you might not realize the role of the chemistry in the changes.

Many of the changes occurred especially in the early times when people had minimal or no knowledge about chemistry, but they empirically developed new materials and uses without knowing that they were made possible by chemical processes. However, understanding or not the science behind the developments, they kept experimenting with modifying the circumstances mostly on a trial-and-error basis.

The cavemen did not understand that fire is a chemical process but figured out that to maintain it, wood had to be used and that

Chemistry: Our Past, Present, and Future
Edited by Choon Ho Do and Attila E. Pavlath
Copyright © 2017 Pan Stanford Publishing Pte. Ltd.
ISBN 978-981-4774-08-6 (Hardcover), 978-1-315-22932-4 (eBook)
www.panstanford.com

the fire was better if the wood was dry and did not contain moisture. Perhaps they found out that if wood was smeared with lard from one of their kills, it could serve as a torch. The ancient ancestor of today's blacksmith discovered that a solid hard material could be formed when certain ores were heated in fire. When the ore was not pure, they "invented" how to make alloys, which were stronger.

The eight chapters of the second part of this book describe how various inventors used chemistry to develop new materials and products for various uses in our life once the principle of chemistry became more and more known. These planned investigations occurred mostly during the past 200–300 years, but even without the understanding of the processes, chemistry had an everlasting impact on our life for thousands of years. Metals, paper, wood, textile just to name a few moved the cavemen to abodes that were heated and lighted; the hide of the saber tooth tiger was replaced by textile garments keeping him warm; extracts of various herbs were used by the medicine man for primitive curing. Once chemists understood the chemistry behind natural processes affecting our life, chemistry helped improving them though scientific research instead of trial-and-error approaches. Chemistry is essential to our everyday lives and the economy. The business of chemistry transforms the natural raw materials of the earth, sea, and air into products that we use every day. The business of chemistry drives innovation and creates jobs and economic growth by creating products that bring major societal benefits to quality of life, health, productivity, convenience, and safety. The science and materials of chemistry make the lives of Americans and others throughout the world healthier, safer, and more sustainable and productive. Indeed, our food, safe water supply, clothing, shelter, health care, computer technology, clean energy sources, transportation, and every other facet of modern life, all depend on the business of chemistry. In each of these areas, new developments and/or improved materials used for cheaper and more easily available existing materials benefiting our life further improved our comfort, providing better transportation, communication, health, and nutrition.

3.2   Construction Materials

3.2.1   Wood

While the first use of wood was for primitive purposes, i.e., heating the cave, creating a weapon to hunt for meat, then searing it, without any modification, it quickly became one of the most important construction material replacing the cave. Consequently, its application expanded to many other areas, e.g., tools, furniture, doors, and paper. While in certain cases it was used without any chemical modifications, generally various empirical and later scientific treatments using chemistry made it more usable. The most important use almost from the beginning was shipbuilding for transportation on rivers and seas. Until the Industrial Revolution, it was the only basis for shipbuilding, but even today it is used for smaller boats. Chemistry was involved in making the ship's lifespan longer. Chemical treatment protected against rots and barnacle. Columbus' ship was copper plated for this purpose. Using glue to create plywood and veneer for furniture can strengthen wood. It is used for erecting prefabricated houses quickly and economically.

Wood, however, has another important application, which contributed to our life: papermaking. Before the discovery of its use for this purpose, writing on surface was limited to silk. Development of papyrus for scrolls had limited application. The pulping process removing lignin from wood and creating high-purity cellulose fibers created a revolution for accessibility of books in an economic way. Depending on the treatment, various types of paper can be manufactured to fit the requirements.

Wood found its way into other areas too. One of it is art. Various musical instruments are made from only wood, and their tone is dependent on the way the wood is prepared. The famous Stradivarius violin owes its unique quality to the treatment of the wood used for its creation. Other area is in the sport equipment. While many of them are replaced by carbon fiber, fiberglass, and metals, but in their initial form, baseball bats, archery bows, golf clubs, and skis were based on wood generally with chemical treatments to improve their properties. Finally, wood is investigated for alternate energy sources other than simple burning. The cellulose in the wood is polymeric carbohydrate, which can be hydrolyzed to glucose, the fermentation of which can provide alcohol for fuel. However, at this time the process is not economic enough to compete with other ways to supplement fossil fuel.

3.2.2 Metals

Wood is the most abundant energy source, which can be renewed utilizing the sun's energy through photosynthesis. The estimated amount of wood for the whole earth is in the range of 10^{12} ton. At the same time, metallic ores, the source of various metals, have to be mined. For strength, metals are far superior to wood, but they have to be converted to elemental metals to be used in various application. Before empirical and scientific metallurgy was developed, only gold, copper, and meteorite iron was available in pure form but in limited quantities. By finding ways to convert metallic ores to metals, such as copper, iron, and various alloys, major advances were made in our life already thousands of years ago, changing history in various ways. The changes were so prominent that history is defined by the development and use of various metals, e.g., Copper, Bronze, and Iron Age.

At the beginning, their use was mostly for weapons for hunting. Hunting weapons and shields improved the efficiency of providing meat for the community while protecting the hunters against dangerous carnivores. Metals also provided tools for improving the cultivation of the land for more productive agriculture. Later, the invention of horseshoes to protect the horses' hooves allowed longer travels either on horseback or in horse-drawn carriages. Unfortunately, at the same time, it made possible more bloody wars between tribes and nations.

Once the empirical smelting processes became more and more scientific, the availability of metals created a revolution. Bridges were built speeding up transportation. Better and better transportation devices, carts with better wheels, bicycles, trains, autos, and finally flying devices such as airplanes and rockets made possible covering never imagined distances in shorter and shorter times. The following is a short description of the benefits copper, bronze, and iron.

3.2.2.1   Copper

While small amount of copper was available in metallic form, its use was rare other than for some jewelry. The pure metal is soft, and in the ancient times, it was not suitable enough for tools and weapons. Its preparation was achieved by smelting copper sulfide as early as 2800 BC in China, but independently other locations also found the way to extract the metal from its ores. Interestingly, since the available copper sulfide was in a mixture of various other ores during the smelting process a mixture of metals was obtained leading to the accidental discovery of alloys. Small amounts of other metals (2–5%) made important changes in the properties of copper.

In modern times when metallurgy became scientific, numerous applications were found for copper, which drastically changed our life. The most important is its use for improved communication through electric wires allowing telegraph and telephone connection. Copper wire is essential for power generation, transmission, electric motors, and various electronic equipment, which we take for granted today. Creation of electricity through generators and its transfer to large distances brought civilization to faraway places. The list is long.

In lesser, but still considerable quantities, copper is used for plumbing and roofing. In architecture, use of copper is both structural (domes and spires) and decorative. Copper has use in etching, engraving, and other artistic procedures.

3.2.2.2   Bronze

It is an interesting situation that as the demand for copper increased, smiths had to use the more abundant and less pure copper ores and the smelting resulted in less pure copper. The inferior appearance of the product was counterbalanced with increased strength of the obtained metal. The resulting alloy had lower melting point (950°C versus 1084°C) and was easier to cast. The result was bronze with 5–15% of tin and arsenic. Lower amount does not strengthen it enough, and higher amount makes it brittle. Without the knowledge of chemistry, the ancient smiths correlated by trial and error the results with the origin of the ores used in smelting.

Once the right combination was achieved, bronze became the most used metal, phasing out pure copper. Further experimenting was done to include other elements such as lead and nickel in the alloys. Armors, spears, and swords were sought-after items by areas that did not have access to the ores. In addition, household items such as knives and sickles became very important. Bronze daggers were ceremonial, indicating high standing, and were buried with the dead. Excavations at various places have discovered bronze cups, bowls, and even large cauldrons. In an interesting way of application, sometimes records were cast in bronze to retain longevity. In Southeast Asia, excavations found even bronze drums.

The smelting process required kilns where high temperature could be reached. Such kilns were used in pottery and porcelain making, and this provided a progress where such work was already done. Many artifacts were found in China, but other places also had the capability. The progress, however, created a problem. To heat the increasing number of kilns, more and more wood was needed as fuel. This resulted in deforestation to such a degree that in some areas, especially in the Mediterranean, smelting came to a halt not because of lack of ores but of wood. It has been estimated that for one mine in the Athens area, the operation and smelting used 1 million tons of charcoal and 2.5 million acres of forest in 300 years. The need for wood was so great that it was often the driving force behind wars.

3.2.2.3 Iron

Pure iron obtained from meteorites breaks easily; however, just as in the case copper, impurities, mostly carbon, within a given range cause strengthening. However, the ancient technology used for bronze was not applicable to obtaining iron from its ores. While some iron artifacts are dated to the Bronze Age, large-scale use of iron did not occur until more complicated technology was discovered, again through trial and error. With the right type of carbon content, iron has a great advantage over bronze in spite of possible rusting. It is harder and long lasting. As soon as the smelting process was sorted out, iron tools, weapons, and other equipment replaced their bronze equivalent just as bronze replaced stone tools. Iron was produced as wrought iron (low carbon content) and cast iron (with high carbon content);

however, it was only in the 17th and 18th centuries that the differences were really understood. Only after this, great progress started toward creating steel, the carbon content of which is in between the two types of iron. However, some iron fragments dated 1800 BC were found to be steel, probably accidentally created by some smith without understanding how it was formed. Excavations in various parts of the world—Iran, India, and China—discovered some steel from 1000 BC and later artifacts were also made of steel.

The development of the iron and steel technology resulted in rapid progress since iron was available in large quantity under better economic conditions and it became a major structural material. The Ironbridge was built in 1779. The large-scale manufacturing of machines, tools, pipes, and various equipment lead to the Industrial Revolution. In 1850, the Bessemer process involving regulated blowing of air through molten iron (4–5% carbon) enabled reduction in the carbon content (2%) to a level required for steel formation. Heating of the furnaces was done first with carbon and later with electric arcs. While reducing the carbon content and eliminating some unrequired impurities, other metals may be added to create stainless steel and other high-resistance steel alloys. Today, skyscrapers, automobiles, rockets, and hulls of large ships are the result of the development of iron metallurgy. While iron is more susceptible to rust than other metals, certain treatments such as galvanization by zinc and covering by lead-oxide-based paints diminish the possibility of rusting.

3.2.2.4 Other metals

Various metals such as chromium, nickel, and zinc are used for specific applications especially as iron alloys mostly because of their strength. Some of the most important metals are aluminum, titanium, tungsten, and lithium.

Aluminum is a very useful metal, which is used in the pure form. It is the most widely used non-ferrous metal. Global production of aluminum is second to that of iron. It is sometimes substituted for copper wire, but its major advantage in addition to corrosion resistance is its light weight. Because of these characteristics, aluminum and its alloys have numerous applications in everyday life:

- Structural materials for automobiles, aircraft, windows, doors, siding, as sheet, tube, and castings
- Various household items such as food containers, cans, foil, and cooking utensils
- Paints and pyrotechnics such as solid rocket fuels and thermite in the powdered form

Titanium itself is highly heath resistant and used in rocket industry. The two most useful properties of the metal are corrosion resistance and the highest strength-to-density ratio of any metallic element. In pure form, titanium is as strong as some steels, but less dense. Under heat and pressure, titanium powder can be used to create strong, lightweight items ranging from armor plating to components for the aerospace, transport. Its alloys with iron, aluminum, and other metals have similar utilization for jet engines, missiles, and spacecraft where strong lightweight materials are needed. Because of these properties, it is the major materials for artificial limbs, orthopedic and dental implants, sporting goods such as baseball bats and golf clubs, and fishing rods.

Tungsten has the highest melting point in the pure form. Since its strength is retained, it is used in many high-temperature applications such as filaments in light bulb, cathode-ray tube, and vacuum tube. These properties also make it an ideal component in rocket engine nozzles. Its most important industrial use (about 60%) is in the form of tungsten carbide for wear-resistant abrasives, and cutting tools such as knives, drills, circular saws, as well as milling and turning tools used by the metalworking industry.

Lithium is not only the lightest metal (with density 0.534 g/cm^3, which is comparable to that of pinewood) but also the least dense solid element. Lithium and its compounds have several industrial applications, including heat-resistant glass and ceramics, as well as high strength-to-weight alloys used in aircraft. Its most promising future application is for these uses, which will consume more than half of the world's lithium production.

3.2.3 Plastic

After Stone Age, Copper Age, Bronze Age, and Iron Age, we are what we can call "Plastic Age." While metals have numerous desirable properties, their increased uses resulted in the demand for developing materials with reasonable strength, easy production, less weight, and other desirable properties.

This is the place where chemistry the most visibly changed our life. The development of various plastics not only decreased the weight in many applications using metal but also made them less expensive. Today there is no household that does not depend on small or large devices made from plastic materials. They are used as structural materials for furniture, building elements, including plastic plumbing. Containers, disposable utensils are essential part of our daily life. Most automobiles have plastic bodies. Casings for electric equipment are made from plastic to prevent electric shocking because of possible short circuits. Wires are insulated by plastic. You would have great difficulty in finding something in your household that does not use plastic. Even some cooking vessels are coated with a fluorine-containing polymer to make their cleaning easier.

3.2.4 Leather and Textile

The cavemen used hides of the animals they killed for food to cover themselves after treating the hides in some experimental way to prevent deterioration. They were not chemists, but through trial and error, they found ways of treating hides. During the progress from being cavemen to civilized humans, various experimental and scientific methods were invented to progress from smelly hides to luxurious leather objects. Today, leather is made through various tanning processes using plant and metallic materials. Earlier leathers were used to be made from the hides of cattle, deer, and pig, but now numerous other skins, even from fish, have applications.

Leather can be found in various parts of our life. Pioneers' wardrobe was based on leather. James Fennimore's famous *Leatherstocking* epic immortalized its role in their life. Its role in saddle and horse rein was important in the life of the American Wild West. Motorcycle riders and racecar drivers frequently wear heavy leather jackets and even pants, which are superior to any practical human-made fabric for abrasion protection in the case of accident. It can form the base for economic and expensive shoes, pants, belts, wallets, and briefcases. Its hydrophilic property and strength are utilized in bookbinding and furniture covering. It is produced in a wide variety of types and styles and is decorated through a wide range of techniques.

At the same time, plant and animal fibers also provided substitute for animal skin clothing, which provided more protection from nature for centuries. The productions of **textile** involve spinning the fibers into yarn and converting them through weaving, knitting, and crocheting to various forms of textile. The basis of these processes has not changed for centuries, but chemistry has helped to speed up the production by treating fibers with various chemicals. This created better yarns and led to more economic production of textiles. Natural textiles are generally finished using various chemical processes to improve their characteristics. The most common technique is applying starch to prevent staining and wrinkles. Advanced technologies further improved machine washability, easy dyeing, and wrinkle resistance.

The fibers used in the modern world are not just of animal and plant origin. Glass fibers are used in space suits. Asbestos fibers provide flame-retardant and protective fabric. Metal fibers reinforce the strength of regular fibers.

However, the greatest improvement chemistry has provided for textiles is the development of synthetic fibers. Many of them were invented to replace natural fibers with certain properties. For example, the invention of nylon provided a strong economic substitute for silk. The replacement of expensive silk with economic nylon in stockings and pantyhose created a revolution in the fashion industry. Nylon is also the major basis for parachutes. Other synthetic fibers widely used are polyesters and acrylic. They are superior in machine washability and permanent press. Carbon fibers are used to reinforce the strength of plastic.

3.2.5   From Pottery to Glass through Ceramic

The earliest starting material for nonmetallic construction materials is clay. It is an aluminum silicate natural rock or soil material with various amounts of metal oxides and possible organic matter. It may have various colors. Depending on its water content, it has various plasticity, which may be increased with additional amount of water and can be formed in the desired shape. It can undergo permanent physical and chemical changes depending on the degree of heat treatment.

Although sun drying is used to provide construction materials for primitive houses, it generally requires a kiln to get solid shapes, which do not lose form upon impact with nature. For pottery, the clay dough is kneaded into the required shape and then heated to high temperature in an oven. It was used in those ancient civilizations that had access to clay and the capability of heating objects to high temperature. Depending on the origin and composition of the clay, the result is earthenware, stoneware, and porcelain. Clay tablets were the first known writing medium. When sintered, it created various items for everyday life: bricks, pots, and dishware. Unusual items made of sintered clay are smoking pipes, art objects, and ocarina. Generally, they are porous and permeable to water; therefore, glaze is applied not just for aesthetics but also for removing porosity.

Porcelain is a special ceramic material, which is made with kaolin clay and heating at 1200–1400°C. It is strong and translucent. Production of porcelain started in China, and for centuries it was the only place where it was created because kaolin clay was not available in Europe and even the production technique was a secret. Porcelain is not only used for tableware but also as building components such as tiles and panels. Because it is a good insulator, it is also used in high-voltage transmission.

Scientifically, glass is defined as a non-crystalline solid material; therefore, porcelain can also be classified as glass. However, in everyday talk, it is applied to products made from silicon dioxide, i.e., sand. For various purposes, including stained glass window, it may contain a wide variety of metals. The physical properties of glass, i.e., transparency, reflection, and refraction of light, provide numerous applications in our everyday life. In addition to windowpanes and tableware, it is an essential part of optical objects, such as lenses, prisms, mirrors, glass wool, and optical fibers using various additives. For example, the addition of boric oxide results in Pyrex glass for cooking glassware, which withstands thermal shocks. Addition of lead oxide results in aesthetic crystal items. Fiberglass, which is used in reinforcing plastic to make boats and fishing rods, is created by the addition of aluminum oxide.

3.3 Conclusion

The answer to the title of this chapter is short. There is no question that chemistry has immensely benefited every aspect of our life. Without its numerous developments, we would be still in the Stone Age. It is interesting to let our fantasies fly to visualize how chemistry will change our world. It took many millennia to get to our present stage from the Stone Age. Even a few centuries will bring many new improvements, some of which cannot be predicted even by our best science fiction writers.

PART II

CONTRIBUTIONS OF CHEMISTRY

Chapter 4

Agriculture

Livia Simon Sarkadi

Szent István University, Faculty of Food Science,
Villányi út 29-43, 1118 Budapest, Hungary

sarkadi@mail.bme.hu

4.1 Introduction

Chemistry has made plenty of contributions to food and agriculture that allow us to raise, harvest, and consume abundant and nutritious food. At the turn of the nineteenth and twentieth centuries, an average kitchen table would be loaded with produce from the root cellar, garden, or local farm; butter from a churn; eggs from hens penned in the backyard; vegetables from the garden; and meat stored in an icebox and cooked over a coal or wood stove. The last century has brought vast changes in how we get food on our tables by making our farms more productive and our food and water supplies readily available. Modern farmers have utilized new chemical advances to improve agricultural production with fertilizers and pesticides and to develop plentiful food supplies. Consumers have benefited from new technologies, which have enhanced the flavor, appearance, availability, and nutritional value of their food. These advances in chemistry help to feed

Chemistry: Our Past, Present, and Future
Edited by Choon Ho Do and Attila E. Pavlath
Copyright © 2017 Pan Stanford Publishing Pte. Ltd.
ISBN 978-981-4774-08-6 (Hardcover), 978-1-315-22932-4 (eBook)
www.panstanford.com

potentially all individuals of the world's rapidly expanding population.

4.2 Development of Agriculture

The word **agriculture** is the English adaptation of Latin agricultura, from ager, "a field," and cultura, "cultivation" in the strict sense of "tillage of the soil." Thus, a literal reading of the word yields "tillage of a field/of fields."

Agriculture has played a key role in the development of human civilization. It was developed at least 10,000 years ago, involving domestication of plants and animals. It has undergone significant developments since the time of the earliest cultivation. The Fertile Crescent of the Middle East was the site of the earliest planned sowing and harvesting of plants that had previously been gathered in the wild. The eight so-called Neolithic founder crops of agriculture appear: first emmer wheat and einkorn wheat, then hulled barley, peas, lentils, bitter vetch, chick peas, and flax.

Until the Industrial Revolution, the vast majority of the human population labored in agriculture. Development of agricultural techniques has steadily increased agricultural productivity, and the widespread diffusion of these techniques during a time period is often called an agricultural revolution.

Agriculture encompasses a wide variety of specialties. Cultivation of crops on arable land and the pastoral herding of livestock on rangeland remain at the foundation of agriculture.

Since the past century, a distinction has been made between sustainable agriculture and intensive farming. Modern agronomy, plant breeding, pesticides and fertilizers, and technological improvements have sharply increased yields from cultivation. Selective breeding and modern practices in animal husbandry, such as intensive pig farming (and similar practices applied to chicken), have similarly increased the output of meat. The more exotic varieties of agriculture include aquaculture and tree farming.

The major agricultural products can be broadly grouped into foods, fibers, fuels, raw materials, pharmaceuticals and illegal drugs, and an assortment of ornamental or exotic products. Since the 2000s, plants were used to grow biofuels, biopharmaceuticals, bioplastics, and pharmaceuticals.

Specific foods include cereals, vegetables, fruits, and meat. Fibers include cotton, wool, hemp, silk, and flax. Raw materials include lumber and bamboo. Drugs include tobacco, alcohol, opium, cocaine, and digitalis. Other useful materials are produced by plants, such as resins. Biofuels include methane from biomass, ethanol, and biodiesel. Cut flowers, nursery plants, tropical fish and birds for the pet trade are some of the ornamental products.

Despite the fact that agriculture employs over one-third of the world's population, agricultural production accounts for less than 5% of the gross world product (an aggregate of all gross domestic products).

Modern agriculture depends on scientific advances to increase the yield of crops and animal products to cope with this rapid population growth. Developments in agricultural chemistry have allowed for crops to be produced in areas that previously were unsuitable for agriculture, and appropriate use of fertilizers has helped to create higher yields to counterbalance the loss of available agricultural land.

4.2.1   Brief Chronology of Main Achievements

Agriculture was developed at least 10,000 years ago, and it has undergone significant developments since the time of the earliest cultivation.

- 9500 BCE: Earliest evidence for domesticated wheat
- 8000 BCE: Evidence for cattle herding
- 7000 BCE: Cultivation of barley; animals are domesticated
- 6500 BCE: Cattle domestication in Turkey
- 6000 BCE: Indus Valley grows from wheat to cotton and sugar
- 5500 BCE: Sumerians start organized agriculture
- 5400 BCE: Archaeological proof for domestication of chicken
- 5400 BCE: Linearbandkeramik Culture in Europe
- 5000 BCE: Africa grows rice, sorghum
- 4000 BCE: Ploughs make an appearance in Mesopotamia
- 3000 BCE: Maize is domesticated in Americas
- 3000 BCE: Turmeric is harvested in Indus Valley
- 2737 BCE: Tea is discovered
- 2000 BCE: First windmill in Babylon

- 1000 BCE: Sugar processing in India
- 500 BCE: Row cultivation in China
- 200: Multi-tube seed drill invented in China
- 700: Arab Agriculture Revolution
- 1000: Coffee originates in Arabia
- 1492: Columbian exchange changes agriculture
- 1599: First Practical Green House is created
- 1700: British Agricultural Revolution
- 1700: British agriculturalist Charles Townshend popularizing a 4-year crop rotation with rotations of wheat, barley, turnips, and clover
- 1794: Cotton gin is invented (The invention was granted a patent on March 14, 1794)
- 1800: Chemical fertilizer began to be used
- 1837: John Deere (1804–86) invents steel plough (Fig. 4.1)

Figure 4.1 Steel plough. *Source*: http://saltofamerica.com/imgArticle/page500/Page500_291_6.jpg.

- 1860: Hay cultivation changes
- 1866: Gregor Johann Mendel (1822–84) describes Mendelian inheritance (Fig. 4.2)
- 1879: Anna Baldwin patented a milking machine that replaced hand milking (Fig. 4.3)
- 1881: The French Louis Pasteur (1822–95) discovered anthrax vaccine for sheep and hogs (Fig. 4.4)

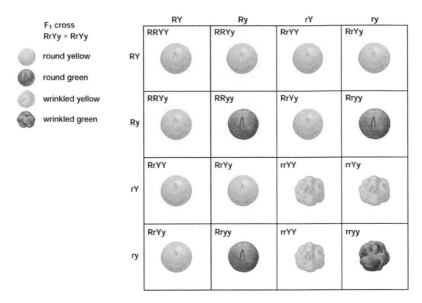

Figure 4.2   Example of Mendelian inheritance. *Source*: http://3201 courtney.weebly.com/uploads/2/3/4/8/23483182/9146682 _orig.jpg.

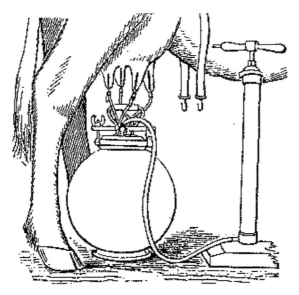

Figure 4.3   Milking machine. *Source*: http://www.americanartifacts.com/ smma/milker/milker.htm.

Figure 4.4   Louis Pasteur vaccinates animals. *Source*: https://fr.wikipedia. org/wiki/Fichier:Louis_Pasteur_in_Pouilly-le-Fort_ (Illustration_-_1881).jpg.

- 1883: Johan Gustav Kjeldahl (1849–1900), Danish chemist, developed a method to analyze the nitrogen content of any organic compound
- 1884: French botanist Pierre Marie Alexis Millardet (1838–1902) invents the Bordeaux mixture to fight vineyard mildew
- 1892: American inventor John Froelich (1849–1933) built the first practical gasoline-powered tractor in Clayton County, Iowa (Fig. 4.5)

Figure 4.5   Gasoline-powered tractor. *Source*: http://historylists.org/ images/froehlichs-tractor.jpg.

- 1900: **Birth of industrial agriculture**
- 1913: Two German chemists, Fritz Haber (1868–1934) and Carl Bosch (1874–1940), developed the process for the production of ammonia at industrial level, which produced nitrogen fertilizers that increased food production at cheaper price
- 1930: First aerial photos for agriculture
- 1931: The first plant patent was issued to Henry F. Bosenberg for a climbing or trailing rose by the United States Patent Office (Fig. 4.6)

Figure 4.6   Copy of the patent for a climbing or trailing rose. *Source*: http://brebru.com/education/plantpa.gif.

- 1939: The Swiss chemist Paul Hermann Müller (1899–1965) discovered the insecticidal qualities of DDT (dichloro-diphenyl-trichloroethane) pesticide

- 1943: The US Department of Agriculture chemists develop aerosol dispersion for insecticides and farm applications
- 1944: Green Revolution begins in Mexico
- 1964: The "Green Revolution": Application of new hybrid plants and liquid soil fertilizers solved the nutrition problem of the Asian population
- 1972: DDT usage banned in the United States (the first country to ban the substance was Hungary in 1968)
- 1974: Monsanto introduced its comprehensive and non-tilling Roundup herbicide
- 1996: Commercial cultivation of genetically modified plants

Agriculture became a science when Justus von Liebig (1803–73) discovered the essential nutrient elements in plants in 1842.

Industrial agriculture (1900) is a form of modern farming that refers to the industrialized production of livestock, poultry, fish, and crops. The methods of industrial agriculture are techno-scientific, economic, and political. They include innovation in agricultural machinery and farming methods, genetic technology, techniques for achieving economies of scale in production, the creation of new markets for consumption, the application of patent protection to genetic information, and global trade. These methods are widespread in developed nations and increasingly prevalent worldwide. Most of the meat, dairy, eggs, fruits, and vegetables available in supermarkets are produced using these methods of industrial agriculture. The birth of industrial agriculture more or less coincides with that of the Industrial Revolution in general.

The **Green Revolution**, occurring between the 1940s and the late 1970s, refers to a series of research, development, and technology transfer initiatives that increased agriculture production around the world, beginning most markedly in the late 1960s.

Norman Ernest Borlaug (1914–2009), American biologist and one of the fathers of the Green Revolution, summed up the role that nitrogen fertilizer played in this grand agricultural transformation by using a memorable kinetic analogy: "If the high yielding dwarf wheat and rice varieties are the catalysts that have ignited the Green Revolution, then chemical fertilizer is the fuel that has powered its forward thrust..." [1]. He was awarded the

Nobel Peace Prize in 1970 in recognition of his contributions to world peace through increasing food supply.

The Green Revolution had a tremendous impact on food production and socio-economic conditions. Applying advanced technologies (including pesticides and synthetic nitrogen) to high-yielding varieties of cereals–crops that account for almost 50% of calories in most diets–caused the marked achievements in world food production. Scientific efforts to remove insect pests, diseases, weeds, and rodents, which are serious constraints to agricultural production, have focused on the breeding of resistant varieties of crop plants, as well as on the development of pesticides, insecticides, and herbicides and integrated pest management strategies.

Despite its achievements, the first Green Revolution did not solve all food and nutrition security issues, partly because the efforts emphasized the food availability component of food security over the food access, food utilization, and sustainability components, resulting in the neglect of core nutrition elements and environmental challenges. It is now critical that chemical science invests and takes a leading role in cross-disciplinary efforts to predict when and where the use of agrochemicals and chemical-based technologies is pushing food production systems over sustainable boundaries and to develop innovative strategies that can enhance social, environmental, and economic sustainability of food systems [2].

The **organic movement** broadly refers to the organizations and individuals involved worldwide in the promotion of organic farming, which they believe to be a more sustainable mode of agriculture. Its history goes back to the first half of the 20th century, when modern large-scale agricultural practices began to appear. The organic movement began in the early 1900s in response to the shift toward synthetic nitrogen fertilizers and pesticides in the early days of industrial agriculture. It lied dormant for many years, kept alive by a relatively small group of ecologically minded farmers. Some governments, including the European Union, have begun to support organic farming through agricultural subsidy reform. Organic production and marketing have grown at a fast pace.

4.3   Chemical Fertilizers and Soil Nutrients

Maintaining good soil structure and fertility is important to ensure high productivity. Using appropriate fertilizers is one of the ways to improve soil nitrogen retention and uptake by plants.

Nitrogen fixation is an essential natural process, in which microorganisms convert the otherwise less-reactive nitrogen into inorganic nitrogen compounds, which in turn first appear in the nutritional chain as organic nitrogen compounds such as proteins in plants.

Improving the efficiency of nutrient uptake and utilization in plants is a major challenge in agricultural productivity. By understanding the metabolic pathways by which nutrients are absorbed and used, biotechnologists can work with chemists to develop crops that deliver higher yields. The use of fertilizers has been pivotal in securing long-term food supplies in the midst of a growing global population.

Fertilizers are natural or artificial substances containing chemical elements that improve growth and productiveness of plants. Fertilizers enhance the natural fertility of soil or replace the chemical elements taken from the soil by previous crops.

Modern chemical fertilizers include one or more of the three elements that are most important in plant nutrition: **nitrogen, phosphorus, and potassium**.

Modern fertilizers stem from a chemical engineering breakthrough pioneered by Fritz Haber in 1908, who discovered the principles of ammonia synthesis, utilizing all the physical and chemical resources available that time. In 1913, Haber (1868–1934) and Carl Bosch (1874–1940) developed the process for producing ammonia at the industrial level. The successful Haber–Bosch process allows ammonia to be produced cost-effectively in commercial quantities for use in nitrogen fertilizers. Haber's original reaction was carried out under high pressures. The improved ammonia synthesis process carries out the reaction at lower pressures and temperatures, which helps save money by reducing the amount of energy required by the process. In 1918, Haber was awarded the Nobel Prize in Chemistry for this discovery. Many people consider the **Haber–Bosch process one of the most monumental chemical engineering achievements of all time**, thanks to its direct impact on global food production [3].

Most nitrogen fertilizers are obtained from synthetic ammonia; this chemical compound (NH_3) is used either as a gas or in a water solution, or it is converted into salts such as ammonium sulfate, ammonium nitrate, and ammonium phosphate.

Phosphorus fertilizers include calcium phosphate derived from phosphate rock or bones. The more soluble superphosphate and triple superphosphate preparations are obtained by the treatment of calcium phosphate with sulfuric and phosphoric acid, respectively.

Potassium fertilizers, namely potassium chloride and potassium sulfate, are mined from potash deposits.

Mixed fertilizers contain more than one of the three major nutrients—nitrogen, phosphorus, and potassium. Mixed fertilizers can be formulated in hundreds of ways.

The achievability of balanced soil and food chemistry was underscored by the 1956 development of the **Kjeldahl method** to analyze automatically the amount of nitrogen in organic compounds. The basic method was developed in 1883 by Johan Kjeldahl (1849–1900). While studying proteins during malt production, he determined that, since nitrogen is a major element in protein, nitrogen analysis could be used to determine the amount of protein in a substance. The Kjeldahl method is a multi-stepped process, including several chemical reactions and finally determination of the amount of protein by calculation from the nitrogen concentration.

Monitoring of protein amount through the estimation of nitrogen must be accurate because proteins are important constituents of foods for a number of different reasons. Proteins contain many essential amino acids that are essential for human health but cannot be synthesized by the body. Therefore, knowing the protein content is essential for determining the nutritional quality of many foods. Proteins are also the major structural components of many natural foods, often determining their overall texture. Many food proteins are enzymes that are capable of enhancing the rate of certain biochemical reactions.

Today the Kjeldahl method is the worldwide standard for calculating the protein content in all kinds of food and beverage samples. The Kjeldahl method is also employed in environmental analysis and the agricultural industry for the determination of nitrates and ammonium.

Innovations in the basic production of chemical fertilizers have been made ever since, including the 1930 marketing of granular fertilizers and the 1965 introduction of suspension fertilizers in the US market. In the 1970s, granulation was further refined to introduce fertilizers suitable for home use blending. The most recent innovations in commercial fertilizers include time-release encapsulation, in which the agent of the synthetic fertilizer gets into the soil during a given time. Thus, it prevents environmentally undesirable overfertilization. The projected demand for nitrogen from chemical fertilizer is estimated to increase to 236 million tonnes in 2050.

Chemical engineers have applied their expertize in chemically synthesizing fertilizers, herbicides, and pesticides that promote crop growth and protect crops from weeds, insects, and other pests. Today, the use of these products is more important than ever to meet the needs of an ever-expanding population.

4.4 Crop Protection and Pest Management

Up to 40% of current agricultural outputs would be lost without effective use of crop protection chemicals. Agriculture is facing emerging and resistant strains of pests. It is, therefore, essential that new crop protection strategies, chemical and non-chemical, are developed.

The use of pesticides coincides with the "chemical age," which has transformed society since the 1950s. In areas where intensive monoculture is practiced, pesticides are used as a standard method for pest control.

Pests are living organisms (weeds, insects, mites) that can cause injury or damage to crops, livestock, and stored food. They can significantly reduce yields, increase the cost of production, and cause complete failure of entire cropping systems.

There are many ways of **controlling pests**. Chemical control is one of the most important methods of pest control. The use of agricultural chemicals offers a number of advantages:

- Products registered for specific pest or crop situations can be selected
- Agricultural chemicals are often fast-acting and produce rapid results

- Most agricultural chemicals can be applied rapidly and efficiently, often over large areas
- Many agricultural chemicals are selective and enable specific organisms to be targeted and controlled

However, agricultural chemicals should always be used with care. Users of agricultural chemicals must ensure that the correct product is applied at the correct dose, at the right time, and with minimum contamination of the environment [4].

Pesticides encompass several different chemical categories. They include **herbicides** (used to kills weeds), **insecticides** (used to kill insects), **fungicides** (used to control molds and fungi), and **rodenticides** (used to kill rodents).

Herbicides are designed to kill or control unwanted plants (or weeds) in crops, pasture, or non-crop situation such as roadsides or recreational areas.

Examples of crop protection chemicals: The herbicidal activity of glyphosate was first discovered in 1970 by Monsanto's John Franz (1929). Glyphosate inhibits a specific growth enzyme called the EPSP synthase (3-phosphoshikimate-1-carboxyvinyltransferase). Glyphosate is rapidly metabolized by weeds, and unlike many other earlier herbicides, it binds tightly to soil so that it does not accumulate in runoff to contaminate surface waters or underground aquifers and does not affect mammals, birds, fish, or insects. Franz received the National Medal of Technology in 1987 for his discovery (the highest honor awarded to America's leading innovators by the US President). In 1990, he also received the Perkin Medal for Applied Chemistry from the Society of Chemical Industry, and in 2007, he was inducted into the National Inventors Hall of Fame.

The herbicide **Mesotrione** is related to the NP leptospermone, produced by the red bottle-brush plant, *Callistemon citrinus*. Leptospermone is only a moderately active herbicide, which gives bleaching symptomology when applied to plants, because it inhibits the enzyme hydroxyphenylpyruvate dioxygenase (HPPD). Mesotrione is an important post-emergence maize-selective herbicide. It is weakly acidic and water soluble, ideal properties for uptake and movement in plants.

The herbicidal **Pseudomonic acid A** is a major metabolite of the bacterium *Pseudomonas fluorescens*. Hydrolysis gives monic acid A, which can then be converted into a variety of semi-synthetic esters and amides, many of which were found to be herbicidal, the result of inhibition of isoleucyl-tRNA synthetase in plants. The 2-(vinyloxy)ethyl ester is particularly active and was taken forward into extensive field trials in cereals.

Insecticides are designed to kill insects and arachnids (spiders, ticks, and mites). Many insecticides work by interfering with the insect's nervous system. Insecticide types are based on their mode of action, the nature of killing agent, the name of the specific target species, or the life cycle stage of the target.

Classification by mode of action:

- Stomach poisons (endosulfan: Thiodan)
- Contact poisons (methomyl: Lannate)
- Lures (fruit fly baits)
- Insect growth regulation (methoprene: Altosid)
- Microbial insecticides (*Bacillus thuringiensis*: Dipel; *Helicoverpa zea nucleopolyhedrovirus*: Gemstar)
- Transgenic insecticidal crops (gene from *Bacillus thuringiensis*)

In 1939, Paul Müller (1899–1965) developed the inexpensive insecticide DDT to control Colorado potato beetles and other insects. DDT (Fig. 4.7) was widely used, appeared to have low toxicity to mammals, and reduced insect-borne diseases such as malaria, yellow fever, and typhus. In 1949, Müller won the Nobel Prize in Medicine for discovering its insecticidal properties. DDT and similar pesticides controlled crop pests and insect-borne diseases for more than 20 years. In the 1960s, public concerns about accumulation of DDT in human body, combined with the increasing resistance in pest species, led to the evolution of new pesticides and the decline of DDT. Today's low-application pesticides provide greater economy for farmers, added worker safety, and are more environmentally friendly than ever before.

Stemofoline is an insecticidal alkaloid produced in the plant *Stemona japonica*. It has a good spectrum of activity, with rapid action, and is a potent agonist of the nicotinic acetylcholine receptor in insects. The downside is its highly complex polycyclic structure, which poses a challenge for the synthetic chemist. A breakthrough came with the recognition of a tropane substructure embedded within the molecular framework of stemofoline. During the exploration of various tropanes, a series of highly active cyanotropanes was discovered with high potency against aphids and whitefly and rapid action both by contact and stomach routes.

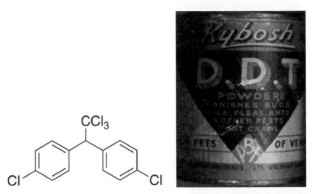

Figure 4.7 Originally packaged DDT. *Source*: http://en.wikipedia.org/wiki/DDT; http://www.vilaglex.hu/Lexikon/Kepek/DDT2.jpg.

Fungicides are agricultural chemicals that protect crops against attack from fungal pathogens. There are two main types of fungicides:

(1) Protectant. These fungicides are deposited as a covering film on the leaf or plant tissue. They act like a protective coat and remain on the plant until a fungal spore lands on the leaf, germinates, and acquires a dose of the agricultural chemical as it tries to penetrate the leaf (e.g., mancozeb: Dithane).

(2) Systemic. These fungicides are able to move within the plant. Some systemic fungicides move in the plant's vascular tissue (e.g., propiconazole: Tilt)

In 1882, French botanist Pierre M.A. Millardet (1838–1902) employed an aqueous solution of copper sulfate and hydrated lime dissolved in water (**Bordeaux mixture**) to effectively combat mildew in French vineyards. The Bordeaux mixture (Fig. 4.8) now controls a number of fungi that attack crops. This also marked the first large-scale fungicide use and revolutionized chemical crop protection.

Chemical innovations in agricultural fungicides continued with the introduction of dithiocarbamate fungicides in 1934 and strobilurin fungicides in 1996. **Azoxystrobin** was discovered during research on *Oudemansiella mucida* and *Strobilurus tenacellus*, which are small white or brown mushrooms commonly found in European forests. Not bigger than a few centimeters, these mushrooms attracted attention of scientists because of their remarkable ability to defend themselves. **Azoxystrobin** is the world's leading agricultural fungicide. It is a fully synthetic compound, whose invention was inspired by the structure and activity of the naturally occurring fungicide strobilurin A.

Figure 4.8   Bordeaux mixture contains copper sulfate and spray lime mixed just before application with water. The mixture sprayed on grapes (right). *Source*: https://pnwhandbooks. org/sites/pnwhandbooks/files/styles/page_thumbnail/ public/plant/images/preparing-tank-mix-bordeaux-mixture/ bordeaux.jpg?itok=GRy-iyTM%20.

Today, the pest management toolbox has expanded to include use of genetically engineered crops designed to produce their own insecticides or exhibit resistance to a broad spectrum of herbicide products or pests. These include herbicide-tolerant crops such as soybeans, corn, canola, and cotton and varieties of corn and cotton resistant to corn borer and bollworm, respectively [5]. In addition, the use of **integrated pest management** (IPM) systems (Fig. 4.9), which discourage the development of pest populations and reduce the use of agrochemicals, has also become more widespread. These changes have altered the nature of pest control and have the potential to reduce and/or change the nature of agrochemicals used.

A highly debated topic and example of a current technology at the crossroads of agriculture and chemistry is genetic engineering (GE), a modern technology for modifying crops and livestock. GE is one of several tools in the modern crop biotechnology kit and allows the introduction of genes from the same species or from any other species, including species that are beyond the normal reproductive range of the plant, into the plant or animal. The need to develop new crop varieties that are adapted to local conditions, conducive to sustainable agriculture, and remain high yielding in the absence of irrigation or large inputs of petrochemicals, is an exceptionally tall and urgent order. Many plant scientists believe that GE can contribute significantly to achieving these goals. However, there are a multitude of concerns about the effects of GE crops on human health, including scientific, social, economic, political, and ethical issues [2].

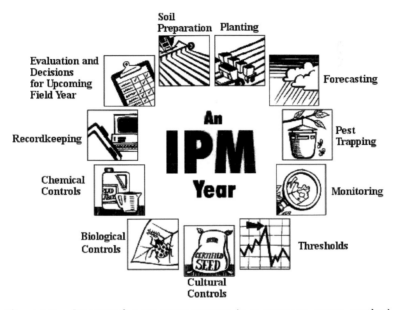

Figure 4.9   Integrated pest management is a common sense method used to eliminate pests. *Source*: http://recreational-turf. wikispaces.com/file/view/ipm1.gif/185871841/ipm1.gif.

4.5   Livestock Production and Protection

In addition to crops, global livestock production faces enormous short-term challenges. The total global meat consumption is predicted to rise to 303 million tonnes by 2020.

4.5.1 Brief History of Livestock Production

The origin of livestock ("livestock" is also used as a synonym for "food-producing animal") production dates back to about 11,000 years ago (after the last glacial period) when humans started to domesticate sheep. This is quite a short period compared to the beginning of the evolution of humanity, which dates back about 14 million years ago. This means that humans and their human-like ancestors survived for millions of years without domesticating animals; plants that were opportunistically collected and animals that were systematically hunted were their most important food source. With the domestication of animals and the cultivation of plants (somewhat less than 11,000 years ago, also addressed as the Neolithic revolution), a fundamental change in the development of humanity happened [6].

Housing of animals in the Middle Ages was mostly reserved for farrowing sows or for other livestock in wintertime when feed supply was limited outdoors. Regularly in autumn, numerous animals were slaughtered to ensure sufficient food supply for the family and to reduce the number of animals, which had to be housed and fed during the cold season.

However, with the growing demand for meat and the increasing crop production, the space for pigs and cattle foraging in fields and forests became limited and animals were increasingly kept indoors over longer periods.

Animal farming did not change very much between the end of the Middle Ages and the 17th century. The transformation of animal farming toward higher productivity happened only when crop rotation was introduced in agriculture, starting first in England.

With the introduction of modern science in agriculture from the middle of the 18th century, scientists started to systematically explore opportunities for further production increases in both plant and animal production. The scientists tried to understand the relationships between soil, plant, weather, and fertilization and recognized the importance of good nutrition and appropriate housing for farmed animals.

The 18th century was the time of devastating waves of rinderpest in Europe. Around 1765, millions of cattle died.

The government decided to kill all sick and suspected animals, 80,000 in total. This drastic measure stopped the disease, and the infectious agent died out for a long period in the United Kingdom. This was the origin of the eradication policy for similar plagues even today. The large plague put veterinary medicine in the focus and many veterinary schools were founded across Europe. In the following years, the number of livestock rose continuously.

The highest increase in farmed animals occurred between 1950s and 1980s, coinciding with the advent of intensive livestock production. Particularly, pig and poultry production became independent of the surrounding land because of feed imports from other parts of the world. Large farming enterprises with as many as 1000 cattle, 10,000 pigs, and 100,000 poultry became the norm in animal production.

The development of animal production in recent decades can be characterized by intensification and specialization.

Intensification means indoor animal housing all year round (non-grazing), high animal densities, a high degree of mechanization and automation (e.g., in feeding, water supply, manure removal, and ventilation), low labor requirement, and often a small air volume in relation to the number of animals in the housing unit.

Specialization means that only one animals species, specially bred for the purpose, is kept in specialized buildings on the farm.

4.5.2 Veterinary Medical Care

The modern animal production practices should also improve animal health and welfare. The use of **veterinary medicines** has been successful in preventing and managing animal diseases such as bluetongue, bovine mastitis, circovirus, foot-and-mouth disease, and salmonella, which just years ago posed serious threats to animal health, food safety, and public health.

Veterinary medical care in food animals consists of the use of:

- Vaccines and prophylactic medication to prevent or minimize infection
- Antibiotics and parasiticides to treat active infection or prevent disease onset in situations that induce high susceptibility

- Antibiotic drugs and hormones for production enhancement, growth promotion, and improved feed efficiency.

Since the 1950s, **antibiotics** have been widely used in food animal production. They are used for many purposes, including the therapeutic treatment of clinically sick animals, for disease prophylaxis during periods of high risk of infection, and for promotion of growth and feed efficiency. Food animals are raised in groups or herds, often in confined conditions that promote the spread of infectious diseases. Antibiotics are frequently used to compensate for poor production practices. Most of the antibiotics used in food animals are the same as or belong to the same classes as those used in humans.

For the therapy of clinical bacterial infections, animals are treated with therapeutic doses of the antibiotic for a period specified on the product label. Therapeutic treatment of individual animals is common practice in dairy cattle production (e.g., treatment of pneumonia or mastitis) but occurs in other species only when it is economically or logistically feasible to handle and treat individual animals (e.g., beef claves in a feedlot, sows, breeding animals). In many cases (e.g., flocks or broiler chickens or pens of salmon), it is impractical to capture, handle, and treat individual animals. In these instances, the entire group is treated, including clinically sick animals, those that may be incubating the disease, and those not infected.

Antibiotics are also used for growth promotion, which is also sometimes called increased feed efficiency. Most antibiotic growth promoters (AGPs) are used in production of pigs, broiler chickens, turkeys, and feedlot beef cattle. The specific physiological basis of the growth-promoting effects of antibiotics is unknown but is hypothesized to involve a nutrient sparing effect in the gut and selective suppression of species of bacteria and clinical expression of infection, i.e., disease prophylaxis. AGPs are typically administered in sub-therapeutic doses for long periods (usually greater than 2 weeks), and sometimes for the entire duration of the production cycle.

AGPs are particularly problematic for resistance, because they are used without veterinary prescription and are administered for long periods at sub-therapeutic concentrations to entire groups or herds of animals.

There are many alternatives to antibiotics. Many of these (e.g., vaccines, health management programs) are already used in good-quality farms and probably reduce the need for antibiotic use.

Prophylactic treatments may be given at therapeutic or sub-therapeutic doses, and the duration of treatment is frequently longer than for therapy. Most commonly, prophylactic treatments are administered to all animals in a group considered to be at risk of infection due to their age or stage of production.

Examples of prophylactic treatments include administration of ceftiofur by injection of hatching eggs or day-old turkey poults to prevent *E. coli* infection; administration of chlortetracycline to feed to beef calves to prevent liver abscess; and administration of tylosin in feed to weaned piglets to prevent diarrhea.

4.6 New Challenges for Agriculture

A rapidly expanding world population, increasing affluence in the developing world, climate volatility, and limited land and water availability mean we have no alternative but to significantly and sustainably increase agricultural productivity to provide food and feed.

- **Population growth.** Global population growth has been the major driving force for growth in world food demand and production. World population is expected to increase from the current 6.7 billion to 9.2 billion by 2050. From 2050, world population will be increasing by 30 million per year [7].
- **Growth in agricultural production slows down.** World agriculture is expected to fall to 1.5% per annum in the next two and a half decades and on to 0.9% per annum in the succeeding 20 years to 2050, compared with 2.1% per annum since 1961. All the major commodity sectors (except for the milk sector) are expected to take part in the deceleration of agricultural growth. The cereals sector has already been in such downward trend for some time now and is expected to continue to have the lowest growth rate of the major commodity sectors during the next 50 years [8]. A doubling of global food production will be required by

2050 to meet the Millennium Development Goals on hunger. This demand for increased food production is exacerbated by economic growth in the emerging economies. As these countries become more affluent, this will translate directly into increased food consumption, particularly for high value-added food such as meat and dairy products. The World Bank estimates that by 2025, one hectare of land will need to feed five people; in 1960, one hectare was required to feed only two people. This needs to be achieved in a world where suitable agricultural land is limited and climate change is predicted to have an adverse impact on food production due to the effects of changing weather patterns on primary agriculture and shifting pests.

- **Increased energy prices.** As the availability of fossil fuels declines, the only renewable carbon resource large enough to substitute for or replace fossil resources for the production of fuels and electricity is biomass. Recent high energy prices are now creating new markets for products that can be used as biomass feedstock for the production of biofuels as substitutes produce fuel ethanol (in this case with subsidies). EU has a target of 5.75% market share of biofuels in the petrol and diesel market by 2010 [8].

  If world agriculture is to become a major source of feedstock for the biofuel industry, this might have implications for food security if feedstock becomes a competitor to food, and for the environment if further deforestation takes place from the eventual expansion of land under the feedstock crops.

- **Water shortage.** It takes 1000 liters of water to grow a kilo of wheat, between 2000 and 5000 liters to grow one kilo of rice, and 16,000 liters to grow the feed it takes to make one kg of beef. Agriculture accounts for 70% of all water use in the world and as much as 95% in many developing countries, almost all of this is for irrigating crops [9].

  Today, more than 1.2 billion people live in areas of physical water scarcity, and by 2025, over 3 billion people are likely to experience water stress. The gap between available water supply and water demand is increasing in many parts of the world, limiting future expansion of irrigation. In areas where water supply is already limited, water scarcity is likely to be the most serious constraint on development, especially in drought-prone areas.

If today's food production and environmental trends continue, we will probably have to face crises in many parts of the world. To avoid this, we need to act to improve water use in agriculture. The hope lies in closing the gap in agricultural productivity in parts of the world through better water management and changes in production techniques, including the use of new drought-resistant crops.

- **Climate change.** The earth has already warmed by 0.7°C since 1900, and based on current trends, the average global temperatures could rise by 2–3°C within the next 50 years [10].

  Food production is particularly sensitive to climate change, since crop yields depend in large part on climate conditions such as temperature and rainfall patterns. There are still large uncertainties as to when, how, and where climate change will affect agriculture production and food security.

The impact of climate change on agriculture depends crucially on the size of the "carbon fertilization" effect. Since carbon dioxide is a basic building block for plant growth, rising concentration in the atmosphere may enhance the initial benefits of warming and even offset reductions in yield due to heat and water stress [11].

Increasing productivity from existing agricultural land represents a significant opportunity because current technologies can be applied to areas where yields are still below average. Historically, increases have come from higher yields as a consequence of improved varieties, better farming practices, and applying new technologies such as agrochemicals and more recently agricultural biotechnology. To meet growing demand for food in the future, existing and new technologies, provided by the chemical sciences, must be applied across the entire food supply chain [12].

4.7 Closing Words

Chemistry is an integral part of agriculture from molecular to organ level. It plays a role from the basics up to the utilization of agricultural products. This is so because chemistry deals with compounds, both organic and inorganic, and agriculture deals with the production of organic products using both organic and inorganic inputs.

References

1. Borlaug, N. (1970). The Green Revolution: Peace and Humanity. A speech on the occasion of the awarding of the Nobel Peace Prize, 1970 Nobel Peace Prize, Oslo, Norway.

2. Fanzo, J., Remans, R., and Sanchez, P. (2011). The role of chemistry in addressing hunger and food security. In: *The Chemical Element: Chemistry's Contribution to Our Global Future*, edited by J. Garcia-Martinez and E. Serrano-Torregrosa. Wiley-VCH Verlag GmbH & Co., KGaA.

3. AIChE. (2009). Chemical engineering innovation in food production. American Institute of Chemical Engineers and Chemical Heritage Foundation.

4. Queensland Government. (2005). *Agricultural Chemical User's Manual: Guidelines and Principles for Responsible Agricultural Chemical Use.* Department of Primary Industries and Fisheries, Brisbane. ISBN 07345 03210.

5. CropLife. (2002). *A History of Crop Protection and Pest Control in Our Society.* CropLife, Canada.

6. Hartung, J. (2013). A short history of livestock production. In: *Livestock Housing: Modern Management to Ensure Optimal Health and Welfare of Farm Animals*, edited by A. Aland and T. Banhazi. Wageningen Academic Publishers, Wageningen, Gelderland, The Netherlands.

7. Department of Economic and Social Affairs. (2007). *World Population Prospects: The 2006 Revision: Highlights.* United Nations, New York.

8. FAO (Food and Agriculture Organization). (2006). *World Agriculture: Towards 2030/2050.* FAO, Rome.

9. Hassan, R., Scholes, R., and Ash, N. (eds.). (2005). *Ecosystems and Human Well-Being: Current State and Trends*, Vol. 1. Island Press, Washington.

10. Stern, N. (2007). *The Economics of Climate Change: The Stern Review.* Cambridge University Press, Cambridge, UK.

11. Wik, M., Pingali, P., and Broca, S. (2008). *Global Agricultural Performance: Past Trends and Future Prospects.* World Bank, Washington, DC.

12. EuCheMS (European Association for Chemical and Molecular Sciences). (2011). *Chemistry: Developing Solutions in a Changing World.* EuCheMS, Brussels.

Chapter 5

Food: Supply and Health

Livia Simon Sarkadi

Szent István University, Faculty of Food Science,
Villányi út 29-43, 1118 Budapest, Hungary

sarkadi@mail.bme.hu

5.1   Introduction

The following main trends will influence the food industry in the next 20 years. Significant demographic and economic developments are increasing the demand for food. Population growth, increasing life expectancy, and economic growth are expanding the demand for food products. Economic growth in emerging economies and global convergence in food consumption are expanding demand for animal protein, dairy products, and processed foods. There is also an increased demand for more healthful and specialized food products. There is an increasing reemphasis on economically priced, safe, quality food [1].

A greater knowledge of the nutritional content of foods is needed to understand fully the food/health interactions, which could facilitate more efficient production of foods tailored to promote human and animal health.

Many opportunities exist for the chemical sciences to supply healthy, safe, and affordable food for all. The main challenges

Chemistry: Our Past, Present, and Future
Edited by Choon Ho Do and Attila E. Pavlath
Copyright © 2017 Pan Stanford Publishing Pte. Ltd.
ISBN 978-981-4774-08-6 (Hardcover), 978-1-315-22932-4 (eBook)
www.panstanford.com

concern agricultural productivity, water, healthy food, food safety, process efficiency, and supply chain waste.

Consumers have benefited from new technologies to be able to consume abundant and nutritious food. Chemistry has enhanced the flavor, appearance, availability, and nutritional value of their food.

Chemical sciences are keys to identifying alternative supplies of "healthier foods" with an improved nutritional profile.

5.2 Brief History of Food Chemistry and Nutrition

Nutritional discoveries from the earliest days of history have had a positive effect on our health and well-being. In 1747, James Lind (1716–94), a physician in the British Navy, performed the first scientific experiment in nutrition. At that time, sailors were sent on long voyages for years and they developed scurvy (a painful, deadly, bleeding disorder). Lind gave some of the sailors sea water, others vinegar, and the rest limes. Those given limes were saved from scurvy. He is well respected today for his work in improving practices in preventive medicine and improved nutrition. British sailors were known as Limeys because they regularly consumed lime juice and enjoyed better health and vigor than sailors in most other navies [2].

In 1770, the French chemist Antoine Laurent Lavoisier (1743–94), the "Father of Nutrition and Chemistry," discovered the actual process by which food is metabolized. He became famous also for the statement *"Life is a chemical process."*

In early 1800s, it was discovered that foods were composed primarily of four elements: carbon, nitrogen, hydrogen, and oxygen, and methods were developed for determining the amounts of these elements. **Chemistry introduced science to analyze the content of food.**

Carl Wilhelm Scheele (1742–86), a Swedish pharmacist, isolated various new chemical compounds from plant and animal substances, which was considered the beginning of accurate analytical research in agricultural and food chemistry [3].

Nicolas Théodore de Saussure (1767–1845), a French chemist, studied CO_2 and O_2 changes during plant respiration (1840) and determined the mineral contents of various plants.

The English chemist Sir Humphrey Davy (1778–1829), in 1807 and 1808, isolated the elements K, Na, Ba, Sr, Ca, and Mg. His contributions to agricultural and food chemistry came largely through his books on agricultural chemistry, of which the first (1813) was *Elements of Agriculture Chemistry, in a Course of Lectures for the Board of Agriculture* [4].

William Beaumont (1785–1853), a US Army surgeon, performed classic experiments on gastric digestion, which destroyed the concept existing from the time of Hippocrates that food contained a single nutritive component. During 1825–33, he experimented on a Canadian, Alexis St. Martin, whose musket wound afforded direct access to the stomach interior, thereby enabling food to be introduced and subsequently examined for digestive changes [5].

In 1840, Justus von Liebig (1803–73), a pioneer in early plant growth studies, was the first to point out the chemical makeup of carbohydrates, fats, and proteins. Carbohydrates were made of sugars, fats were fatty acids, and proteins were made up of amino acids. He published the first book on food chemistry in 1847 [6].

In 1871, Jean Baptiste Duman (1800–84), suggested that a diet consisting of only protein, carbohydrate, and fat was inadequate to support life. In 1862, the Congress of the United States passed the Land-Grant College Act, authored by Justin Smith Morrill. This act helped to establish colleges of agriculture in the United States and provided considerable impetus for the training of agricultural and food chemists. Also in 1862, the United States Department of Agriculture was established and Isaac Newton (1642–1727) was appointed the first commissioner. In 1863, Harvey Washington Wiley (1844–1930) became the chief chemist of the US Department of Agriculture, from which office he led the campaign against misbranded and adulterated food, culminating in the passage of the first Pure Food and Drug Act in the United States (1906).

Franz Ritter von Soxhlet (1848–1926) invented the Soxhlet extractor (Fig. 5.1) in 1879, and in 1886, he proposed pasteurization be applied to milk and other beverages. Soxhlet is also known as the first scientist who fractionated the milk proteins in casein, albumin, globulin, and lactoprotein. Furthermore, he described for the first time the sugar present in milk, lactose.

Johan Gustav Christoffer Thorsager Kjeldahl (1849–1900), Danish chemist, developed a method in 1883 for determining the amount of nitrogen in certain organic compounds using a laboratory technique that was named the Kjeldahl method after him (Fig. 5.2).

Since the late 18th century until the first half of the 20th century, most of the essential chemical elements and dietary substances were discovered and characterized.

The discovery of vitamins beginning in the early 20th century contributed significantly to our knowledge of proper nutrition and to the fight against diseases of malnutrition. In 1913, the first vitamin found in butter and egg yolks was discovered by Elmer Verner McCollum (1879–1967) and Marguerite Davis (1887–1967). Vitamin A (precursore beta-carotene) is an essential nutrient for vision and the protection of epithelia. Its chemical structure was determined in 1931, and it was first synthesized in 1947 (Fig. 5.3).

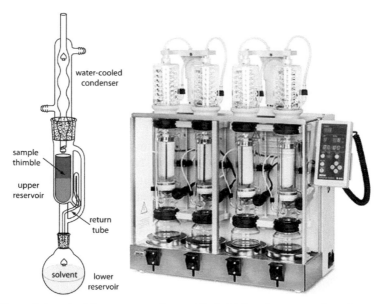

Figure 5.1   Soxhlet extractor. Manual (left side) and automated equipment (right side). *Source*: http://chem.libretexts.org/@api/deki/files/12596/Figure7.24.jpg?revision=1&size=bestfit&width=225&height=543 http://donaulab.hu/img/ProductLine/1191.max.jpg.

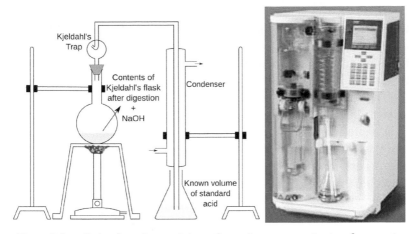

Figure 5.2   Unit for determining the nitrogen content of organic materials by the Kjeldahl method. Manual (left side) and automated equipment (right side). Source: https://commons.wikimedia.org/w/index.php?curid=18937792; https://buchicorp.files.wordpress.com/2010/12/buchi-kjelflex-k3601.jpg?w=129&h=150.

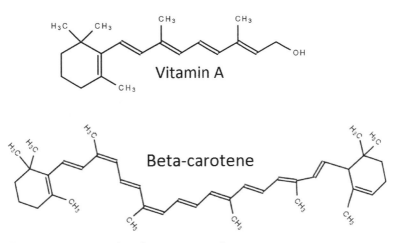

Figure 5.3   Vitamin A and its precursore beta-carotene.

There are some other important facts in the history of nutrition. Elsie Widdowson (1906–2000) and Robert McCance (1898–1993) coauthored *The Chemical Composition of Foods* in 1940, which became the basis for modern nutritional thinking.

In 1941, The National Research Council (USA) set up the first Recommended Dietary Allowances (RDAs), the estimated amount of a nutrient (or calories) per day considered necessary for the maintenance of good health. The RDA is updated periodically to reflect new knowledge.

In 1968, Linus Pauling (1901–94) coined the term *orthomolecular nutrition*. He proposed that by giving the body the right molecules in the right concentration (optimum nutrition), these nutrients would be better utilized and provide superior health and contribute toward longer lives. Pauling was awarded the Nobel Prize in Chemistry in 1954.

The first **food pyramid** was published in Sweden in 1974. A food pyramid or diet pyramid is a pyramid-shaped diagram representing the optimal number of servings to be eaten each day from each of the basic food groups. The food pyramid introduced by the US Department of Agriculture in 1992 was called the *Food Guide Pyramid*. It was updated in 2005 and then replaced by *MyPlate* in 2011 (Fig. 5.4).

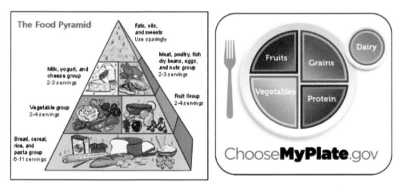

Figure 5.4   The Food Pyramid and MyPlate. *Source*: http://www.chiro.org/ nutrition/ABSTRACTS/food.shtml; https://choosemyplate-prod.azureedge.net/sites/default/files/printablematerials/ myplate_white.jpg.

Chemical components are keys to understanding our modern-day food and nutritional needs. Chemistry not only identified these critical elements for human health but defined what is considered essential for sustaining life.

The life of people with diabetes became easier with the discovery of artificial sweeteners, which help to maintain proper blood sugar level. In modern days, they are also very important for proper diets to limit calorie intake.

Modern food chemistry focuses on the chemistry of foods; reactions and interactions of food components; and the changes occurring in food during production, processing, packaging, preserving, storing, and cooking. Food chemistry encompasses everything from agricultural raw materials to consumer end-use products.

5.3 Food Availability

Population growth, increasing life expectancy, and economic growth are expanding the demand for food products. The world's population has been estimated to be more than 8 billion by 2050. The increase in human population has led to a sustained pressure to produce more food from limited resources. To match global food demands with limited natural resources requires sustainability to be optimized across the whole food production chain. A doubling of grain production will be needed to meet the population's demands in 2050; however, yield increases of the world's cereals have begun to stagnate [7].

Developed agricultural systems are required to improve agricultural yields to meet growing demand for food. Soil chemistry and applications play a critical role in developing soil and crop management practices through enhanced understanding of soil processes, plant nutrition, fertilizer production, development of improved crop varieties, and methods for controlling pests and diseases [8].

In addition to crops, global livestock production faces enormous short-term challenges. Total global meat consumption is predicted to rise to 303 million tonnes by 2020 [9]. Technologies are needed to counter the significant environmental impact and waste associated with rearing livestock. Livestock production currently accounts for one-fifth of greenhouse gas emissions worldwide. Most wild fisheries are at or near their maximum sustainable exploitation level [10].

Many opportunities exist for the chemical sciences to supply healthy, safe, and affordable food for all. The main challenges concern agricultural productivity, water, healthy food, food safety, process efficiency, and supply chain waste.

5.4 Food and Health

Nutrition is a major, modifiable, and powerful factor in promoting health, preventing and treating diseases, and improving the quality of life. The first nutritional experiment is recorded in the Book of Daniel in the Bible. Daniel was among the finest young men captured by the King of Babylon when the Babylonians overran Israel and was to serve in the King's court. He was to be fed from the King's table of fine foods and wine. Daniel objected and preferred his own choices, which included vegetables (pulses) and water. The chief steward was afraid for his head but agreed to a trial. Daniel and his friends received Daniel's own diet for 10 days and were then compared to the King's men. As they appeared fitter and healthier, they were allowed to continue with their own foods, not defiling themselves with those of the King [2].

The tenet "Let food be thy medicine and medicine be thy food" espoused by Hippocrates (c. 460–370 BC) nearly 2500 years ago is receiving renewed interest.

The global challenges are to provide sufficient high-quality, nutritious, and affordable food for the rapidly increasing global population. In addition, there are also global problems of diet-related diseases, such as obesity, which is an established significant risk factor for many adverse health conditions, and there is a pressing need to ensure that our ageing society maintains its health and well-being in order to reduce spiraling health and social costs.

Chemical sciences are the keys to identifying alternative supplies of "healthier foods" possessing an improved nutritional profile. One of the main challenges is to produce food with less fat, salt, and sugar that can be detrimental to health, while maintaining consumer perception and satisfaction from the products. Another challenge will be to develop improved food sources and fortifying foodstuffs to combat malnutrition and target immune health [11].

5.4.1 Food and Its Classification

Almost all foods are of plant or animal origin. The majority of foods consumed by humans are seed-based foods. **Edible seeds** include cereals (such as maize, wheat, and rice), legumes (such as beans, peas, and lentils), and nuts. Oilseeds are often pressed to produce rich oils, such as sunflower, flaxseed, rapeseed (including canola oil), and sesame. Cereal grain is a staple food that provides more food energy worldwide than any other type of crop. Maize, wheat, and rice together account for 87% of all grain production worldwide [12].

Vegetables are the second type of plant matter that are commonly eaten as food. These include root vegetables (such as potatoes and carrots), leafy vegetables (such as spinach and lettuce), stem vegetables (such as bamboo shoots and asparagus), and inflorescence vegetables (such as globe artichokes and broccoli).

Some other classifications of food are as follows:

- **Fast food** is the term given to food that can be prepared and served very quickly. The term "fast food" was recognized in a dictionary by Merriam-Webster (1951).

- **Junk food (Fig. 5.5)** is an informal term applied to some foods that are perceived to have little or no nutritional value, or to products with nutritional value but which also have ingredients considered unhealthy when regularly eaten, or to those considered unhealthy to consume at all. The term was coined by Michael Jacobson (1943–), Director of the Center for Science in the Public Interest, in 1972. Junk food includes foods such as soft drinks, hamburgers, hot dogs, ice cream, cake, French fries, chocolate, confectionery, pizza, cookies, fried chicken, onion rings, and doughnuts.

- **Whole foods (Fig. 5.6)** are foods that are unprocessed and unrefined, or processed and refined as little as possible before being consumed. Whole foods typically do not contain added ingredients, such as sugar, salt, or fat. Examples of whole foods include unpolished grains; fruits and vegetables; unprocessed meat, poultry, and fish; and non-homogenized milk. The term is often confused with organic food, but whole foods are not necessarily organic, nor are organic foods necessarily whole.

- **Organic foods (Fig. 5.7)** are made in a way that complies with organic standards set by national governments and international organizations. Since 1990, the market for organic food and other products has grown rapidly. The principal methods of organic farming include crop rotation, green manures and compost, biological pest control, and mechanical cultivation. Organic farmers are restricted by regulations to using natural pesticides and fertilizers.

Figure 5.5   Junk foods. *Source*: http://previews.123rf.com/images/egal/ egal1012/egal101200015/8394658-Junk-food-collection-isolated-on-white-background-Stock-Photo-unhealthy.jpg.

Figure 5.6   Whole foods. *Source*: http://www.cornichon.org/cgi/mt4/mt-search.cgi?blog_id=2&tag=Whole%20Foods&limit=20.

Figure 5.7   Organic foods. *Source*: https://coffeechalk.wordpress.com/ 2014/12/15/ten-for-tuesday-11/.

- **Functional food** is defined as any food or food ingredient that may provide a health benefit beyond that conferred by the nutrients the food contains.

Chemistry is one of the tools to develop new technologies to enhance the healthful functionality of foods through selective breeding or enhancement of the plant substrate or by optimizing specific chemical constituents.

Consumer interest in the relationship between diet and health has increased the demand for specific foods or physiologically active food components, so-called functional foods. The term functional food was first introduced in Japan in the mid-1980s [13].

All foods are functional to some extent because they provide taste, aroma, and nutritive value. Functional foods have no universally accepted definition. Within the last decade, however, the term functional as it applies to food has adopted a different connotation—providing an additional physiological benefit beyond that of meeting basic nutritional needs.

Functional foods include the following:
- **Conventional foods** such as grains, fruits, vegetables, and nuts.
- **Modified foods** such as yogurt, cereals, and orange juice.
- **Medical foods** such as special formulations of foods and beverages for certain health conditions.

- **Foods for special dietary use** such as infant formula and hypoallergenic foods.

The most common functional foods are as follows:

- **Cold-water fish—sardines and salmon**. These protein-packed fish have high amounts of omega-3 fatty acids, which can lower the overall risk of heart disease, reduce joint pain, and improve brain development and function.
- **Nuts**. Nuts, including cashews and almonds, are high in magnesium, which can lower blood pressure. Almonds, pecans, and walnuts can help lower cholesterol. They can also help to control blood sugar levels.
- **Whole grains**. These are a good source of fiber, help lower cholesterol, and assist with blood sugar control, making them a good choice for people with diabetes.
- **Beans**. Beans are another source of fiber, as well as protein, potassium, and folate.
- **Berries**. Strawberries, cranberries, blueberries, raspberries, or blackberries are beneficial functional foods. Not only are they low in calories, their anthocyanin pigments, which give them color, offer health-promoting benefits.

5.4.1.1 Functional foods from plant sources

Domesticated food plant species were derived from wild plant species decades or centuries ago and have been highly cultivated and bred for specific characteristics that appeal to the culture where they are sold. Food plants are primarily of interest for their nutritive properties. Food plants can be viewed as chemical entities in which we are interested for their abilities to provide specific chemicals for growth, development, and health. Overwhelming evidence from epidemiological, in vivo, in vitro, and clinical trial data indicates that a plant-based diet can reduce the risk of chronic disease, particularly cancer [13].

- **Oats**. Oat products are a widely studied dietary source of the cholesterol-lowering soluble fiber beta-glucan. There is now significant scientific agreement that consumption of this particular plant food can reduce total and low-density lipoprotein (LDL) cholesterol, thereby reducing the risk of coronary heart disease (CHD).

- **Soy**. Soy was in the spotlight during the 1990s. The cholesterol-lowering effect of soy is the most well-documented physiological effect. Several classes of anti-carcinogens have been identified in soybeans, including protease inhibitors, phytosterols, saponins, phenolic acids, phytic acid, and isoflavones.
- **Flaxseed (also called linseeds).** Among the major seed oils, flaxseed oil contains the most (57%) of the omega-3 fatty acid, alfa-linolenic acid.
- **Tomatoes**. Tomatoes have received significant attention because of interest in lycopene, the primary carotenoid found in this fruit, and its role in cancer risk reduction.
- **Garlic (*Allium sativum*)**. Garlic has been used for thousands of years for a wide variety of medicinal purposes. Its effects are likely attributable to the presence of numerous physiologically active organosulfur components (e.g., allicin, allylic sulfides). The purported health benefits of garlic are numerous, including cancer chemopreventive, antibiotic, anti-hypertensive, and cholesterol-lowering properties.
- **Broccoli and other cruciferous vegetables**. Epidemiological evidence has associated the frequent consumption of cruciferous vegetables with decreased cancer risk.
- **Citrus fruits**. Several epidemiological studies have shown that citrus fruits are protective against a variety of human cancers. Citrus fruits are particularly high in a class of phytochemicals known as the limonoids.
- **Cranberry**. Cranberry juice has been recognized as efficacious in the treatment of urinary tract infections since 1914, when Blatherwick reported that this benzoic acid-rich fruit caused acidification of urine [14].
- **Tea**. Tea is second only to water as the most widely consumed beverage in the world. A great deal of attention has been directed to the polyphenolic constituents of tea, particularly green tea. Polyphenols comprise up to 30% of the total dry weight of fresh tea leaves. Catechins are the predominant and most significant of all tea polyphenols. The four major green tea catechins are epigallocatechin-3-gallate, epigallocatechin, epicatechin-3-gallate, and epicatechin.

- **Wine and grapes**. There is growing evidence that wine, particularly red wine, can reduce the risk of cardiovascular disease (CVD). France, in particular, has a relatively low rate of CVD despite diets high in dairy fat. Although this "French Paradox" can be partly explained by the ability of alcohol to increase HDL cholesterol, more recent investigations have focused on the non-alcohol components of wine, particularly flavonoids.

5.4.1.2   Functional foods from animal sources

Although the vast number of naturally occurring health-enhancing substances are of plant origin, there are a number of physiologically active components in animal products that deserve attention for their potential role in optimal health [13].

- **Fish**. Omega-3 (n-3) fatty acids are an essential class of polyunsaturated fatty acids (PUFAs) derived primarily from fish oil. It has been suggested that the Western-type diet is currently deficient in n-3 fatty acids.
- **Dairy products**. Dairy products are one of the best sources of calcium, an essential nutrient that can prevent osteoporosis and possibly colon cancer. In fermented dairy products, there are other beneficial component known as probiotics.

  Probiotics are defined as "live microbial feed supplements which beneficially affect the host animal by improving its intestinal microbial balance." Of the beneficial microorganisms traditionally used in food fermentation, lactic acid bacteria have attracted the most attention. In addition to probiotics, there is growing interest in fermentable carbohydrates that feed the good microflora of the gut.

  These **prebiotics**, defined by G.R. Gibson and M.B. Roberfroid in 1995 as "nondigestible food ingredients that beneficially affect the host by selectively stimulating the growth and/or activity of one or a limited number of bacteria in the colon and thus improves host health," may include starches, dietary fibers, other non-absorbable sugars, sugar alcohols, and oligosaccharides [15]. Oligosaccharides consist of short-chain polysaccharides composed of three and 10 simple sugars linked together. They are found naturally in many fruits and vegetables (including banana, garlic, onions, milk, honey, artichokes). The prebiotic concept has been further extended to encompass the concept of synbiotics, a mixture of pro- and prebiotics. Many synbiotic products are currently on the market in Europe.

- **Beef.** An anticarcinogenic fatty acid known as conjugated linoleic acid (CLA) was first isolated from grilled beef in 1987 [16]. CLA refers to a mixture of positional and geometric isomers of linoleic acid (18:2, n-6) in which the double bonds are conjugated instead of existing in the typical methylene interrupted configuration. Nine different isomers of CLA have been reported as occurring naturally in food. CLA is unique in that it is found in highest concentrations in fat from ruminant animals (e.g., beef, dairy, and lamb).

Diet is only one component of an overall lifestyle that can have an impact on health. Functional foods for health are an important part of an overall healthful lifestyle that includes a balanced diet and physical activity.

5.5   Food Preservation

Food preservation is any of a number of methods by which food is kept from spoilage after harvest or slaughter. In ancient time, salt was the only available choice for food preservation. When meat or fish is immersed in salt, the salt gradually removes the water from the organic material by the process of osmosis, thus killing the bacteria liable to cause the foodstuff to rot. Food can be preserved for long periods of time when salted, and so salted food can be transported long distances without fear of spoilage [17].

In the middle of the 19th century, the German Justus Liebig was the first who made meat extract (Fig. 5.8). Advances in preservation and manufacturing techniques have allowed creating processed foods.

As food chemistry fueled industrial practices, a wide variety of processed food was developed. New technologies extended the shelf life of food, including freeze drying (liophilization) (1906), deep freezing (1920), and precooking and freezing foodstuff (1939), and making concentrates from fluids (1946), furthermore, high-temperature pasteurization and canning, refrigeration and freezing, chemical preservatives, and irradiation. Advances in packaging materials have played an important role in modern food preservation [18].

Figure 5.8   The one-time placard of Liebig's meat extract. *Source*: http://fotos.verwaltungsportal.de/gallery/5/9/2/4/gross/ 4103965773.jpg.

Several non-thermal processing methods and technologies are also available or currently in development to deactivate microorganisms and extend microbiological shelf life in foods. Those techniques include ultra-high-pressure techniques, ionizing radiation such as pulsed X-ray, ultrasound, pulsed light and pulsed electric fields, high-voltage arc discharge, magnetic fields, dense phase carbon dioxide, and hurdle technologies.

- **Traditional canning**. In the late 1790s, French confectioner Nicolas Appert (1749–1841) discovered that many foods (including meats, fruits, vegetables, and milk) will resist spoilage for extended periods once they are sufficiently heated and sealed in airtight glass containers. Building on his discoveries, England's Peter Durand developed a way to seal food in lightweight, easy-to-seal, unbreakable, airtight containers made from tin-plated iron, for which he received a patent in 1810. Bryan Dorkin (1768–1855) and John Hall set up the first commercial canning factory in England in 1813. By the 1920s, mechanized factories became widespread. In addition to automation processes that enable the commercial-scale production of cans, the container has evolved from steel to aluminum and then to a variety of metal alloys. These advances have significantly reduced the weight and material cost of cans.

- **Pasteurization.** In the early years, no one knew how Nicolas Appert's process preserved foods successfully. More than 50 years later, Louis Pasteur (1822–95) solved the mystery by demonstrating that the growth of microorganisms is the primary cause of food spoilage and food-borne illnesses and that a high percentage of them could be killed by heating liquids to about 130°F (55°C) or higher, for relatively short periods, without altering the chemical makeup of the food. This simple process became known as pasteurization and was quickly and widely adopted (Fig. 5.9).

Figure 5.9   The one-time placard of pasteurization. *Source*: http://scienceblogs.com/ethicsandscience/wp-content/blogs.dir/377/files/2012/04/i-530ed977bea1a9bcd8a2b7a8d7c02474-Pasteur.jpg.

- **Frozen foods**. Freeze-dried foods have a long shelf life and quickly regain their original flavor, aroma, size, shape, and texture, once water is added back to them, which reduces spoilage and the weight of the food, making it easier and cheaper to transport. In freeze-drying processes, food material is introduced to a vacuum chamber where it is frozen, and then most of the water is removed by sublimation by reducing the chamber pressure to a few millibars, which is finally subjected to control heating to remove any remaining water.

- **Quick-freezing**. This process for foods was first commercialized in the 1920s by Clarence Birdseye (1886–1956), whose name has become practically synonymous with frozen foods. Birdseye found that by blanching vegetables (cooking them briefly in boiling water) just before freezing, the process could deactivate certain enzymes that cause off-colors and off-flavors, thereby enhancing the quality of the thawed vegetables. The first commercial use of "puffing" to produce such cereals as Cheerios and puffed rice also began.

- **Freeze drying**. This process was pioneered in the 1930s. Frozen foods are dried after deep freezing, in which the entrained water is removed by a process known as sublimation by heating the frozen product in a vacuum chamber. The removal of water slows spoilage, thus providing longer shelf life and reducing the weight of the food, which makes it cheaper and easier to transport.

- **Refrigerants**. The first compressor-assisted refrigerator made for home use was introduced in 1918 by the firm Kelvinator (Fig. 5.10). Since being introduced for home use, refrigeration has altered food preservation by providing the ability to transport and store fresh foods safely. In the early 1920s, refrigerators were unpopular since the sulfur dioxide coolant proved to be toxic. The solution was Freon 12, a chlorofluorocarbon (also referred to as a CFC, CCl_2F_2) compound made into a refrigerant gas by Thomas Midgley (1889–1944) and Charles Franklin Kettering (1876–1958) in 1931. Refrigerators quickly became standard in homes, restaurants, and grocery stores.

Figure 5.10 Household refrigerator by Kelvinator. *Source*: http://timerime. com/upload/resized/180352/2316243/resized_image2_ 3dac4c0bf04664a0964d721256015271.jpg.

- **Chemical preservatives**. Chemical preservatives are applied to foods as direct additives during processing, or develop by themselves during processes such as fermentation. Food preservatives can be classified as follows:
  - **Natural preservatives** such as salt, sugar, vinegar, syrup, spices, honey, and edible oil.
  - **Chemical preservatives** such as benzoates, sorbates, nitrites and nitrates of sodium or potassium, sulfites, glutamates, glycerides.

    Both natural and chemical preservatives are categorized into three types:

- **Antimicrobials** that destroy or delay the growth of bacteria, yeast, and molds. For example, nitrites and nitrates prevent botulism in meat products. Sulfur dioxide prevents further degradation in fruits, wine, and beer. Benzoates and sorbates are anti-fungals used in jams, salads, cheese, and pickles.
- **Antioxidants** that slow or stop the breakdown of fats and oils in food in the presence of oxygen (oxidation) leading to rancidity. Examples of antioxidants include BHT (butylated hydroxytolune), BHA (butylated hydroxyanisole), TBHQ (tertiary butylhydroquinone), and propyl gallate.
- **Antienzymatic preservatives** that block enzymatic processes such as ripening occurring in foodstuffs even after harvest. For instance, erythorbic acid and citric acid stop the action of enzyme phenolase that leads to a brown color on the exposed surface of cut fruits or potato.

- **Irradiating food.** In 1997, the US Food and Drug Administration (FDA) first approved the use of a technique known as irradiation to kill disease-causing bacteria and parasites and spoilage microorganisms in fresh and frozen poultry and meats. Today, irradiation of food and agricultural products is allowed by about 40 countries. The approval by US regulators validated what food scientists had known for a long time through extensive research that low doses of ionizing radiation absorbed by various foods could effectively kill disease-causing bacteria such as *Salmonella*, *Escherichia coli*, and *Listeria*, and delay food spoilage. Extensive testing of food irradiation shows that treated foods are not at risk of "becoming radioactive" because approved energy levels are too low to induce radioactivity.
- **High-pressure processing (Fig. 5.11).** Pressed inside a vessel exerting 70,000 pounds per square inch or more, food can be processed so that it retains its fresh appearance, flavor, texture, and nutrients while disabling harmful microorganism and slowing spoilage. By 2001, adequate commercial equipment was developed so that by 2005, the process was being used for products ranging from orange juice to guacamole to deli meats and widely sold [19].

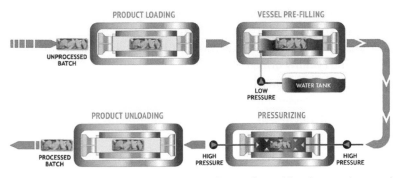

Figure 5.11   High-pressure processing. *Source*: http://hpplosangeles.com/
wp-content/uploads/2016/08/diagram.jpg.

5.6   Food Additives and Ingredients

Food additives are defined by the FDA as "any substance, the intended use of which results or may reasonably be expected to result, directly or indirectly, in its becoming a component or otherwise affecting the characteristics of any food." In other words, an additive is any substance that is added to food. Direct additives are those that are intentionally added to foods for a specific purpose, while indirect additives are those to which the food is exposed during processing, packaging, or storing [20].

Additives and preservatives maintain the quality and consistency of foods. They also maintain palatability and wholesomeness of the food, improve or maintain its nutritional value, control appropriate pH, provide leavening and color, and enhance its flavor.

Food additives can be divided into several groups, although there is some overlap between food preservatives [12]:

- **Antimicrobial agents**, which prevent spoilage of food by mold or microorganisms. These include vinegar and salt, and also compounds such as calcium propionate and sorbic acid, which are used in products such as baked goods, salad dressings, cheeses, margarines, and pickled foods.
- **Antioxidants**, which prevent rancidity in foods containing fats and damage to foods caused by oxygen.
- **Artificial colors**, which are intended to make food more appealing and to provide certain foods with a color that

humans associate with a particular flavor. These coloring substances are erythrosine (red), cantaxanthin (orange), amaranth (Azoic red), tartrazine (Azoic yellow), and annatto bixine (yellow orange).

- **Natural and artificial flavors**, which are contained in many processed foods.

  Natural flavors are essential oils, oleoresins, and essences that are typically extracted or distilled from such products as spices, fruits or fruit juices, vegetables or vegetable juices, plant materials, meat, seafood, poultry, eggs, dairy products, or fermentation products of such foods using roasting, heating, enzyme-based processes, and other techniques.

  Artificial flavors are chemical compounds synthesized in the lab by chemical engineers and other technical professionals to mimic the flavors found in nature and then manufactured in commercial-scale quantities. Synthetic flavoring agents, such as benzaldehyde for cherry or almond flavor, may be used to simulate natural flavors.

  Artificial flavor enhancers make food taste better or give it a specific taste. Flavor enhancers, such as monosodium glutamate (MSG), intensify the flavor of other compounds in a food (Fig. 5.12).

  There are only **four basic categories of taste** our tongue can differentiate (sweet, salty, sour, and bitter), yet the nose can differentiate thousands of distinct odors. As a result, most artificial flavors require a blend of both taste and smell components to create the total signature flavor experience.

- **Bleaching agents**, such as peroxides, are used to whiten foods such as wheat flour and cheese.

- **Chelating agents**, such as citric acid, malic acid, and tartaric acid, are used to prevent discoloration, flavor changes, and rancidity that might occur during the processing of foods.

- **Nutrient additives**, including vitamins and minerals, are added to foods during enrichment or fortification. For example, milk is fortified with vitamin D, and rice is enriched with thiamin, riboflavin, and niacin.

- **Natural and artificial sweeteners. Natural sweeteners** such as refined sugar (sucrose, in the form of crystallized sugar or sugary syrups) are produced most often from sugar cane or sugar beets and have long dominated the natural sweetener market.

Figure 5.12 Main food groups containing monosodium glutamate. *Source*: http://www.heatherparisi.com/download/ac69204_msgfood. jpg.

Artificial sweeteners (Fig. 5.13) with a sweetness level of 500–600 times greater than those of traditionally refined sugar are valued by calorie-conscious consumers and diabetics. While their early discovery was accidental, chemical engineering expertize perfected the synthesis of these chemical compounds and commercial production.

The earliest artificial sweetener, **saccharine**, was discovered in 1901 by John Francis Queeny (1859–1933), and its widespread commercial availability as "Sweet N Low" helped usher in a diet revolution in the United States in the 1950s.

Recently, **aspartame** (made of two amino acids phenylalanine and aspartic acid) has emerged as a leading artificial sweetener. Aspartame was first sold in 1985 in the United States; this low-calorie intense sweetener marketed as NutraSweet was developed in 1955 as a possible anti-ulcer drug. It is now used in thousands of processed foods, including diet sodas, desserts, yogurt, and candy.

Another artificial sweetener **(sucralose)** sold as Splenda is made from sugar unlike the other artificial sweeteners. Using a patented process, its makers produce a no-calorie, chlorinated derivative that tastes like sugar with virtually no bitter aftertaste. The process selectively replaces three hydrogen–oxygen groups on the sugar molecule with three chlorine atoms. This change in molecular structure converts sucrose to sucralose, which, according to Splenda's manufacturer, is an inert (nonreactive) molecule. Once the product is consumed, it passes through the body without being used for energy, so it has no calories and the body does not recognize it as a carbohydrate [18].

- **Thickening and stabilizing agents**, which function to alter the texture of food. For example, emulsifier lecithin keeps oil and vinegar blended in salad dressings, and carrageen is used as a thickener in ice creams and low-calorie jellies.

Figure 5.13 Sacharine, aspartame, and sucralose. *Source*: https://scienceandfooducla.files.wordpress.com/2013/11/baking3.png?w=610&h=396; http://media.pennlive.com/bodyandmind_impact/photo/10996948-large.jpg; http://www.seriouseats.com/images/20080923-splenda.jpg.

Starches from various sources are a major source of carbohydrates in processed foods and often added as an ingredient to achieve a desired function. Starches are added to many foods, ranging from soups, stews, and gravies to pie filling, sauces, and custards. Since starches absorb water and become a gel when cooked, starch additives help thicken many foods and improve the stability of the food during heating, intense mixing, or long shelf life. They also help give many foods a good "mouth feel."

Chemical engineers have played a key role in isolating the desired starch from various cereal grains (such as corn, wheat, rice, and sorghum), roots and tubers (such as potatoes, sweet potatoes, cassava, and arrowroot), and other sources, and then in engineering systems to produce easy-to-use, free-flowing powdered starches in commercial-scale quantities. Such systems use a variety of chemical engineering processes, including wet grinding or milling, screening or sieving, washing, centrifuging, dewatering, and drying [18].

Food additives play a vital role in today's food supply. Food additives help to assure the availability of wholesome, appetizing, and affordable foods that meet consumer's demands from season to season.

5.7 Food Packaging

Packaging food with plastics, metal, glass, and ceramic technologies helps to preserve food during sale, shipping, and preparation. Ralph Wiley invented industrial saran polymer in the 1930s, and household saran wrap was introduced in 1953 to provide an excellent barrier to oxygen, moisture, aroma, and chemicals under extreme humidity and temperature conditions [18].

The past 50 years brought dramatic advances in the field of plastic products, including food packaging. PVC, polyethylene, and polypropylene wrappings, multilayer wrappings, relating systems (plastic and metal laminates), and PET (polyethylene terephthalate) bottles enable the long-term storage of raw meat, sausages, dairies, candies, processed foodstuff, water, and soft drinks in shops and at home.

Vacuum packaging was invented in the 1900s to prolong the shelf life of foods by moving the oxygen from inside the food package. The widespread practice of freezing foods began with fruit and fish.

Controlled-atmosphere packaging was invented in the 1950s. The process controls oxygen and carbon dioxide levels inside the packaging environment to limit respiration by fruits and vegetables and reduces the amount of off-gas ethylene produced, which delays ripening and spoilage.

Aseptic packaging provides major advantages over traditional canning in the 1960s. During aseptic packaging, both the food and packaging are sterilized at high temperatures for very short periods. Foods processed using aseptic packaging retain their vitamins, minerals, and desired textures, colors, and flavors more effectively than those processed with traditional canning. In 1989, aseptic packaging technology was voted the food industry's top innovation of the last 50 years by the Institute of Food Technologists.

Modified-atmosphere packaging began to be used widely in the 1980s. It is a more advanced variation of controlled-atmosphere packaging, in which the "head space" atmosphere within a food package or the transportation/storage vessel is modified by flushing it with a blend of inert (nonreactive) gases.

Improvements on traditional vacuum packaging, controlled-atmosphere packaging (CAP), and modified-atmosphere packaging (MAP) represent chemical engineering breakthroughs [18].

The invention of chemically sterilized "brick packs" was another important chemical engineering contribution to food safety. The multilayer packages are widely used to package juice, milk, tomato sauce, and countless other products. Brick packs protect contents from spoilage and provide extended shelf life without the need for refrigeration. The ingenious, brick-shaped package is constructed from high-quality paperboard, plastic, and aluminum with each layer playing a specific role.

Recently, flexible, laminated "retort pouches" are used as an alternative to traditional metal cans or glass jars. The multilayered retort pouches are filled with wet foods, sealed, and then heat-treated to sterilize the contents so that the food within the pouch is never exposed to air until eaten and stored without refrigeration until opened. This thin, multilayer retort pouches allow high-temperature sterilization using shorter heating times than for traditional canning. This process technology is widely used to supply sterile, prepackaged foods for soldiers, astronauts, hikers, campers, and consumers [18].

5.8 Closing Words

Chemistry has been pivotal to food production from farm to table. Malnutrition is still a condition that affects vast numbers of people worldwide; chemists will be essential in formulating fortified food products to help combat malnutrition and improve health. Much of the innovation and technology within chemistry can be earmarked and in motion to ensure that food security is achieved for all during the next decades.

References

1. Wahlstrom, D., and Dahl, S. (2009). *Destination 2025. Focus on the Future of the Food Industry*. Deloitte Development LLC.

2. History of Nutrition, visited 03.01.2015. http://www.nutritionbreak-throughs.com/html/a_history_of_nutrition.html.

3. Owen, R. (1996). *Fennema: Food Chemistry*. Marcel Dekker, Inc., New York, Basel, Hong Kong.

4. Davy, H. (1813). *Elements of Agricultural Chemistry in a Course of Lectures for the Board of Agriculture*. Richard Grifein and Company, London.

5. Beaumont, W. (1833). *Experiments and Observations of the Gastric Juice and the Physiology of Digestion*. F. P. Allen, Plattsburgh, New York.

6. von Liebig, J. (1847). *Researches on the Chemistry of Food,* edited from the author's manuscript by W. Gregory. Londson, Taylor and Walton, London.

7. Cassman, K.G., Dobermann, A., Walters, D.T., and Yang, H. (2003). Meeting cereal demand while protecting natural resources and improving environmental quality. *Annual Review of Environment and Resources*, **28**, 315–358.

8. Fanzo, J., Remans, R., and Sanchez, P. (2011). The role of chemistry in addressing hunger and food security. In: *The Chemical Element: Chemistry's Contribution to Our Global Future*, edited by J. Garcia-

Martinez and E. Serrano-Torregrosa. Wiley-VCH Verlag GmbH & Co. KGaA.

9. Delgado, C., Rosegrant, M., Steinfeld, H., Ehui, S., and Courbois, C. (1999). *Livestock to 2020: The Next Food Revolution.* International Food Policy Research Institute, Washington, DC.

10. Barange, M. (2005). *Science for Sustainable Marine Bioresources.* Plymouth Marine Laboratory, Plymouth.

11. Simon-Sarkadi, L. (2014). Global challenges require global cooperation. In: *Vision 2025: How to Succeed in the Global Chemistry Enterprise,* edited by H.N. Cheng, S. Shah, and M. Li Wu. American Chemical Society, ACS Symposium Series, 1157, pp. 69–75.

12. Abdulmumeen, H.A., Risikat, A.N., and Sururah, A.R. (2012). Food: Its preservatives, additives and applications. *International Journal of Chemical and Biochemical Sciences,* **1**, 36–47.

13. Hasler, C.M. (1998). Functional foods: Their role in disease prevention and health promotion. *Food Technology,* **52**(2), 57–62.

14. Blatherwick, N.R. (1914). The specific role of foods in relation to the composition of the urine. *Archives of Internal Medicine,* **14**, 409–450.

15. Gibson, G., and Roberfroid, M.B. (1995). Dietary modulation of the human colonic mibrobiota: Introducing the concept of prebiotics. *Journal of Nutrition,* **125**, 1401–1412.

16. Ha, Y.L., Grimm, N.K., and Pariza, M.W. (1987). Anticarcinogens from fried ground beef: Health-altered derivatives of linoleic acid. *Carcinogenesis,* 8, 1881–1887.

17. Kurlanski, M. (2002). *Salt: A world history.* Walker & Co., New York, pp. 41–46; 291–317.

18. AIChE. (2009). Chemical engineering innovation in food production. American Institute of Chemical Engineers and Chemical Heritage Foundation.

19. Riddervold, A. (2008). High pressure food preservation. In *Food Conservation: Papers Given to the International Conference on Ethnological,* pp. 12–16. ISBN 9780907325406.

20. US Food and Drug Administration. (1993). *Everything Added to Food in the United States.* C.K. Smoley (c/o CRC Press, Inc.), Boca Raton, FL.

Chapter 6

Energy

James Wei

School of Engineering and Applied Science, Princeton University,
Princeton, New Jersey 08544-5263, USA

jameswei@princeton.edu

We lavishly use energy at home and work and in moving people and goods around. Table 6.1 gives the consumption and sources of energy in the United States.

Table 6.1   Consumption and sources of energy in the United States

Consumption	%	Source	%
Residence, commerce	36	Natural gas	25
Industry	38	Oil	39
Transportation	26	Coal	21
		Nuclear	9
		Wind, hydro, geo	4
		Biofuels	3
Total	100	Total	100

We have already covered transportation in Chapter 10, so in this chapter, we concentrate on the use of energy in residence,

Chemistry: Our Past, Present, and Future
Edited by Choon Ho Do and Attila E. Pavlath
Copyright © 2017 Pan Stanford Publishing Pte. Ltd.
ISBN 978-981-4774-08-6 (Hardcover), 978-1-315-22932-4 (eBook)
www.panstanford.com

commerce, and industry. In the source column in Table 6.1, we can find that fire or combustion consumes all the fossil fuels and biofuels, which is about 88% of the total. Nuclear energy makes the next biggest contribution. The role of chemistry is critical to the extraction and refining of fuels from nature, to the processes in releasing energy, and to manage safety and environmental consequences.

6.1 Fossil Fuel: Fire and Combustion

Fire always had a central role in human affairs and considered divine and magical in many civilizations and cultures. In the Greek mythology, humans were weak and unable to take care of themselves, till the titan Prometheus took pity on them and stole fire from the gods and gave it to men. With the help of fire, humans became powerful and prosperous. Hephaestus (or his Roman equivalent Vulcan) was the Greek god of fire and forge, who made all the tools and weapons. The four elements of nature are air, fire, water, and earth. The Roman Vestal virgins keep the sacred fire, which must be burning at all times. Agni is the god of fire in the Hindu religion.

Archaeological evidence points to the first use of fire about a million years ago, perhaps to provide light and heat at night. The *Homo habilis* of East Africa lived in a warm climate. But when they migrated to the Middle East and further north to Europe and Asia, especially during the numerous ice ages, they needed fire to endure the intense cold. Fire made it possible for them to walk from East Asia across the Bering Strait to Alaska, and then all the way down to Central America, and finally to frozen Tierra del Fuego. In homes, fire made possible light, heating, cooking; in workshops, fire made possible materials such as metals and ceramics. In modern homes, the main source for electricity is from power stations that are mainly powered by combustion, which made possible machines for cleaning, washing, air conditioning, refrigeration, radio, television, and computers.

6.1.1 Fuel Mining, Refining

Ancient plants and animals got buried under the ground and transformed over many million years into fossil fuels such as coal, oil, natural gas, shale, and tar sand. About 60% of the current sources are from oil and gas deposits. Natural seepages from underground reservoirs occurred in many places in the world and became popular in the Middle East and the Persian Gulf. To increase the flow of gas and oil, wells were drilled. The earliest known natural gas wells were drilled in China in 347 BC, at a depth of up to 800 feet with bits attached to bamboo poles (Fig. 6.1). The natural gas produced was burned to evaporate brine, which produced salt. By the 9th century, Persian alchemists distilled petroleum to produce products such as kerosene in a device called the alembic, which were used in lamps. Marco Polo reported the collection of oil from seeps when he visited the Azerbaijani city of Baku on the shores of the Caspian Sea in 1264. "Colonel" Edwin Drake is popularly credited with the first drilled oil well in the United States, which took place in Titusville, Pennsylvania, in 1858. The method was patterned after the salt well drillers, using a steam engine to power the penetration of the soil. But the sides of the walls began to collapse at 16 feet, so he used a series of 10 feet pipes to support the well. He struck bedrock at 32 feet. At the depth of 70 feet, he struck oil, which gushed out and was collected. That led to a rush, and thousands of drilling rigs were all over Pennsylvania. The crude oil is subsequently refined to produce a range of products for the market, and the most important early product was kerosene for the lamp, which replaced the need to hunt whales for oil.

Figure 6.1   Chinese drilling.

Chemistry makes many contributions to the extraction of oil and gas. Primary recovery is the passive collection of oil and gas from a drilled well, which is driven by a number of natural pressures. These include natural water displacing oil, expansion of the natural gas at the top of the reservoir as well as initially dissolved in the crude oil, and gravity drainage. The amount of oil recovered in this phase is typically 5–15%. When there is insufficient underground pressure to force the oil to the surface, secondary recovery methods are applied to recover more oil. They rely on the supply of external pressure into the reservoir in the form of injecting fluids to increase reservoir pressure. The methods used include water injection, natural gas reinjection and gas lift, which injects air, carbon dioxide, or some other gases into the bottom of a well that needs rejuvenation.

When water is used to push the oil out of the rock, an ideal outcome is a uniform broad front from the injection hole to the receiving hole, which pushes out all the oil, shown in Fig. 6.2(left). However, when the oil is very viscous, a phenomenon called "fingering" often takes place where the water moves in a narrow path and bypasses pockets of oil left behind, shown in Fig. 6.2(right).

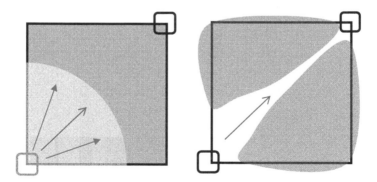

Figure 6.2   Fluid flow.

The solution is to make the water more viscous and to make the oil less viscous. Many chemicals such as polymers are added to the water to increase its viscosity and to recover more of the oil. Guar gum is extracted from the seeds of a plant and is a long chain polymer of sugar molecules. Synthetic modifications of natural gum are also used to improve its performance. Other methods include adding detergents to make the oil and water more miscible, and injecting carbon dioxide instead of water. On average, the recovery factor after primary and secondary oil-recovery operations is between 35% and 45%.

Tertiary recovery or enhanced oil recovery (EOR) methods try to make the oil more mobile in order to increase extraction. One method is thermally enhanced oil recovery (TEOR), which heats the oil and reduces its viscosity, making it easier to extract. Steam injection is the most common form of TEOR. A gas turbine is used to generate electricity, and the waste heat is used to produce steam, which is then injected into the reservoir. Fire flooding (in situ burning) injects air underground to burn some of the oil, which heats the surrounding oil. Tertiary recovery makes possible the recovery of another 5% to 15% of the reservoir's oil.

6.1.1.1 Massive Hydraulic Fracturing

The underground reservoirs of oil and gas may be in relatively high-porosity rocks such as sandstone, so it is relatively easy for the oil and gas to flow to the drill holes for recovery above ground. There are also low-porosity rocks such as shale where the flow rate of oil and gas is too small for satisfactory recovery. The unit of permeability is millidarcy (md). An impervious rock such as granite and limestone has permeability less than 0.1 md. Clay and sandstone may have around 1 md, and good oil reservoir rocks have 10–10,000 md. Hydraulic fracture is a method to pump water at high pressure into a rock layer to break it up. The hydraulic fluid is actually a solution of polymers to thicken the fluid, plus sand to keep the fractures apart after the pressure is withdrawn.

Figure 6.3   Fracking.

In normal drilling, the drill bit goes downward in the vertical direction. It is easy to collect the oil that is near the drill hole but is difficult to collect the oil that is far from the drill hole. When the oil-bearing rock layer is long and horizontal, numerous vertical wells are needed so that no oil rock is very far from a drill hole, which increases the costs of drilling. Directional drilling is one of the key technologies involved, where the drilling does not simply go straight down but can turn corners in other direction. Directional drilling may have been developed to drill slanted wells in order to steal oil from deposits that belong to other owners. This technique is applied to the task to fracture a long and thin horizontal layer of shale, by first drilling to the desired depth and then drilling sideways. After the well is drilled, a device perforates the horizontal part of the production pipe to make small holes in the casing, exposing the wellbore to the shale. Then a mixture of water (90%), sand (9.5%), and chemicals is pumped into the well under high pressure to create small fractures in the shale, which increases the mobility of natural gas or oil. Sand keeps the fractures open after the pressure is released. Water is not thick enough to carry sand, so thickeners have to be added. This led to the most dramatic recent story of how the United States changed from an importer of oil and gas to self-sufficiency, and further into an exporter of gas to friendly nations. This is called massive hydraulic fracturing, especially for natural gas trapped in shale. The new technology has helped the United States in a single decade to improve its international balance of payment and create many hundred thousands of well-paying jobs! The export of energy gives the United States the clout to influence foreign relations. In the 2014 Crimea dispute between Russia and Ukraine, the United States came to the rescue of Eastern European nations from dependence on Russian gas by exporting its surplus natural gas.

6.1.1.2 Refining

When crude oil is brought into an oil refinery, it is fed into a distillation unit that separates out fractions according to their boiling points. Leaving from the top of a distillation column are the lower boiling fractions, which consist of light gases and gasoline. Leaving from the middle are the middle boiling fractions of kerosene and distillate gas oil, and leaving at the bottom are the residual heavy oil and asphalt. At the beginning of the oil industry, the original market for petroleum products was kerosene for lamps, which is the portion that boils between 150 and 275°C. There were no ready markets for gasoline, which was discarded and poured into creeks. After the rise of the Otto engine and automobiles, gasoline became the premium product. The diesel engine and the turbo engine led to demands for diesel fuel and jet fuel. The lighter distillate fuel oil is used for homes and commerce, but the heavier and dirtier residue fuel oil can be used only in industry for electric power generation and ocean-going ships. Sulfur is an important impurity in some crude oils and burns to make sulfur oxide, which is an acid pollutant responsible for acid rain and must be removed.

The most important properties of a fuel are the melting and boiling points, which determine whether the fuel is a solid, liquid, or gas in storage and in use; the heating value in BTU/pound, which determines the density of combustion energy; and impurities such as sulfur and nitrogen, which determine corrosion, safety, and environmental pollution.

Coal is vegetable matter buried underground under heat and pressure. It goes through ranks from peat to lignite, sub-bituminous coal, bituminous coal, and finally anthracite. The proximate analysis is a rough analysis of the contents of coal: moisture; volatile organic matter, which is expelled as gas and liquid droplets upon heating; fixed carbon, which turns into combustible char; and ash, which contains inorganic elements such as silicon and aluminum. Peat is the accumulation of partially decayed vegetation in wetland or bogs, which is high in moisture and low in fixed carbon. Lignite or brown coal is derived from peat. It has less moisture and is considered the lowest ranking in coal. It has a low value in heat generated per ton, so it is usually strip mined in open bits and burned in a nearby power plant. Bituminous is black coal, which is of higher quality than lignite and contains bitumen, a form of tar that creates smoke on burning. Anthracite is a hard and shiny coal, which has the most fixed carbon and is suitable for burning at home, as it is smokeless.

6.1.2 Combustion Process

A fire occurs when oxygen meets fuel at sufficiently high temperatures. The simplest to describe is a pre-mixed gaseous flame, such as between methane and air. The chemical equation for methane combustion is as follows:

$$CH_4 + 2O_2 \rightarrow CO_2 + 2H_2O + heat$$

One pound of methane will combine with 4 pounds of oxygen to make 2.75 pounds of carbon dioxide and 2.25 pounds of water. If the original mixture is 1 pound methane with 6 pounds of oxygen, then after complete combustion, there would be 2 pounds of oxygen left. Since air is about 21% oxygen, combustion without residue would require a mixture of 1 pound methane with 19 pounds of air, which we call a stoichiometric mixture. If the air-to-fuel ratio is higher than the stoichiometric value, the mixture is called lean, but if the air-to-fuel ratio is lower, it is called rich. All fuels can be consumed in a lean mixture, but some fuels cannot be burned in a rich mixture and become soot or smoke. The rate of combustion is controlled by chemical kinetics, which is the speed of the reaction between oxygen and fuel molecules.

In an appropriate mixture of air and fuel, combustion can take place when there is a spark or flame around, and the temperature is above the flash point. When the air–fuel mixture is heated above the auto-ignition temperature, it will spontaneously ignite even if a spark or flame is not around. Let us look at some values (Table 6.2).

Table 6.2   Flash point and auto-ignition temperature of some materials

Material	Flash point (°C)	Auto-ignition (°C)
Methane	−188	—
Gasoline	−43	280
Ethanol	16	363
Diesel	52	256
Newspaper	—	240
Vegetable oil	327	—

Gaseous combustion can take place when fuel vapor and oxygen are mixed together. When gasoline is poured on the floor, there is enough vaporization that a match sets it on fire. But if diesel fuel or vegetable oil is poured on the floor, there is not enough vaporization for a match to produce a flame. When brandy is poured over a bowl of cherries, it will not ignite unless it is pre-heated to create sufficient vapor. The Bunsen burner has a vertical column of methane in the center, which is separated from air by cylinders of air–fuel mixtures, where combustion takes place.

Besides gaseous burning, there is another form of burning that takes place at the interface between air and a char or oil droplet surface. Between air and the solid surface, there is a thin film where oxygen must travel from air to the solid surface. The rate of combustion is controlled by the rate of this mass transport, which is proportional to the surface area where oxygen meets the solid. A solid cube with a width equal to d has a volume of d^3, as well as a surface area of $6d^2$. So the surface-to-volume ratio is given by $6/d$. A large cube has a smaller surface-to-volume ratio than a small cube. Since the combustion rate is controlled by the surface area, it is advantageous to pulverize the coal into small granules and to atomize liquid into smaller droplets.

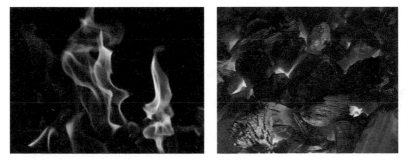

Figure 6.4   Combustion.

When solid coal is introduced into an oven with air, it goes through several phases. In phase 1, coal particles are heated to remove water at below its boiling point. In phase 2, the dry coal particle undergoes pyrolysis or de-volatilization to remove gases such as carbon dioxide and nitrogen, and also volatile matters that are small organic molecules such as methane and propane. The volatile matter mixes with oxygen and burns in the gas phase and achieves the highest temperature. Char is left behind, together with ashes. In phase 3, a smoldering surface combustion takes place, which is controlled by the rate of diffusion of oxygen and at a temperature lower than that in phase 2. The yellow, orange, and red colors of a flame come from solid particles, which are soot or smoke caused by inadequate supply of oxygen. A gaseous blue flame is the hottest and does not contain any solid particles.

The direct burning of coal usually takes place in three types of burners: Pulverized burner uses the smallest particles of 10–100 microns; fluidized bed uses medium size particles of 1–5 mm; and fixed bed uses large particles of 10–50 mm. In the pulverized and fixed bed burners, the processes of de-volatilization, gaseous flame, and char burning take place in a series of zones, and the maximum temperature can be 1500°C, which is high enough to cause a reaction between nitrogen and oxygen to form NO_x (oxides of nitrogen, which are pollutants). In the well-mixed fluidized bed, the entire oven is more uniform and the maximum temperature can be lowered to 1000°C, which avoids the formation of NO_x. The best design and operation of a furnace involve good mixing of fuel with oxygen and quickly raising the mixture to combustion temperature. A minimum amount of smoke and fume should be discharged to the atmosphere, and a method should be in place to remove the ashes.

6.1.3   Safety, Health, and Environment

Coal combustion is a very large-scale operation, and it creates massive problems and challenges from mining to refining, burning in boilers and engines to disposal of exhaust gases and ashes. Underground coal mines are notorious as scenes of explosion and collapse; open pit mines create scars on earth and acid runoffs, which poison fish and wildlife. Stationary combustion is responsible for 90% of SO_x (sulfur oxide pollutants) and 50% of NO_x and particulates emitted to the atmosphere. Most of them are released by burning coal at electric power plants. The concept of "Cradle-to-Grave Stewardship" must be supported to be aware of the hazards posed in every step of the process and to take responsibility to minimize damages to the public and to nature.

Combustion creates many pollutants, some of which are gases such as sulfur dioxide, nitrogen oxides, and carbon monoxide, and some of which are particles such as smoke and soot. They are objectionable because they are responsible for visibility reduction, fog formation, reduction in solar radiation, changes in temperature and wind, acid rain, harm to materials such as ozone cracking of rubber tires, as well as harm to plants, animals, and human health. Global warming has become the most important long-term effect of combustion, which threatens to change the climate pattern on earth. It is strongly affected by the emission of carbon dioxide from fossil fuel combustion. The 2014 report of the Intergovernmental Panel on Climate Change (IPCC) mentioned eight major climate risks: coastal flooding causing death and harm; inland flooding causing harm and economic losses; extreme weather disrupting electrical system; extreme heat, especially for the urban and rural poor; food insecurity linked to warming, drought, or flooding; water shortage causing agricultural or economic losses; loss of marine ecosystem, essential to fishing; and loss of terrestrial and inland water ecosystems.

What actions can a power plant take to reduce harmful emissions? A number of approaches are available to control air pollution:

- **Pre-combustion control**: This involves changing to lighter fuels that contain less harmful elements, or better cleaning heavier fuels before putting them in the boiler. Hydrogen is the perfect fuel as it contains no carbon, sulfur, or nitrogen. Natural gas is mostly methane and an excellent fuel as it produces much heat and little emission. The sulfur content of US coal ranges from 0.2% for western coal from Montana and Wyoming, to 5.5% from eastern coal from Illinois and West Virginia. Some of the sulfur exists in the form of inorganic pyrite FeS_2, which is many times heavier than coal and can be separated by pulverizing the coal, followed by washing and settling.

- **Combustion control**: There are two sources of NO_x in the flue gas: One comes from nitrogen in the fuel, and the other comes from high-temperature combustion, which combines atmospheric nitrogen with oxygen. A combustion process that lowers the flame temperature would thus lower the emission of NO_x. Conventional pulverized bed and fixed bed combustions take place in ovens with distinct zones for de-volatilization to char burning, and the flaming zone can reach 1500°C. On the other hand, the fluidized bed is more uniform, and the combustion can be run at 1000°C. Limestone can be pulverized and mixed with pulverized coal and injected to a chamber that is lifted by air from below, to react with the fuel nitrogen and sulfur.

- **Post-combustion control**: If an even higher degree of sulfur removal is desired, the exit flue gas can be treated by a mixture of limestone ($CaCO_3$) and water in a scrubber. A dry scrubber would require the more expensive lime (CaO). This process would produce a large quantity of calcium sulfate, which is gypsum and can be used to construct dry wall boards. Ash particles can be removed from the flue gas by the cyclone or electrostatic precipitator. In a cyclone precipitator, the flue gas with particles are injected into the shell of a cylinder to promote spinning. As a result, the particles are pushed against the wall due to centrifugal force and flow down as dust, while the clean gas is removed from the top center. In an electrostatic precipitator, a source of electron is injected to the flue gas, which tends to be attached to the particles, which are attracted by plates having positive charges.

The combustion of coal in London led to a type of pea-soup fog, which is familiar in Sherlock Holmes stories. The abundant sunshine and automobile exhausts in Southern California gave rise to another type of serious problem of photochemical smog. Sunlight falling on these automobile exhaust pollutants leads to the formation of ozone, which is particularly harmful to senior citizens, children, and people with heart and lung conditions such as emphysema and asthma. It can cause eye and nose irritation and harm the body's ability to fight infections. In 1970, the Clean Air Act Amendments posed a challenge to the automobile industry, as it set an arbitrary 90% reduction in the emission of carbon monoxide (CO), hydrocarbons, and NO_x, from the 1970 emission to the 1975 emission. Since there were no proven methods to achieve these requirements, the automobile industry was forced to innovate rapidly. This is called "technology-forcing legislation," with no assurance that an adequate and affordable technology will emerge on time. Fortunately, the effective technology of catalytic converters was put together in time and has been a fixture in all the automobiles sold around the world since 1975.

Catalytic reactors have been used to great success in the petroleum and chemical industries for many decades, under the watchful eyes of professional plant engineers equipped with numerous monitoring instruments. There were many skeptics when catalytic reactors were proposed to solve the automobile pollution problem, as it would require an unprecedented launching of 100 million small reactors placed in the hands of ordinary citizens with no technical knowledge and little incentive to maintain the reactor in good working order. The main automobile pollutants from the tailpipe are carbon monoxide, hydrocarbons, and nitrogen oxides. The federal law specifies a test cycle that simulates a California urban car, starting a cold engine in the morning, and driving in a stop-and-go fashion through the city for 23 min. An unregulated car would emit 3.4 g/mile of hydrocarbons and 35.0 g/mile of CO, as well as 5 g/mile of NO_x.

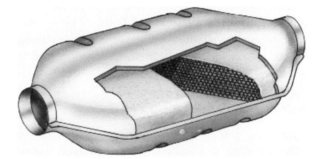

Figure 6.5   Catalytic muffler.

The catalytic converter is a ceramic monolith, containing usually cylinders with diameter from 3 to 6 inches and of length 5 inches. They contain many parallel channels extending in a longitudinal direction along the entire length of the cylinder and are extruded and then fired. The channels are spaced at around 12 per inch and arranged to allow the rapid passage of exhaust gases with very little pressure drop. A wash coat of porous alumina is then deposited on the inside of the channels to provide support for the precious metals platinum, palladium, and rhodium. The exhaust gas would go through a monolith in less than 20 ms. At a sufficiently high temperature of around 500°F, the precious metals have the ability to convert oxygen and carbon monoxide by more than 90%. However, from a cold start in the morning, the initial temperature of the monolith may be as low as 32°F, and the conversion rate would be negligible. The operation of the catalytic converter depends on the front of the monolith being warmed up by the exhaust gases in a short enough time to begin the catalytic reactions, which would help in warming up the rest of the monolith. It has proven to be a great success, operates without big problems, and is adopted around the world.

6.2 Biofuel

Twigs and branches of wood were the oldest fuels. The forests and farms produce valuable food and wood, as well as a large quantity of biomass, which have significant heating values. Such solid fuels have low energy density and are not economically valuable for transport to users far away. We can burn them locally or convert them into liquid fuel for transportation by biochemical fermentation or thermochemical methods. There are many opponents to biofuels for different reasons. Some biofuels are converted from corn, which can reduce food supply and increase food prices.

6.2.1 Biochemical Fermentation of Food

Beer and wine are among the greatest inventions of mankind, adding joy to lives. What is a party without wine and song? Scenes of wine making were found in Egyptian tombs as early as 2000 BCE. After dealing with the great flood and all the animals on the Ark, Noah got drunk on wine. The Greeks thought that a good life should have a balance between Apollo, who is cool and intellectual, and Dionysius or Bacchus, who is the passionate and dynamic god of wine. Beer and wine are produced by fermentation with yeasts, and the alcohol content does not rise much above 15%.

The fermentation of wine from grapes depends on the action of the yeast *Saccharomyces cerevisiae*, which is a microbe about 0.010 mm in diameter. It secrets an enzyme *zymase*, which has the power to split a molecule of sugar with six carbons into two alcohol and two carbon dioxide molecules:

$$C_6H_{12}O_6 \rightarrow 2C_2H_5OH + 2CO_2$$

Figure 6.6 Fermentation.

The fermentation of beer from starch requires a two-step process. Starch is a polymer of sugar, but it has to be split into six-carbon sugars, first with the help of another enzyme *amylase,* which is also used to make bread rise. So to obtain beer from rice or rye, the first step is to convert starch to sugar, involving another yeast in a process also called malting, into a sugary liquid called wort. Then the wort is fermented again to convert sugar to alcohol.

The more fiery whiskey, rum, gin, and brandy with 50% alcohol had to wait till the invention of distillation, which was associated with the practice of alchemy. The process begins by placing beer or wine in a vessel, heating the vessel from below with a flame, and collecting and chilling the rising vapor to condense into liquid. Pure alcohol boils at 78°C, which is lower than the boiling point of water at 100°C. Since alcohol is more volatile than water, the vapor above a water–alcohol mixture has more alcohol. In one stage of distillation, the distillate has approximately 40% alcohol. If a higher percent of alcohol is desired, the distillate can be distilled in two or more stages. In an additional degree of sophistication, the distillate can be stored in a charred oak barrel to make "aged whiskey."

Figure 6.7   Distillation.

The concept of biofuel involves obtaining fuel not from fossils of ancient vegetation under the ground but from contemporary vegetation. The easiest process is to convert sugar or corn into alcohol by fermentation, and then to use distillation to increase the alcohol concentration to the point that it can be burned or blended into gasoline. Common ethanol fuel mixtures include E15, which is 15% ethanol and 85% gasoline, and E85, which is 85% ethanol and 15% gasoline. The higher alcohol mixtures have little volatility and are difficult to start in cold weather. It is necessary to make seasonal changes to lower alcohol fuels during the winter. E100 can be used in Brazil as the winters are not very cold. However, the biggest objection to biofuel from starch is that the conversion of food to fuel has the unfortunate effect of making food scarcer and more expensive. Many people have realized that this competition creates a tension that is difficult to resolve. Another objection is that distillation requires the input of a great deal of energy, resulting in a much lower net energy production.

6.2.2   Biochemical Fermentation of Biomass

When you look at a Brazilian sugarcane field, you might find that of the total weight of the plant biomass, 35% is left in the field as leaves and stem; the harvest removed can be pressed to yield 30% sugar, as well as 35% bagasse. We can convert leaves and stem, plus bagasse, without interfering with the food supply. This is often called biomass fermentation, which involves working with wood, waste, straw, manure, cane, and byproducts of agriculture and forestry. However, this is a much more difficult and energy-consuming technology, as these materials are very fibrous and not soluble in water. Another obstacle is the cost of collection of these low-value materials from many forests and farms and the cost of transportation to a processing plant.

The material in wood can be classified as cellulose, hemicellulose, and lignin. Cellulose is a fibrous material made of sugar molecules connected together to form long straight chains, and the bonds between the sugar molecules are exceedingly strong and difficult to break. Hemicellulose are chains of sugars with numerous side branches, and lignin is an aromatic material of smaller molecules.

Methods to break down biomass into smaller molecules to make fuel include pyrolysis and fermentation. They are divided into biological methods at low temperature with microbes and enzymes, and thermochemical routes, which use high-temperature heat to break down cellulose by gasification and pyrolysis. A good deal of contemporary research is concentrated on the fermentation route, which is difficult and consumes much energy. The fuels derived by these routes are called second-generation biofuels, which do not use food as the starting material.

6.2.3   Thermochemical Reactions of Biomass

Pyrolysis is heating biomass without oxygen or steam to degrade the fibers and turn them into gases and liquids, and leaving behind a solid char. Wood begins to char at around 200–300°C and produces a number of gases and liquids, which can be further processed to make fuels or chemicals. The most studied feedstock is switch grass, which is native to North American prairie and has no food value.

Gasification is heating fibers in the presence of water vapor and some oxygen at a temperature as high as 700°C. The output is a syngas, which is primarily hydrogen, carbon monoxide, and carbon dioxide. A typical volume ratio of H_2/CO is 0.7, and this syngas can be burned at a local power plant. When the goal is producing a transportation fuel with a ratio of 2.0, we can follow with a water-shift reaction to convert some of the carbon monoxide into hydrogen and carbon dioxide:

$$H_2O + CO \rightarrow H_2 + CO_2$$

This is followed by further synthesis with the Fischer–Tropsch process to produce a transportation fuel:

$$(2n + 1)H_2 + nCO \rightarrow C_nH_{2n+2} + nH_2O$$

In the current state of technology, the economics of these processes must receive government subsidies to be competitive with fossil fuel processes.

6.3   Batteries

A battery changes chemical energy into electrical energy. The history of the battery began with Alessandro Volta in 1791 when he made a pile of alternating pairs of copper and zinc discs on top of each other, separated by layers of cloth or cardboard soaked in brine. This pile produced a steady and stable current over a period of time. Many other types of batteries have been invented since then, involving two different materials as the cathode and the anode, separated by a solution called an electrolyte.

There are two types of batteries. The primary battery generates electricity by consuming electrodes in chemical reactions, and we throw them away when they are spent. A flashlight uses A or D batteries that have zinc as the shell and negative electrode, carbon in the center as positive electrode, and an electrolyte of solution of manganese oxide and ammonium chloride. When a flashlight is connected to this battery, zinc is consumed to make zinc chloride and electricity flows at the familiar 1.5 V.

The secondary battery releases electricity upon demand but can be recharged to store electricity generated elsewhere. We depend on lead–acid batteries for our cars, which have lead for electrodes and sulfuric acid for the electrolyte. Lead is 11 times denser than water, so a lead battery is too heavy for hand-held devices. We turn to the lightest of all metal elements, lithium, with a density of 0.53 g/ml, which can float on water. The early lithium-ion batteries posed significant safety concerns as they overheated and burst into flames. Lithium-ion batteries are widely used for portable electronic devices such as phone, laptop computers, cameras, as well as power drills and saws, and wheelchairs. Electrified automobiles such as the Tesla also use lithium-ion batteries.

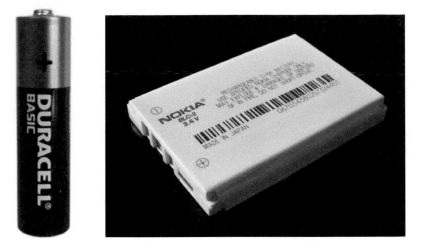

Figure 6.8   Batteries.

Solar and wind energies are highly desirable as these are renewable, very abundant, and do not emit any harmful gases. However, these do not supply a steady flow of energy. A solar generator delivers electricity during a sunny day but makes no delivery during cloudy days and at night. Wind energy is also notoriously unpredictable, from gusty moments to calm moments.

What is needed is a temporary storage of energy during sunny days to be released on cloudy nights. The performance rating of a battery depends on a number of factors: energy density or how much energy can be stored in a given volume, such as watt-hour/liter; how much energy can be stored in a given weight, such as watt-hour/kilogram; power density measures how fast can the energy be delivered, such as watt/kilogram; cycles measure how many times can it be recharged and discharged and continue to function.

6.4   Nuclear Power

Nuclear energy has superb advantages as well as towering disadvantages. Earth has a tremendous supply of nuclear fuels, which is much bigger than fossil fuels or renewables, and nuclear energy does not produce any greenhouse gases to contribute to global warming. Nuclear power also has very negative consequences that threaten health, safety, and the environment. A number of nuclear plant accidents have occurred, such as the Three Mile Island accident in the United States, the Chernobyl disaster in Ukraine, and the Fukushima meltdown in Japan. An additional concern is the fear of terrorists or rogue states generating or seizing enough nuclear materials to make an atomic bomb.

Nuclear energy involves either fission, which is the splitting of a heavy atom such as uranium into two or more fragments, or fusion, which is the joining of two hydrogen atoms into helium. Fission power is descended from the atomic bombs dropped on Hiroshima and Nagasaki and has been powering a large amount of electricity generations around the world for many decades. Fusion power is descended from the hydrogen bomb, which was never used in war, and its peaceful use is still in the experimental stage.

6.4.1   Mining and Fuel Enrichment

Currently, we know three nuclear fuels suitable for generating power and electricity: uranium-235, plutonium-239, and uranium-233. Uranium-235, also called ^{235}U, is obtained from mines in places such as Congo, Kazakhstan, Canada, and Australia. Plutonium-239 is produced by the absorption of neutrons by uranium-238, and uranium-233 is produced by the absorption of neutrons by natural thorium-232.

The ores of uranium are oxides and contain approximately 0.7% of uranium-235; the rest is mostly uranium-238. A nuclear power plant needs an enriched uranium-235 of about 3%, and a weapons-grade plant needs concentrations of at least 80%. Enrichment is the process of increasing the content of uranium-235, which is a very difficult and energy-consuming process. The technology of separation between molecules A and B depends on differences in their properties. We pan gold from sand based on their weights as gold sinks to the bottom of the mixture; we distill alcohol from wine to make brandy as alcohol is more volatile than water; and we sieve sand from pebbles based on their sizes. But the physical difference between uranium-235 and uranium-238 is very small, and they have identical chemical properties. There are two effective methods of separation, both based on the fact that their weights are about 1% apart. First, we have to turn uranium into a gas, which begins with treating uranium oxide with hydrofluoric acid (HF), which is a very corrosive gas that can even etch glass. This produces uranium hexafluoride (UF_6), which is solid at room temperature but turns into a gas at 56°C. So a shipment of UF_6 consists of 0.7% $^{235}UF_6$ and 99.3% $^{238}UF_6$.

The next move is to place this gaseous mixture into a tube under high pressure, with porous walls so that the gas can go through the pores in the tube walls to reach the outside, which is kept at lower pressure. The tube wall is sintered nickel or aluminum, with very small pore size 0.010–0.025 mm. The uranium hexafluoride gas is exceedingly corrosive and attacks almost any metal. The only method that can stand this attack is putting a protective coating with the polymer Teflon, which is a solid fluorocarbon also used to coat non-sticky frying pans. This ratio of diffusion rate is equal to merely 1.0043! This is a very small difference of 0.4%, so the degree of separation for a single stage of operation is very small. In one stage of separation, we can upgrade a concentration originally at 0.7000% to stage one at 0.7028%! If we feed the slightly enriched stream of stage one to a second separation stage, we can upgrade it further to stage two at 0.7056%. So if we keep on adding stages after stages, we would get richer and richer streams. How many stages would it take to go from 0.7% to 3.3%? The answer is at least 1428, even when everything is working perfectly under ideal conditions. The Atomic Energy Commission built a plant in Oak Ridge in Tennessee with a gaseous diffusion separation unit with 4000 stages, providing insurance just in case something is not working perfectly. This plant also produces depleted tailings that contain 0.3% U-235. Higher pressure on the inside of the tubes leads to faster flow, so an enormous amount of electricity is needed to compress the gases between stages. It also requires an enormous amount of cooling water to cool down the compressed gases.

The location of this plant at Oak Ridge in Tennessee was chosen because of a series of dams that belong to the Tennessee Valley Authority, which provides the electricity and cooling water needed for this operation. The material for the first atomic bomb dropped in Hiroshima of 1945 was produced in this plant. Gaseous diffusion was the foundation of nuclear weapons and peaceful use for many decades all over the world.

In recent years, the most economical method to produce enriched uranium is gaseous centrifugation, which is affordable by countries that are not industrially advanced, such as Iran, Pakistan, and North Korea. A centrifuge can be a familiar kitchen appliance with a cylinder that spins on an axis; it can be used to separate milk into cream and skimmed milk. The heavier milk is pushed toward the edge, and the lighter cream moves toward the center. The separation factor is proportional to the radius of the cylinder and the square of the angular velocity. When uranium hexafluoride gas is spun in a centrifuge, the heavier U-238 is more concentrated at the edge and the lighter U-235 is more concentrated at the center. The centrifuge was considered in the American war efforts in 1940 but was discarded as the gaseous diffusion technology showed more promise. A number of German bomb researchers were captured by the Russians, who put them to work on separation methods. Some of them made significant advances, which continued after they returned to Germany in the 1950s. Their work was followed by the Pakistani Abdul Q. Khan in the 1970s, who moved decisively to improve the centrifuge. He built slender tubes of 10 cm diameter and several meters long, which spin at up to 1500 revolutions per second. It is estimated that it is possible to achieve a separation factor of 1.3, in comparison with the 1.004 for the gaseous diffusion. At this separation factor, it would take only six stages to go from 0.7% to 3.3%, so that the capital investment and energy cost is much smaller than in the gaseous diffusion. This led to a nuclear proliferation so that atomic bombs are available not only to major powers of the United States, Russia, and China, but also to Pakistan, North Korea, and Iran.

Fusion power depends on the fusing of deuterium and tritium, both isotopes of hydrogen. Deuterium is usually represented as D or ^{2}H, which has one electron whirling around a nucleus with one proton and one neutron, thus twice as heavy as ordinary hydrogen (H). Tritium is represented as either T or ^{3}H and has one electron in addition to one proton and two neutrons in the nucleus. The sources of deuterium is ordinary water, which contains 99.98% ordinary hydrogen and 0.02% deuterium. Ordinary water is H_2O, and heavy water is D_2O. Heavy water is 10% heavier than ordinary water, freezes at 3.8°C, and boils at 101.4°C. The separation can be carried out by distillation or electrolysis.

The source of tritium is by proton bombardment of lithium-6. Lithium is also the principal ingredient of lithium and lithium-ion batteries, which are crucial to portable electronic devices and electrical automobiles. Lithium ores are found in many countries, particularly in Chile, China, Australia, and Argentina. The reaction to generate tritium is as follows:

$$^6Li + {}^1n \rightarrow {}^4He + {}^3H$$

Tritium is radioactive and decays to ^{3}He, with a half-life of 12.3 years. It was used in watches to make the hands visible at night, till their harmful effects were demonstrated.

Figure 6.9   Nuclear power plant.

6.4.2 Nuclear Reactor Process

Everything in the world is made of elements, such as oxygen and carbon. There are something like 100 elements, each with a nucleus and a number of electrons, like a sun with many planets whirling around. The number of electrons determines the atomic number of the element, such as one electron is hydrogen and six electrons is carbon. The nucleus at the center has a number of protons and neutrons. The number of protons is equal to the number of electrons around, and the sum of the protons and neutrons determines the atomic weight. The three forms of hydrogen are called isotopes. The natural abundance of ^{1}H is 99.98%, ^{2}H is 0.02%, but negligible ^{3}H. Carbon has six protons in the nucleus and usually has six neutrons in the nucleus, for an atomic weight of 12. But carbon can also have eight neutrons in the nucleus, which results in an atomic weight of 14, designated as ^{14}C. This is the famous carbon-14, which is used to date ancient bones and manuscripts, as it undergoes radioactive decay to half its abundance in 5100 years.

Let us consider uranium, which has 92 electrons and can have atomic weights 233, 235, and 238 plus many others. The natural abundance of U-235 is 0.72% and U-238 is 99.27%. U-235 has the remarkable property that when it is bombarded with a neutron, it can lead to a reaction called fission to produce barium and krypton and three neutrons, plus a great deal of energy:

$$^{235}U + {}^1n \rightarrow {}^{141}Ba + {}^{92}Kr + 3{}^1n$$

This nuclear fission produces more neutrons than it consumes. Now let us put 100 pounds of natural uranium in a reactor. When the three neutrons produced hit other atoms of U-235, they have the potential of causing three more fissions and producing nine neutrons and even more energy. So the number of neutrons and energy produced can escalate at the rate of (1, 3, 9, 27, 81, ...), which is a chain reaction leading to an explosion. The multiplication factor k is the ratio 3. However, not all the neutrons manage to collide with U-235 atoms. Some of them fly off to the surroundings, and some hit other atoms and are absorbed. A smaller reactor would have a larger fraction lost to the surrounding, and thus smaller value for k. A neutron can be absorbed by U-238 to produce plutonium and two electrons:

$$^{238}\text{U} + {}^1\text{n} \rightarrow {}^{239}\text{Pu} + 2e^-$$

Thus, instead of an acceleration of neutrons and energy, there can be a decrease in the concentration of neutrons and energy produced to quenching (1, 1/2, 1/4, 1/8, 1/16, ...). Here we have $k = 1/2$, and we have a sub-critical reaction. The plutonium produced can be recovered later and can be used to make the second type of atomic bomb, which was dropped on Nagasaki.

To make a bomb, we need to increase the concentration of U-235 and increase the size of the pile to reach the "critical size," so that the value of k is greater than 1. To make a nuclear reactor that sends out a steady stream of heat, we need to moderate the concentration of U-235 so that k is slightly greater than 1; then we can put in "control rods," which are neutron absorbers such as boron and cadmium, so that $k = 1$. Then the rate of nuclear reaction would be at a steady level such as (1, 1, 1, 1, ...). An atomic bomb needs k to be much higher than 1, and an atomic reactor needs k to be around 1.

There is a far more powerful method of generating nuclear power, which uses resources that are truly inexhaustible and promises to have no negative environmental impacts. This method is thermonuclear fusion, which is used by the sun and all stars to create heat and light. Instead of splitting an atom that has a very high atomic weight, nuclear fusion combines two light atoms together to make one atom of higher weight. For instance, the main reaction in the sun is

$$^1\text{H} + {}^2\text{H} \rightarrow {}^3\text{He} + \text{energy}$$

This reaction takes place in the interior of the sun, which is at 15 million degrees Celsius and under a pressure of billions of atmosphere. The hydrogen bomb was developed by Edward Teller and the Atomic Energy Commission, which uses an equal mixture of deuterium and tritium in the reaction

$$^2\text{H} + {}^3\text{H} + \rightarrow {}^4\text{He} + {}^1\text{n} + \text{energy}$$

Peaceful use of nuclear fusion to generate power was started by Lyman Spitzer, a professor of astronomy at Princeton University in 1950. He did not want a rapid burst of reaction and power as in a bomb, but a moderate and steady release of power.

The nuclear fusion reaction is so powerful that it would blow all the material apart. The sun can contain the expansive forces because it has a tremendous pull of gravity to prevent large-scale escape. For contained fusion power on earth, there are three principles that must be accomplished: a *very high density* of nuclei in a confined space, an *extremely high temperature*, and a sufficiently *long time interval* of containment.

Two large-scale international research projects are under way: the International Thermonuclear Experimental Reactor (ITER), which is located in France near Marseilles, and the National Ignition Facility (NIF), which is located in Livermore, California. The ITER uses the principle of magnetic confinement by building a torus for the hot gases or plasma to circulate under a power magnetic field to keep it contained. The NIF uses the principle of inertial confinement by firing concentric laser beams on a central sphere where capsules of nuclear fuel are contained, which compress to heat them up and keep them confined. These are gigantic experiments financed by all the advanced nations and have been plagued by many unexpected instabilities, leading to delays and budget overruns. The latest forecast is that these may be ready by 2027.

6.4.3 Safety, Health, and Environment

Nuclear power creates another challenge: What to do with the radioactive wastes from the mines, factories, and power plants? When a radioactive element decays, it can emit three types of radiation. The alpha particle is identical to a helium nucleus with two protons and two neutrons. The beta particle is the electron, and gamma is a radiation without any particle. The danger posed by a radioactive isotope depends on the penetrating power of the ray, the energy of the decay, and on the pharmacokinetics or where this isotope would travel and remain in the human body, and when it would be excreted. The alpha particle is large and cannot penetrate deep, whereas the beta particle is an electron, which can penetrate much deeper. High-energy gamma rays have the most penetrating power. The energy of decay is usually measured in MeV, which is million electron-volts. Strontium-90 has a beta decay with an energy of 0.546 MeV. It has chemical properties similar to calcium, so it can enter the body and remain in the bones. It has a half-life of 28.8 years, so its concentration would lower to half in 28.8 years, and to one fourth in 57.6 years, and so forth. Cobalt-60 has both beta and gamma decay, with the energy of 2.8 MeV and half-life of 5.3 years. The gamma radiation with such a high energy can be very damaging. The shorter half-life means a much more intense ray for a shorter period of time. Iodine-131 is a short-lived beta and gamma emitter, but it concentrates in the thyroid gland and can cause more injury than cesium-137, which is water soluble and is rapidly excreted in urine. Radium is very harmful as it has long residence in the body, and its radiation is far more damaging to tissues.

A nuclear power plant creates thousands of wastes, and many of them are radioactive with a wide range of half-lives. The ones with a short half-life of less than a year can be stored in a warehouse to cool off, as the decay concentration is of the order of (1, 1/2, 1/4, 1/8, 1/16, 1/32) and reduced to 1/32 in 5 years. During this period, there would be intense activities leading to high temperature and would require frequent inspections for corrosion and leakage. The longer living wastes have to be stored for long periods in places such as salt mines. High-level waste (HLW) is produced by nuclear reactors. It contains fission products and transuranic elements generated in the reactor core. It is highly radioactive and often hot. HLW accounts for over 95% of the total radioactivity produced in the process of nuclear electricity generation. The amount of HLW worldwide is currently increasing by about 12,000 metric tons every year. A 1000 MW nuclear power plant produces about 27 tons of spent nuclear fuel every year. The ongoing controversy over high-level nuclear waste disposal is a major constraint on the nuclear power's global expansion. Most scientists agree that the main proposed long-term solution is deep geological burial, either in mines or deep boreholes. However, almost six decades after commercial nuclear energy began, not a single government has succeeded in opening such a repository for civilian high-level nuclear waste.

Nuclear fusion power does not seem to generate any radioactive waste. It would be a great contribution to human welfare, if we can develop the appropriate technology in a reasonable time frame.

References

Allen, David T., and D. A. Shonnard, "Green Engineering," Prentice-Hall, 2002.

Bartok, William, and A. F. Sarofim, "Fossil Fuel Combustion," John Wiley, New York, 1991.

Benedict, Manson, T. H. Pigford, and H. W. Levi, "Nuclear Chemical Engineering," McGraw-Hill, 1981.

Davis, MacKenzie L., and D. A. Cornwell, "Introduction to Environmental Engineering," McGraw-Hill, 1991.

Donaldson, Erle, W. Alam, and N. Begum, "Hydraulic Fracturing Explained," Gulf Publishing Company, Houston, 2013.

Faraday, Michael, "The Chemical History of a Candle," Oxford University Press, 2001.

Flagan, Richard C., and J. H. Seinfeld, "Fundamentals of Air Pollution Engineering," Prentice-Hall, 1988.

Glassman, Irvin, and R. A. Yetter, "Combustion," Fourth edition, Elsevier, Amsterdam, 2008.

Glasstone, Samuel, "Principles of Nuclear Reactor Engineering," D. van Nostrand, Toronto, 1955.

Harrison, Roger G., P. Todd, S. R. Rudge, and D. P. Petrides, "Bioseparations Science and Engineering," Oxford University Press, 2003.

Lackner, Maximilian, A. B. Palotas, and F. Winter, "Combustion: From Basics to Applications," Wiley-VCH, Weinheim, Germany, 2013.

Lake, Larry W., "Enhanced Oil Recovery," University Co-op, Austin, Texas, 2011.

Miller, Bruce G., and D. A. Tillman, "Combustion Engineering Issues: For Solid Fuel Systems," Elsevier-Academic Press, Amsterdam, 2008.

Shell International Petroleum Company, "The Petroleum handbook," Elsevier, Amsterdam, 1983.

Wei, James, "Catalysis for Motor Vehicle Emissions," pp. 57–129, in "Advances in Catalysis," volume 24, Academic Press, 1975. Edited by D. D. Eley, H. Pines, and P. B. Weisz.

Chapter 7

Medication: Diagnosis

Veronica Németh

University of Szeged, Szeged, Dugonics tér 13, 6720 Hungary

nemethv@chem.u-szeged.hu

7.1 Introduction

In the past, "a great surgeon made a big cut", though there were endeavors to ensure that both the diagnosis (finding disease) and treatment procedures had the minimal negative consequences for the patient.

One important factor was the huge development of medical imaging procedures, which led to the popularity of "keyhole" surgery (minimally invasive surgery). The surgeon can no longer be proud of the large surgical scar. The open surgical procedures were increasingly pushed back, avoiding many complications and adverse consequences. The new procedures led to shorter length of hospital stays, less human suffering, and sooner return of the patient to family and work. Ultimately, the cost was reduced.

Another important factor in effective disease finding was the development of evidence-based clinical chemistry (laboratory diagnostics), which also had a prominent role in chemistry. The chapter describes these two disciplines.

Chemistry: Our Past, Present, and Future
Edited by Choon Ho Do and Attila E. Pavlath
Copyright © 2017 Pan Stanford Publishing Pte. Ltd.
ISBN 978-981-4774-08-6 (Hardcover), 978-1-315-22932-4 (eBook)
www.panstanford.com

7.2 Diagnostic in Clinical Laboratory

7.2.1 Historical Perspective

A significant milestone of the long road leading to the development of scientific chemistry is the age of iatrochemistry that replaced the thousand-year-long existence of alchemy in the 16th century. Alchemists found a more secure way to make a living by producing medications. The doctors took the knowledge of chemistry and made their medication, while the alchemists began to examine patients. The transitional period is vividly expressed in the painting *Science* (17–18th century, Fisher Collection) after Gerard Thomas (Fig. 7.1).

Figure 7.1 *Science*, after Gerard Thomas (17–18th century, oil on canvas). *Source*: http://uploads.neatorama.com/wp-content/uploads/2011/01/lastpic-500x328.jpg.

In the center of the painting is the master holding a glass filled with urine to light. The young, apparently pregnant woman watches him closely. The foreground and the background, however, show his assistants performing real alchemy by crushing and melting ores. In the painting, distinctive pieces of the alchemist's workshop appear: distillation column to the left, the assistant holding bellows, stuffed lizard hanging from the ceiling, etc.

Scenes depicting "doctor's visit" were very frequently portrayed by the 17th century Dutch painters. The main characters did not resemble real doctors; rather, they appeared more like the beloved "Dottore" of "commedia dell'arte," who was a pompous, know-it-all character rather than a real scientist, representing fraud, quackery, and pseudoscience.

Jan Steen, a prominent figure of the Dutch Golden Age, in his *The Doctor's Visit* (Wellington Museum, London) painting depicts a young woman collapsing on a chair, lovesick or may be pregnant (Fig. 7.2). The clay pot with burning charcoal and a small piece of wick beside it can also be discovered at the feet of the lady. These seemingly ordinary objects functioned as "pregnancy test" in the 17th century.

Figure 7.2   Jan Steen, *The Doctor's Visit* (oil on panel, 1661–62). *Source*: https://en.wikipedia.org/wiki/Pregnancy_test#/media/File: JanSteen-Doctor%27sVisit%281658-1662%29.jpg.

At that time, one of the "sure" methods to diagnose pregnancy used a piece of burning wick tucked under the young woman's nose. If the unpleasant smell made her faint, it had a very high probability that she was with child. A much clearer indication is the flask in the hands of an older woman chatting with the doctor who will examine the patient's urine after taking her pulse. Examining urine is a basic and ancient medical diagnostic tool. Starting from the end of the 15th century, there were printed books with pictures showing shades of urine. In the Middle Ages and in the early modern period, observing urine was the physician's most typical activity. The patient became convinced of the doctor's great professional skills as he held the bottle up to the light, smelling or occasionally tasting the liquid. Indeed, they tasted the urine if it was sweet (diabetes), hence probably this was the first laboratory diagnostic test.

7.2.2 We Prefer Testing to Tasting

The analysis of human bodily fluids as an opportunity for making a diagnosis is as old as the development of modern medicine. Until the mid-19th century, doctors were chemists too. Urine remained the primary subject of examination, and the light microscope became the tool for observing sediments. From 1780, the accumulated knowledge in chemistry enabled the detection of sugar in urine. The alcoholic fermentation test made tasting urine unnecessary. In the middle of the 19th century, a German chemist, Hermann Fehling, developed the classical chemical sugar test, as well as established the method to measure the amount of sugar. This chemical reaction has since entered into the high school education.

In the second half of the 19th century, rapid developments in chemistry made a major impact on medical diagnosis. The outstanding chemists of the age had greatly contributed to the development of the science-based bodily fluid analysis and to the training of specialists in this field. Methods and tools for the qualitative and quantitative detection of certain materials were instantly applied for the analysis of the composition of bodily fluids, leading to the birth of clinical chemistry, which is the application of analytical chemistry for human health. Even today, its development is closely linked to the development of chemical sciences.

Clinical laboratory diagnostics is incredibly diverse. A well-equipped laboratory is prepared for the examination of hundreds of parameters. Here we could only present some of them to demonstrate how the contribution of chemists increased significantly the chances of successfully curing a disease.

7.2.3   Enzymes in Blood Serum: Enzyme Diagnostics

Enzymes are present in very small ($\sim 10^{-9}$ mol/l) concentrations in the blood serum. Their direct measurement would be very complicated, but the products that the enzymes generate in a short time (1–5 min) at a high concentration can be easily measured. Therefore, not the enzyme concentration but its activity is measured, which is proportional to the reaction rate.

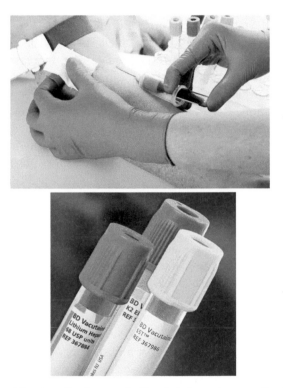

Figure 7.3   Blood collection and blood tubes. *Source*: (top) http://www.rainbow-project.org/wp-content/uploads/2014/12/bigstock-Drawing-Blood-6001048-1024x683.jpg;   (bottom) http://extww02a.cardinal.com/us/en/distributedproducts/images/B/B2953-33.jpg.

A good third of routine clinical analysis determines enzyme activity from body fluids (serum, plasma, urine, spinal fluid) for diagnostic purposes (Fig. 7.3). Although biochemists and chemists have isolated 12,000–13,000 enzymes, only 15 or 20 enzyme-activity measurements are used for diagnostic purposes.

Until 1948, routine clinical laboratories analyzed only four enzyme activities for diagnostic purposes (amylase, lipase, acid phosphatase, and alkaline phosphatase).

7.2.4 Laboratory Diagnosis of Myocardial Infarction

Cardiovascular diseases (coronary artery disease) had first emerged in the United States (later also in Europe) as the leading cause of death. They belong to the so-called diseases of civilization since the main risk factors are inadequate diet, increased calorie intake, sedentary lifestyle, and stress.

Initially, when the clinical analysis was restricted to determining amylase, lipase, and phosphatase enzyme activity, the diagnosis of heart attack (acute myocardial infarction) was limited to a few, less specific methods. These were the markers of inflammation (increase in white blood cell count, the C-reactive protein) and a stress marker (blood glucose level increase). The first specific clinical assay measured the activity of glutamate oxaloacetate transaminase (GOT), which is now known as aspartate aminotransferase.

The serum creatine-kinase (CK) activity assay offers an excellent opportunity to determine if the chest pain is caused by heart attack because after heart attack, CK is released in large quantities from the damaged heart muscle cells to the blood. The advantage of this method, since the 1970s, over the GOT assay is that the blood CK level increases sooner after heart attack, and it is more sensitive and specific.

This new method served as a basis for the development of CK isoenzyme assays. If the chemists could measure only the total CK level in the blood, the diagnosis of a heart attack would not be specific because CK is also released by the much larger mass of skeletal muscles.

However, there the enzyme has a form (CK-MB) that is specific to cardiac muscle. Using the CK-MB test from 1975, a more accurate diagnosis can be obtained and even the size of the infarction-affected area can be deduced.

With the help of the CK-MB test, the diagnosis time of heart attack shortened from 3 days to 12–24 h, and this allowed an earlier therapeutic intervention, reducing the death rate from the previous 30% to 10%. During treatment, the CK-MB test is repeated at certain intervals to track the effectiveness of the therapy.

However, the enzyme markers have some weaknesses, so research was initiated for further cardiac markers, which led to the diagnostic application of troponins.

Troponins are proteins but are not enzymes; they regulate the functions of muscle proteins. They have been successfully applied as cardiac markers since 1989. The test is highly sensitive and specific for the heart, and it is an invaluable resource in mild heart attack that no other test could diagnose. As early as 3–4 h following myocardial infarction, serum troponin levels increase and remain high for as long as 1–2 weeks. Nowadays, cardiac troponin became the standard for myocardial necrosis because it is more specific and sensitive than the classic cardiac enzymes.

Thanks to the work of chemists, the number of cardiac markers detected in serum and their importance in the clinical chemistry continues to evolve dynamically.

Now let us look at an example of technical design development of the measurements, which led to the possibility of self-tests at home.

7.2.5   Application of Dry Chemistry in Clinical Chemistry

In 1970 in Rochester (NY, United States), Eastman Kodak Co., known for its imaging products, was thinking about a development (military and space technology) that could provide liquid-free (dry chemical) methods for the determination of blood or urine components under special circumstances (e.g., submarines, spacecraft). The dry film technology seemed fit for this purpose in which the company had great traditions and experience. The idea was that thin layer analysis could be applied in routine clinical chemistry. The dry film includes all reagents that are dried onto a plastic film, and the sample (blood, urine) contains the water required for the reaction. The reagents are dissolved, the reaction occurs, and the resulting color can be measured by a photometer.

However, many problems occurred during the technical realization. For example, the serum did not spread out on the film due to its high surface tension. To overcome this problem, chemists impregnated the cellulose acetate film with TiO_2, causing the serum layer to spread and filling the capillaries. About 10 μl of serum is capable of filling the capillaries of about 1 cm^2 surface. The capillaries cannot absorb and react with more; this is favorable for the assay because the measured signal does not depend on the sample volume.

The reagents were applied on small slides (chip) layer by layer. The sample is pipetted onto this slide, and after incubation, the reflection is measured in reflection photometer.

7.2.6 The Strip Technique

The initial dry technique was further developed, which led to the currently used strip technology. As a result, a significant portion of the urine tests may be carried out by non-professionals.

A variable number of filter papers or glass fiber pads, affixed to thin plastic strips, contain the required dry reagents. The strip is dipped into the urine, and the color that appears is compared to the color scale on the side of the box to determine the corresponding value.

Inside the plug of the boxes, a moisture-adsorbing silica gel prevents the strips from getting wet. The reagent impregnated into the pads may expire, especially those used for glucose detection (Fig. 7.4) because it contains sensitive enzymes (glucose oxidase, peroxidase).

Simplified at-home test kits facilitate the personal monitoring of health. For example, diabetes patients used to have to visit a laboratory to determine if sugar was present in their urine. In the 1960s, the first portable, battery-operated blood glucose meter was introduced, which works with chemical sticks to detect glucose, improving the life quality of patients considerably. In the 1970s and 1980s, home-use diagnostic kits were introduced for fecal occult blood, ovulation, pregnancy, and strep (Fig. 7.5).

The role and importance of clinical chemistry continue to grow in medical practice. The number of tests increases rapidly. Today an average hospital laboratory performs millions of analysis a year (Fig. 7.6). To meet this demand, powerful automatic laboratory machines were installed (1000–2000 tests per hour).

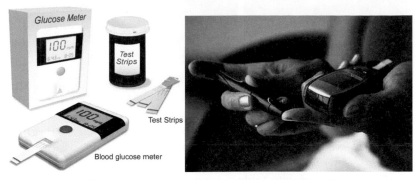

Figure 7.4 Blood glucose detection at home makes life easier for diabetes patients. *Source*: (left) https://www.diabeticpick. com/blog/wp-content/uploads/2015/02/Glucose-Testing-Machine.png; (right) http://img.webmd.com/dtmcms/live/ webmd/consumer_assets/site_images/articles/health_tools/ control_your_diabetes_slideshow/getty_rf_photo_of_person_ checking_blood_sugar.jpg.

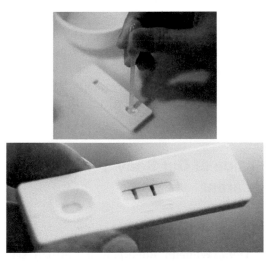

Figure 7.5 Urine test for pregnancy can be done at home simply. *Source*: (top) http://www.healthowealth.com/wp-content/uploads/2014/01/529.jpg; (bottom) http://www. akbirthcenter.org/wp-content/uploads/2011/07/free-pregnancy-test-home.jpg.

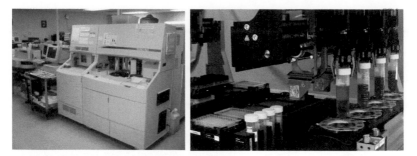

Figure 7.6   Clinical laboratory in a hospital setting showing several automated analyzers. *Source*: https://upload.wikimedia.org/wikipedia/commons/9/9e/LabMachines.jpg.

The number of detectable compounds also increased by order of magnitudes. A well-equipped clinical laboratory has the capacity to analyze 150–300 parameters. The analytical characteristics (sensitivity, selectivity, specificity, reliability) continue to improve.

Earlier clinical chemistry results were used almost exclusively to confirm or rule out a suspected diagnosis. Currently, the fields of applications have considerably expanded. For example, the analysis of risk factors can be very effective in preventing cardiovascular diseases.

The value of laboratory results is increasing because, thanks to the work of chemists and clinical biochemists, there are more and more parameters (e.g., Troponin, S100 protein) with 100% specificity to a given disease, allowing the physician to make a more accurate diagnosis than before.

7.3   The Inside of Our Body Becomes Transparent

7.3.1   X-Rays Triumph

In November 1895, Wilhelm Roentgen, a German scientist and professor at the University of Wuerzburg, Germany, made a discovery that changed the course of medicine. He discovered how to see through the skin and soft tissues to visualize the skeletal system inside the body.

At the end of the 19th century, examination of cathode rays was a popular trend among experts around the world. Roentgen, too, repeated the experiments of other researchers: He passed high-voltage electricity through a glass tube partially saturated with air or other gases. As early as 1858, it was noted that phosphorescent glow appeared on the walls of the glass tube when electric current passed through it. This phenomenon was explained in 1878. The phosphorescence-inducing "particle flow" was named "cathode ray" by William Crookes. We now know that cathode rays are indeed a flow of electrons emitted from the negative electrode, the cathode, and the phosphorescence is created by the electrons hitting the glass. In 1892, Heinrich Hertz, who discovered radio waves, showed that cathode rays could penetrate through a thin metal foil. In 1894, Philipp Lenard modified the tube, who opened the sealed glass and closed it with a thin aluminum foil, allowing the rays to leave the vacuum tube, and they could be captured by a fluorescent screen placed in their way. However, it became quickly apparent that the cathode rays exiting the low-pressure tubes could pass only 2–3 cm under normal air pressure. Roentgen had the idea of applying the glass tube that Crookes originally used, the one that had no thin aluminum window, and tried to detect cathode rays as they were exiting the glass tube. Under these conditions, no one had ever detected cathode rays before. On one occasion, the scientist noticed a strange phenomenon. He wrapped the cathode ray tube in black paper and darkened the room to better observe the tube's weak, bluish fluorescing light. He noticed that every time he turned on the device, an object on the table a meter away from the tube began to glow green, although the rays from the tube could not get there to trigger the phenomenon. (This object was a paper screen that he pasted with barium platinum cyanide salt, which is a type of fluorescent salt.) He concluded that the tube emitted something else besides the cathode rays, some kind of invisible radiation. Roentgen named the mysterious radiation that could penetrate black paper X-rays, as X usually represents unknown in calculations. We now know that an X-ray is a short-wavelength, high-energy electromagnetic radiation. X-rays arise when electrons of the cathode ray collide at the anode. They slow down to interact with the atoms of the anode, and the interaction forms X-rays. In the weeks following the discovery, he did not leave his laboratory, and ate and slept there. His most striking discovery was recognizing that the rays could penetrate human body, making bones and internal organs visible. He also realized that images of the skeleton can be recorded on photosensitive film. On December 22, 1895, he made the first picture of the hand of his wife, Bertha (Fig. 7.7).

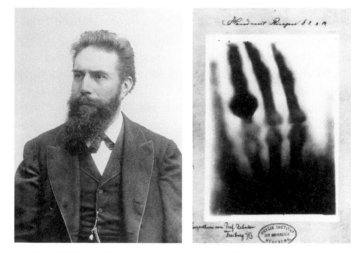

Figure 7.7   Wilhelm Roentgen and the first X-ray picture. *Source*: (left) https://en.wikipedia.org/wiki/Wilhelm_R%C3%B6ntgen#/ media/File:Roentgen2.jpg; (right) https://upload.wikimedia. org/wikipedia/commons/e/e3/First_medical_X-ray_ by_Wilhelm_R%C3%B6ntgen_of_his_wife_Anna_Bertha_ Ludwig%27s_hand_-_18951222.gif.

His first publication on the mysterious X-ray radiation was published in Wuerzburg, Germany, on December 28, 1895. The article summarized the results of 3 weeks of hard work and contained all observations, findings, and considerations that the knowledge in that era made possible.

In the history of science, there are only a few discoveries with impact as robust as Roentgen's. The news about the miraculous X-ray was widely broadcasted by the world's press because it was immediately recognized that an amazing diagnostic tool became available for medicine. Within a few months, X-ray images were used to realign fractured bones in several capitals across Europe. In 1896, more than 1000 articles and 49 books were published on the subject, demonstrating how intensely X-rays were studied. In 1901, the very first Nobel Prize in Physics was awarded to Roentgen, underlining the significance of the discovery.

Modern medicine routinely applies the amazing visualizing ability of X-rays for examinations. The physician can gain insight into the inner parts of body, which otherwise would not be possible without a risky surgery. It is an extraordinary tool in particular in the field of internal medicine and surgery because it allows the early detection of diseases of the stomach, lungs, and other organs. It helps the surgeon to plan and prepare in advance for surgery. It shows the position of broken bones and also helps to accurately determine the location and position of a foreign object invading the human body (bullet, piece of metal), making it possible to decide the best way to remove it.

The discovery of X-rays made a huge impact on science. Henri Becquerel read Roentgen's report and immediately began to study this new type of radiation. In 1896, he discovered artificial radioactivity. This, however, is another story, whose key figure is a Polish girl, who was known to the world as Marie Curie. Her chemical studies and laboratory skills greatly helped her to achieve the separation of the first radionuclide produced by the radioactive decay of uranium. But let us return to the screening of the human body and learn about an uplifting story.

7.3.2 Petites Curies

In 1914, at the beginning of World War I, Marie Curie, who was already a two-time Nobel laureate, recognized that there were only a few X-ray machines available for the use of physicians in France. She temporarily stopped her scientific work and organized the collection and distribution of laboratory X-ray machines to hospitals. She ordered the first X-ray car. In an ordinary automotive, an X-ray machine and a generator were installed while the car's engine provided the power. Madame Curie put into service 20 such mobile X-ray units, popularly known as petites Curies (the "little Curie"), and she set up 200 X-ray stations.

She often took images of wounded soldiers, but it also happened that doctors performed surgery with her assistance under the X-ray machine. The surgeon could see the shadow of his tweezers entering into the wound, bypassing the bones, and seize the shrapnel. However, there were a lot of wounded soldiers and there was a great shortage of professionals who could have been experts of operating the X-ray machine and developing the film. Then Marie Curie organized radiological courses (Fig. 7.8), worked tirelessly, and trained over 250 nurses to become X-ray specialists. Using the images, they could determine where exactly the bullets or shrapnel were in the mangled, suffering body. As a result, thousands of French soldiers escaped amputation or death.

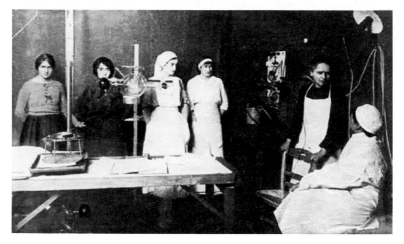

Figure 7.8   Marie Curie organized radiological courses. *Source*: http://www.1914-1918.be/photos2/photo_sans_album/marie_curie_003.jpg.

7.3.3 X-Ray Detection

There are several options for detecting X-rays. The oldest method uses a sensitive surface similar to photographic films, the ones used in cameras. This is still a commonly used technique today. Another, much faster but less accurate method is the use of a fluorescent screen that results in a higher radiation dose. Here the physician can immediately see (without developing a film) the studied area. The fluorescent screen contains a substance (e.g., ZnS) that emits photons in the visible spectrum upon X-radiation. The number of photons emitted is proportional to the intensity of the incident radiation, so a silhouette appears on the screen. However, to detect it with a naked eye, a relatively high-intensity X-ray beam should be used. Therefore, this method of analysis is applied only in urgent cases.

With the advent of digital tools and photon detectors, it is now possible to process a radiograph directly using a computer. In such systems, a two-dimensional, position-sensitive photodetector replaces the photographic film, which means that the detector emits a pulse when a photon is received, and the spatial position of the photon is also recorded. This is stored in the computer, and the image can be displayed at any time. The advantage of such recordings is that it involves less exposure to radiation. Increased sensitivity can be achieved by phase-contrast imaging. The key here is to take advantage of the wave nature of X-radiation. When such a radiation passes through the material, not only the size of a wave, but also phase will change.

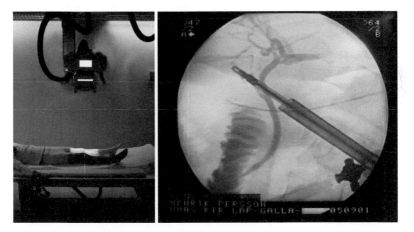

Figure 7.9 X-ray machine and a keyhole surgery (laparoscopy) aided by X-ray technology. *Source*: (left) http://www.diagnoscan.hu/wp-content/uploads/2014/06/mellkasrontgen2.jpg; (right) https://upload.wikimedia.org/wikipedia/commons/c/c7/Laprascopy-Roentgen.jpg.

Often, when the wave passes through different parts of the body while the size of the wave alters slightly only, the phase will change significantly. In this case, the conventional method serves with poor images. However, the phase where the changes are greater can be measured with a little trick. The point is that not the absolute phase of the wave, but the change relative to another wave is measured. In this case, sharp boundaries will appear on those areas where the phase changes. So we see the boundaries of various parts intensified. With the aid of this method, the spatial resolution can be increased compared to conventional techniques (Fig. 7.9).

7.3.4 Further Development of X-Ray Technology

In many pathological cases, it is important to find out the spatial position of the alteration, which is often impossible to reveal by traditional X-ray imaging that produces a silhouette of the examined body part from a particular projection angle. However, our body extends in all three spatial dimensions; therefore, a silhouette does not describe it sufficiently. If we look at our bodies from different directions, we can get different silhouettes, but we still do not know the three-dimensional position and shape of the organs. To do so, often images from two different angles are taken, which helps a great deal, but this approach does not result in an accurate three-dimensional image of the object of study. The tomographic methods remedied this deficiency by providing a detailed three-dimensional volumetric image of the studied area.

7.3.5 CT Scan

Everybody knows as much about the CT scan that we have to lie down on a bed, stay still, and something in the ring surrounding our body begins to move. Less clear is what moves and why. What is C and what is T? "T" is the abbreviation of tomography, which is derived from the Greek word "cut" and "section." The term refers to the imaging procedure in which the X-ray source and the detector, which is mounted on the opposite side, are rotated around the body, producing a sectional image of the body in a single rotation, and then stacking those two-dimensional slices together, a three-dimensional image is formed as the ring moves along the axis of the body.

This sounds very simple, but if you think about it more, it is not so obvious how to come up with the optical sections. Here comes into the picture the letter C, referring to the word computer or computed. CT optical sections are produced as follows: different areas of the body absorb the X-ray beam to different degrees, and this effect is summed in the detector, after the beam has passed through the body. The size of absorption of the X-ray beam by the various parts can be computed by a mathematical technique using these sums, and this will reveal a section of the body. To amplify the contrast of certain areas, the patient can be injected with contrast agents that better absorb the X-rays.

We will tell you later about these agents, but for now let us get back to the beginning of the CT technique.

In 1972, a British computer engineer at EMI Ltd, Godfrey Hounsfield, and his neuroradiologist associate, for the first time in the history of medicine, made an area of the brain transparent that had not been visible before (Fig. 7.10). They called the device computerized axial tomography. Hounsfield designed a device emitting narrow X-ray beams at different angles. This machine generated a series of thin sections, so-called tomograms, and then these data were digitalized and turned into a series of special mathematical instructions (algorithm) by the sensors. Finally, a high-speed computer created a three-dimensional X-ray image from the algorithm. The method has some similarities to human vision, the two-eye vision, that views an object from two angles, or from two points of view, and the brain reconstructs the three-dimensional relationships.

When they presented their X-ray images of human heads to experts, radiologists immediately realized that the new method, the computer tomography (CT), can scan not only the tissues of the brain but all the soft tissues and can reveal their pathologic changes. It took less than 5 years to have a thousand CT devices around the world. With the help of such scans, the size and shape of the lesions can be made visible in three dimensions without invasive surgery; therefore, interventions can be designed well in advance. Between 1973 and 1983, the use of CTs spread worldwide in the clinic, and from 1974, thanks to the US scientist Robert Ledley, not only the head but also the entire body could be scanned.

Figure 7.10 Godfrey Hounsfield and the prototype CT scanner. *Source*: (left) http://stat.ameba.jp/user_images/20130906/07/ igatakeru/34/f7/j/o0320024012674242562.jpg; (right) https://upload.wikimedia.org/wikipedia/commons/3/39/ RIMG0277.JPG.

7.3.6 Development of CT Apparatus

The first-generation of CT scanners was slow and bulky, with a mere two optical sections per minute processing ability. In 1969, the first experiments carried out by Hounsfield on phantoms took 9 days. In 1974, the fourth-generation CT scanners came to the market with a row of detectors forming a complete circle, thus only the X-ray source needs to be rotated. The development is well demonstrated by the improvement in the resolution and in the scanning time. While in 1972, they could reconstruct an 80 × 80 pixels image, the resolution reached 1024 × 1024 pixels in 1993. In parallel, the scanning time was reduced below 1 s from several minutes. The development was partly due to the application of computers and their speed. In the early 1990s, 30 optical sections could be produced in 30 s. At this rate, a snapshot could be produced of nearly every organ. Newer and newer developments followed, and the devices became obsolete at an alarming rate. Today, the sensor of the new CT scanner made by General Electric rotates in just 0.28 s, so even the not-so-slow heart beat can be studied. The greater speed has another advantage; patients spend less time embraced by the ring, so they are less likely to panic and stir from the uncomfortable position.

The new machine reduces the radiation dose, which is also very important, especially when examining children (Fig. 7.11). The basic expectation for any from the new imaging method is the lowest possible exposure to radiation while producing images with the maximum amount of information. Several methods are known or being developed to minimize the dose. For example, 25% reduction became available by the ultrafast ceramic (UFC) detectors, which were invented in part by the contribution of chemists.

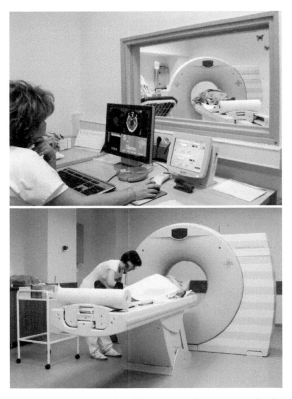

Figure 7.11   CT scanning. *Source*: Diagnoscan Magyarország. http://www. diagnoscan.hu/sajtoszoba/.

Computerized processing of conventional X-ray scans reformed medical imaging. X-ray films with silver emulsion became outdated, and they were replaced by images using digital signals. The modern "image" is a set of data stored in the computer that can be recollected in a multi-dimensional layout and also can be easily transmitted anywhere in the world. The prospective surgery can be planned based on an image, and the prostheses can be conformed to the patient's exact anatomy.

Although the CT apparatus is developed primarily for medical applications, it is successfully used in other fields as a non-invading imaging. Anthropologists can examine Egyptian mummies without harming them. By the aid of portable and mobile CT scanners, cross-sectional images of live trees can be taken without cutting down trees.

Thus, the main point of the tomographic method is determining the structures of the individual two-dimensional optical sections by measuring the intensity in different directions and compiling them into a complete three-dimensional picture. Of course, this method works not only with X-rays but with any such radiation where a direction (vector, straight line) and a respective intensity can be assigned to the passing beams and the intensity along the line is proportional to the absorption or the emission of the beam. An example of the first case is X-ray tomography (CT), while positron emission tomography (PET) is an example of the second one. Now we proceed to the diagnostic applications of radioactive isotopes.

7.3.7   Isotopes in Diagnostics

As we mentioned in relation to the work of Marie Curie, chemistry played a decisive role in the establishment and further development of nuclear science. The nuclei of some isotopes are stable, but others are converted into new nucleus (unstable). Biologically and chemically, they behave equally. However, unstable or radioactive isotopes emit radiation during their radioactive decay, hence the name radioactive isotopes. They may occur naturally or may be the product of nuclear fission, and also it can be produced artificially. The emitted rays reveal their presence, so these isotopes can be used as "tracer" to track hidden processes; all you need is a suitable instrument. In 1913, a Hungarian-born scientist George de Hevesy was the first to suggest using radioactive isotopes as tracers (Fig. 7.12). (In 1943, he received the Nobel Prize in Chemistry for the development of the method.)

Figure 7.12 George de Hevesy and a cyclotron. *Source*: (left) http://nuclphys.sinp.msu.ru/persons/images/hevesy_george.jpg; (right) http://source.pet.hu/images/m10_i95_ciklotron_1.jpg.

According to an urban legend, the idea of radioactive tracers was conceived in the following way. Young Hevesy was told that the housekeeper collected food scraps to make meatloaf to serve the next day. One Sunday, Hevesy intentionally left half of a steak on the plate after injecting the steak with a weak radioisotope. The next day, meatloaf was served for dinner. The instrument detecting radiation began to tick.

7.3.8 Scintigraphic Procedures

Nowadays, the radioactive tracer technique is widely used in medicine. The radioactive nutrient or drug absorbed by the body provides information on the path, speed, and location where the body stores the test substance by measuring radiation. Even complex physiological processes can be tracked or analyzed. Radioactive isotopes, so-called radiopharmaceuticals (a complex formed by a suitable carrier molecule, a radioactive isotope), that are introduced in the body for medical purposes accumulate mainly on the sites of rapid metabolism, like cancer cells. The gamma radiation emitted by the isotope is detected by the camera above the patient's body. Then the incoming signals are sent to a computer to create an image. These methods are the basis for scintigraphic imaging. In contrast to CT, their advantage is that they provide information about the function of the examined organ and the speed of their operation. For example, the metabolism of the thyroid changes in certain thyroid diseases (Fig. 7.13); it either accelerates or slows down (hyper- or hypofunction). The faster or slower incorporation rate of radioactively labeled materials correlates with the altered metabolism.

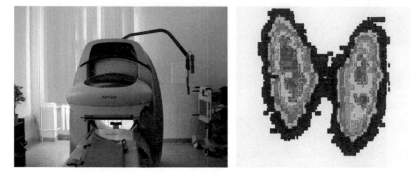

Figure 7.13   Apparatus for making scintigraphy and the scintigraphic image of thyroid. *Source*: (left) https://en.wikipedia.org/wiki/Scintigraphy#/media/File:Scyntygrafia.JPG;   (right) http://www.thyrolink.com/merck_serono_thyrolink/en/images/2.1.2_cts-normal-th_180_tcm1553_62394.gif.

Chemists have to develop substances for isotopic examinations that are nontoxic compounds and have a short half-life. The importance of this is to keep to a minimum the exposure of the patient or the animal to the harmful effects of ionizing radiation (Fig. 7.14).

The optimal isotope emits gamma rays only (alpha and beta rays do not exit from the body and, therefore, cannot be detected, but they are damaging). The best fit to these strict criteria is technetium (atomic number 43), which is one of the most potent radiation sources used in medicine. It is the first artificially produced element (1937), hence its name. It is very unique in the periodic table; while its neighbors are stable elements, technetium is radioactive. Technetium is generated during the radioactive decay of molybdenum-99. The short-lived, metastable technetium-99 isotope is produced in technetium generators for medical purposes, linked to carrier molecules, and the radiopharmaceutical is administered intravenously to the patient. It preferentially accumulates in the growing osseous tissue, and it can be monitored by its gamma radiation. Its half-life is 6.01 h, so 94% is decayed in the human body within 24 h. Nowadays, it has replaced iodine-131 because iodine-131 emits harmful beta radiation. This iodine isotope was previously widely used for thyroid function tests. However, technetium must be linked with a carrier molecule to get into the thyroid gland.

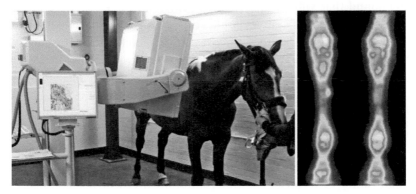

Figure 7.14   Veterinary scintigraphy and bone scan. *Source*: (left) https://dps2008.files.wordpress.com/2013/06/rossendales-1.png; (right) http://www.horsedoc.de/eng/docs/images/Zehenap.jpg.

Due to their ability to destroy cells, certain radioactive isotopes can kill cancer cells; thus, they also have therapeutic effects. However, one should proceed with caution since radiation can also damage healthy cells and that can cause cancer.

7.3.9   Positron Emission Tomography

In this method, an unstable isotope undergoes decay and emits a positron, which interacts with an electron, producing the photons that are detected. The positron is an antimatter particle; in fact, it is antiparticle to the electron. The encounter between a particle and its antimatter counterpart annihilates both particles, producing two high-energy gamma photons moving in opposite directions (Fig. 7.15).

The PET device counts the number of these photons, but how these processes result in a tomogram? An isotope is injected into the patient's body that emits a positron during decay.

These are short-lived isotopes such as fluorine-18 isotope with a half-life of 120 min, oxygen-15 isotope with 2 min half-life, or carbon-11 isotope with 20 min half-life. The radioisotope should, therefore, be prepared on site (in a cyclotron) by a chemist just before the application. (Of course, this greatly increases the price of the device.) Then the positron-emitting isotope is coupled to a compound that will be delivered to the desired sites in the body. The fluorine atoms, for example, are coupled to a compound similar to sugar (6-fluoro-d-glucose). In the body, this substance then travels to the places where sugar would go (Fig. 7.16). The radioactive tracers are administered by intravenous injection.

The number of emitted positrons and, therefore, the number of photons generated in the secondary process are proportional to the number of fluorine atoms. In other words, a tomographic intensity map correlates the spatial and temporal distribution of the metabolic processes. Thus, this method provides a unique opportunity to map, for example, different areas of the brain or to localize tumors. PET scan shows not the anatomical changes but the functional characteristics of the organs and tissues in a given moment. This is a very significant step forward in medical diagnostic imaging because a disease may first cause alterations in the functions of organs and tissues before anatomical changes can be seen.

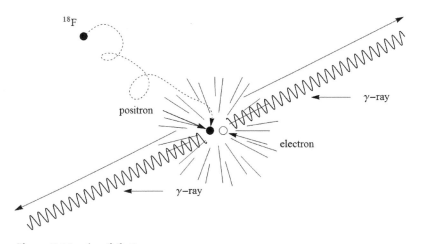

Figure 7.15   Annihilation.

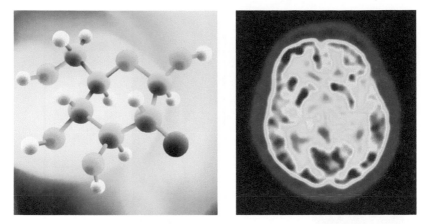

Figure 7.16 Model of 6-fluoro-d-glucose and PET scan of the human brain. *Source*: (left) http://radiography.0catch.com/pet-2-06.gif; (right) https://upload.wikimedia.org/wikipedia/commons/c/c6/PET-image.jpg.

Doctors use PET to diagnose malignant cancers, heart disease, and a variety of central nervous system disorders at their early stage and to determine treatments and track the effectiveness of the applied therapies. The method is fast, accurate, and safe and does not place risk or stress on the patient. PET scan can help in designing more targeted and appropriate treatment. It can spare patients from unnecessary suffering and greatly improve the chances of survival. At the same time, with the help of more accurate information on the extent of the disease, procedures that would not improve the patient's condition can be avoided. Radical surgical interventions can be avoided since PET provides a more accurate picture of the extent of tumors.

State-of-the-art diagnostic tools now combine PET scan and CT into a single integrated imaging technology (Fig. 7.17). The PET scan detects the abnormal, metabolically altered areas whose position can be accurately identified by the anatomical image generated by CT.

Additionally, a PET scan can indirectly help medicine because it provides more cost-effective and faster technique to scientists by allowing them to continuously track the radioactively labeled experimental drug administered to laboratory animals.

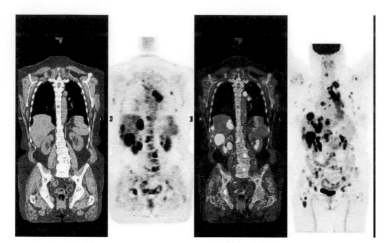

Figure 7.17 Abnormal whole-body PET/CT scan with multiple metastases from a cancer. *Source*: https://en.wikipedia.org/wiki/User: Myohan/Nuclear_Medicine_Sandbox#/media/File:Abnl_petct. jpg.

7.3.10 Diagnosis in a Magnetic Field

Finally, we introduce the nuclear magnetic resonance (NMR) or magnetic resonance imaging (MRI). Although these two terms cover methods based on the same principles, in practice they are distinguished from one another. These methods are based on the detection of a response by the hydrogen atom. MRI is sensitive only to hydrogen in water molecules, while NMR is sensitive to all the hydrogen in the body. (From now on, we will call both methods MRI for brevity.) Hydrogen is suitable for this technique because its nucleus is a single proton, and it is very abundant in the human body in chemical bonds since 70% of our body is water.

To perform the study, the patient is positioned in strong, artificially generated magnetic field. This field tilts the axis of protons in hydrogen atoms. During the procedure, these are "bombarded," layer by layer, with extra energy, which alters the tilting of the axis (Fig. 7.18). Then the proton, while trying to restore its original angle, emits the absorbed energy. It is this emitted energy that the instrument measures, and the computer uses to generate a three-dimensional image.

The image informs about the water content, density, and ultimately the composition of the tissue. Since different tissues behave differently, detailed structural information can be obtained about the organs. A variety of disorders can be visualized because the disorders change the structures of the tissues in the organs. Examples include tumors and their metastases, inflammatory processes, and blood supply disruptions. MRI is superior when compared to CT; it provides images not only in the transverse direction, but in any arbitrary direction. In addition, it provides a better contrast of the soft tissues.

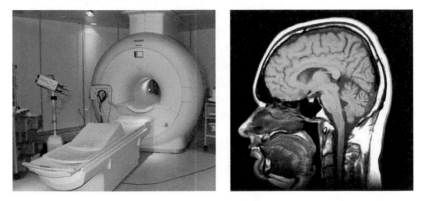

Figure 7.18 MRI apparatus and an MRI scan of head. *Source*: (left) https://en.wikipedia.org/?title=Magnetic_resonance_imaging#/media/File:MRI-Philips.JPG; (right) https://upload.wikimedia.org/wikipedia/commons/3/36/MRI_head_side.jpg.

In the presence of medically implanted foreign bodies (pacemakers, magnetic metals), MRI cannot be performed. Chemists and manufacturers are constantly developing novel implant materials (e.g., nanocoating for shielding) to minimize the risks associated with MRI.

7.3.11 Contrast Materials

For better visibility, to enhance the contrast between body parts and their environment, so-called contrast agents were developed by chemists. Roentgen himself realized that the denser the material, the less it is penetrable by X-rays.

Although there are differences in composition and density among organs, sometimes this is not enough to get a detailed picture. In this case, the contrast can be increased in the studied organ by administering a substance that strongly absorbs X-rays.

The simplest example is the examination of the gastrointestinal tract, when a barium compound is most commonly used. The water-soluble barium compounds are toxic, but barium sulfate dispersed in water or in the gastric or intestinal juice is insoluble, i.e., it does not dissociate into ions, so it cannot be absorbed by the vasculature. Barium sulfate is a white powdery substance, which is suspended in water and given to the patient. The suspension is either swallowed or administered rectally, depending on which section of the gastrointestinal tract is examined. X-rays do not pass through barium sulfate, so all curvatures of the contrast-coated tract are superbly outlined on each image (Fig. 7.19).

The use of contrast agents has allowed the study of blood vessels and vascular diseases (angiography). These studies can be performed by both CT and MRI techniques.

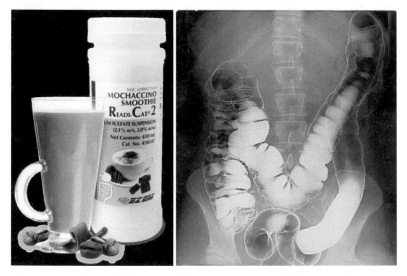

Figure 7.19 Flavored barium sulfate suspension and the use of contrast agents in colon. *Source*: (left) https://www.jeffmed.com/ItemPictures/Detail/450307.jpg; (right) https://en.wikipedia.org/wiki/Radiocontrast_agent#/media/File:Human_intestinal_tract,_as_imaged_via_double-contrast_barium_enema.jpg.

In the case of CT, intravenous iodine is injected into the body, but the physician should use these drugs with the greatest caution because serious complications may occur. He or she should consider the benefits of the expected information and the risks associated with the probable side effects. However, with an accurate diagnosis, the patient could be more efficiently treated or could even fully recover.

Here is an example to understand its significance: In the United States, 100,000 limb amputations are carried out a year. Chemists continue to develop new contrast agents to minimize the occurrence of complications and side effects.

The gadolinium element (atomic number 64) was discovered by the Swiss chemist Jean Charles Marignac in 1880. It belongs to the rare-earth metals. Because of its highly paramagnetic properties, it is often administered to patients as contrast agent during MRI examinations based on magnetic resonance. The principle essentially is the same as the use of barium sulfate for X-ray examinations of the gastrointestinal tract. Gadolinium functions similarly but in the magnetic field of MRI scanners. The gadolinium compound is chelated to minimize the gadolinium-induced unwanted side effects. The medical practitioner then intravenously administers this compound to the patient. There are also iron-oxide-containing MRI contrast agents. These are dextran (a polysaccharide) coated, suspended colloids of iron oxide nanoparticles. They are mainly used to diagnose tumors in liver, spleen, and lymph nodes.

7.4 A Small Detour: Crystal Structure Determination

The discovery of X-rays has helped sciences and the development of medicine in other aspects as well, namely, by characterizing the structure of crystals. The X-ray diffraction method (X-ray crystallography) is based on the fact that X-ray wavelength falls in the interatomic distances of atoms and chemical bonds (Fig. 7.20). During X-ray irradiation of crystals, the beam collides with the electrons of the atoms and the direction of the beam changes.

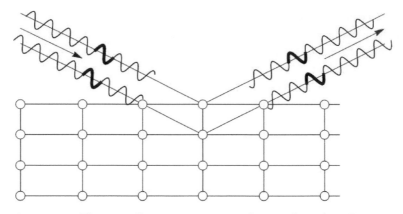

Figure 7.20 The crystalline atoms cause a beam of incident X-rays to diffract into many specific directions.

The scattering and reflection are characteristic for the crystal structure. The reflected beam is detected on a perpendicularly positioned screen coated by photoemulsion. Lighter and darker spots appear on the screen in different directions in a regular pattern (diffractogram, see Fig. 7.21).

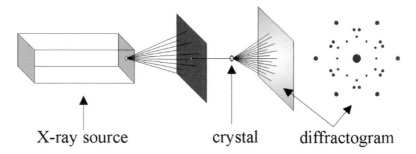

X-ray source   crystal   diffractogram

Figure 7.21 Measuring arrangement and diffractogram.

By the distance and location of the spots, chemists can calculate the distance between atoms of the studied crystals, the bond angles, and ultimately the complete spatial structure of the material.

The type of the atoms can be deduced from the size of the spots. The full structure of very large molecules can also be determined. X-rays had been used very early for exploring the inner structure of crystalline materials (the first X-ray images were taken in 1912). Among others, a precise description of the structure of a material is important as this knowledge will allow artificial production of the material and to reveal the relationship between the structure and the physicochemical properties of the material. Let us look at some examples!

When in the early 1940s, the first antibiotic, penicillin, was used in an attempt to cure bacterial infections, despite initial success, the first attempts ultimately failed as the penicillin stock ran out during the treatment and the patient died. The mass production of penicillin started in 1944, but the technology used at the time did not result sufficient amounts. During the World War II, penicillin significantly reduced the number of deaths and amputation caused by infected wounds in the troops of the Allied forces. However, the drug was still difficult to produce. So at that time, it was common practice that the urine of penicillin-treated patients was collected and chemists re-isolated penicillin for reuse. The development of semi-synthetic penicillin at industrial scale solved the supply problems. The chemical structure of penicillin was defined in 1945 by the British biochemist Dorothy Crowfoot Hodgkin by X-ray diffraction. Thus, she contributed to the production of synthetic drug (Fig. 7.22). She received the Nobel Prize in 1964 for this and other structure determinations.

In 1953, James Watson and Francis Crick in Cambridge, England, developed the model of the DNA double helix structure, which relied largely on the X-ray results of Rosalind Franklin, a British chemist at King's College in London, examining DNA strands (Fig. 7.23).

It is difficult to assess the impact of the discovery of the DNA structure on the further development of science. Here we only mention the Human Genome Project, completed in 2000, deciphering the complete human genetic material. It served as a new foundation of medicine and opened avenues toward personalized, genetics-based medicine.

Of course, as with any technology, X-ray crystallography also has limitations. One problem is that macromolecules (e.g., proteins) must be crystallized, which in most cases is not an easy task for the chemist. Another problem is whether the observed crystal structure corresponds to the structure of naturally occurring molecule. In addition to X-ray diffraction, chemists use magnetic resonance to determine the structure of substances. The two technologies complement each other well.

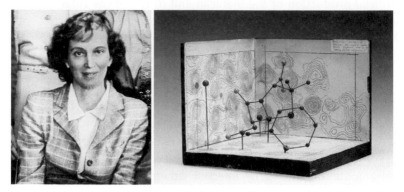

Figure 7.22 Dorothy Crowfoot Hodgkin and the model of penicillin made by her. *Source*: (left) http://blog.sciencemuseum.org.uk/insight/files/2014/12/Dorothy-Hodgkin-credit-Science-Museum-SSPL-low-res.jpg; (right) https://en.wikipedia.org/wiki/Dorothy_Hodgkin#/media/File:Molecular_model_of_Penicillin_by_Dorothy_Hodgkin_%289663803982%29.jpg.

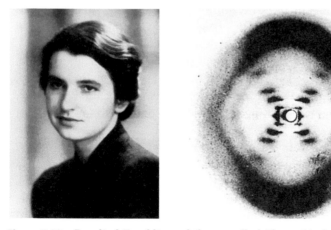

Figure 7.23 Rosalind Franklin and the so-called Photo 51 showing X-ray diffraction pattern of DNA. *Source*: https://zonageekuc.files.wordpress.com/2014/10/2caad-16_06_xraydiffraction-l.jpgQ.

The road completed in slightly more than a century from Roentgen till the Human Genome Project is long, complicated, and arduous. But convincing examples show that scientific discoveries, like the links of a chain, are related to each other. Do not forget the computer tomograph; this incredibly complex machine has its roots in a Wuerzburg laboratory where, in an autumn afternoon in 1895, a barium-salt-soaked piece of paper, left accidentally on the table, began to glow.

Chapter 8

Medication: Curing

Erika Godor and Dorottya Godor

Semmelweiss Medical University, Budapest, Üllői út 26, 1085 Hungary

8.1 Introduction

It is without any shadow of a doubt that modern medicine would be unimaginable without the countless contributions of chemistry. Throughout much of human history, medicine and health care were primitive. If people became sick or injured, doctors could do little more than comfort them or give them folk remedies that often turned out to cause more harm than good.

Hard-working resourceful chemists and chemical engineers have helped to found modern medicine by developing novel pharmaceuticals, creating new medical equipment, and refining diagnostic procedures. Millions of human lives have been saved and improved by the health advances developed through chemistry.

As you read this chapter, you will realize that medicinal chemistry is usually not about producing one drug for one particular disease. It often happens that a single compound proves to have therapeutic potential in a number of conditions

Chemistry: Our Past, Present, and Future
Edited by Choon Ho Do and Attila E. Pavlath
Copyright © 2017 Pan Stanford Publishing Pte. Ltd.
ISBN 978-981-4774-08-6 (Hardcover), 978-1-315-22932-4 (eBook)
www.panstanford.com

that might even seem to be unrelated. It is also frequently seen that one drug paves the way for the development of treatment for all kinds of diseases by way of its congeners and derivatives.

8.1.1 Examples of the Impact of Chemistry on Medicine—in Raw Numbers and in Social Terms

Just think of the impact of the development of penicillin and subsequent antibiotics by looking at the number of deaths at the beginning and the end of the 20th century.

In 1900, pneumonia, tuberculosis (TB), and diarrhea of infectious origin accounted for one-third of all deaths in the United States, 40% being children. As of 1997, only 4–5% of all deaths were due to pneumonia, influenza, and human immunodeficiency virus (HIV) infection [1]. It is important to recognize that developments in medicinal chemistry went hand in hand with improvements in sanitation and childhood vaccination program in achieving these results.

Certain drugs benefited humankind not only in terms of mortality rates, but also had a social impact by removing the stigma and false beliefs associated with certain diseases.

For thousands of years, the strange behavior of people suffering from what we now know as schizophrenia was attributed to the influence of evil spirits. These people were either burned at the stake for witchcraft or shut up in "lunatic asylums" or "bedlams" under inhumane conditions for treatment that we would now regard as torture. The discovery and elucidation of the mechanism of action of antipsychotics dissolved many common misconceptions about schizophrenia. Generally speaking, if a drug is effective for a certain condition, then it must have a site of action somewhere in our body where it fixes a problem. That is, the disease has a biological basis on a molecular level. In the case of schizophrenia, we now know that not demons, but an excess of dopamine lies behind this debilitating condition. This chemical imbalance can be influenced with the appropriate drugs, and many patients can be managed outside of mental institutions.

Figure 8.1 *Philippe Pinel at the Salpêtrière* by Tony Robert-Fleury. The painting depicts chief physician Dr. Pinel ordering the removal of chains from patients at the Paris women's asylum in 1795. Women are chained to the wooden posts of a building. *Source*: Wikimedia Commons.

8.1.2   Chapter Aim

In this chapter, we would like to show you how pharmaceutical chemistry has made a difference to the lives of millions of people by way of creating drugs used in some common conditions. Some of these medications are household names. The stories behind these drugs are fascinating and full of twists of chance. One neglected molecule revisited decades later can turn out to be able to provide a better quality of life, as was the case with the antianxiety agents.

It would fill an entire book if we tried to present chemistry's role in the drug therapy of all major diseases such as heart and vascular disease, cancer, diabetes, and infectious diseases. Instead, we selected three health issues, namely, the management of pain, inflammation, and depression. Even only these three topics provide sufficient examples of the numerous ways chemistry can contribute to medicine. Chemistry's role in medicine is not simply about discovering molecules with a therapeutic potential. The development of new formulations in order to avoid organs that account for side effects or for the breakdown of much of the dose administered and cost-effective methods of synthesis, are equally important challenges.

8.1.3 Versatility of Medicinal Chemistry

Before we embark on our journey, we would like to elaborate on a number of common traits that can be recognized in the history of drug development and that represent the versatility of medicinal chemistry.

- First, many of the medications we use today (including certain antibiotics, painkillers, and antipsychotics) would be nonexistent without the 19th century German dye industry.
- Serendipity is ever present: sometimes compounds discarded as useless and put on the shelf for decades turn out to be useful when rediscovered by adventurous scientists.
- Perseverance in spite of successive failures is also needed. An example of this is the Salvarsan story. Paul Ehrlich and his colleagues screened hundreds of organic compounds before finding the one that showed activity against the bacterium causing syphilis.
- Even small changes in a molecule can make a big difference by improving activity and selectivity and reducing side effects.
- Certain drugs can not only fix one problem, but also work with the same mechanism of action in various cells with different results. Pain relief and prevention of blood clotting might seem to be unrelated, but prostaglandin synthesis is the common denominator, a process that can be disrupted with aspirin.
- The history of drugs for different conditions is intertwined. Certain sulfonamide compounds have antibacterial action, and it was only later discovered that the ones lacking this property can be used as diuretic agents, e.g., in heart disease.
- A side effect can help us understand how a drug works and the mechanisms that result in disease. Parkinson's disease is caused by an inadequate amount of dopamine in the brain. Schizophrenia results from excessive amounts of this neurotransmitter. So it is no wonder that some antipsychotic agents can cause side effects similar to Parkinson's syndrome, such as tremor and movement disorders while alleviating the psychotic symptoms of schizophrenia.

- It is also important to recognize that not only the discovery of new active agents matters but also designing formulations other than conventional tablets can improve patient compliance. Depot (i.e., slow release) injections can be given to schizophrenic patients who are reluctant to comply with their doctor's instructions and take pills every day. Cancer patients experiencing severe pain are given fentanyl-releasing skin patches. These patches provide constant pain relief for 72 h, whereas the effect of a dose of intravenous morphine wanes off in 3–4 h. We can now achieve hormonal contraception not just with "the pill," whose efficacy can decrease in the case of a diarrheal illness or starting a course of antibiotics. Hormone-releasing intrauterine devices are also available that provide superior efficacy and can remain in place for extended periods.
- Sometimes it takes not only wits but also a great deal of courage (or should we call it recklessness) to make groundbreaking discoveries. The evolution of pharmacology and the research into naturally occurring compounds would have followed a longer and more tortuous course without the self-experimentations of Sertürner and his teenage friends with morphine.
- Finally, you do not necessarily have to be a chemist working arduously in a research laboratory to make a contribution to drug development and people's well-being. Sometimes even a rural family doctor can make observations that are clinically relevant, as exemplified by the recognition of aspirin's antithrombotic effect.

8.1.4 Current Trends in Drug Development

We have to mention that drug development today is no longer about taking a brown glass that has been collecting dust for years off the laboratory shelf and using poor animals and humans to experiment with it on a trial-and-error basis. We have methods to determine the exact chemical structure of the molecule of interest, follow its route and transformation in the human body, and describe its interaction with the biological target.

This works the other way around as well, and it is referred to as *rational drug design*. That is, if we have a specific biological target (e.g., an enzyme) and know its three-dimensional structure, we can design molecules using computer modeling that are complementary to our target in shape and charge. These molecules can interact specifically with the target, thus fewer side effects will be created. This approach can reduce the money and time spent on the development of new drugs [2].

8.1.5 Drug Safety: A Lesson from the Past

Drug development and use is also supported by randomized controlled trials (RCTs), which include thousands of people in order to establish the effectiveness of therapy and to find out more about adverse drug reactions. The importance of this is exemplified by the thalidomide catastrophe. This drug entered the European market in the 1950s, at a time when there were no strictly regulated clinical trials assessing long-term safety. By 1961, this resulted in more than 8000 children born with severe limb abnormalities and organ damage to mothers who took the drug during pregnancy [3].

8.1.6 Where Next?

The RCTs are one of the cornerstones of "evidence-based medicine." The term causes controversy in the medical profession as some argue that it is a kind of "one size fits all" policy. Important efforts are being made to establish personalized medicine, which is the customization of healthcare for individuals based on genetic and other information.

The following sections discuss some drugs that made a difference to medicine.

8.2 Pain Management

Even if we do not have some serious underlying disease, we all have headaches, a sore throat, a toothache, and menstrual cramps at times. There is, of course, more persistent pain that can even be excruciating, such as that experienced by patients suffering from cancer or sciatica. Folk medicine has used the bark of the willow tree, the poppy plant and other herbs for millennia for these conditions. In this section of our chapter, we would like to show you what further improvements chemistry has brought for the relief of pain.

8.2.1   Aspirin

8.2.1.1   A household name

If there is one drug that most of us keep in our medicine cabinet, it must be aspirin, which has become a household name. This is the best-selling drug in the world in terms of the total number of tablets sold [4]. Besides, it also played a remarkable role in the development of the pharmaceutical industry, which originally branched off from the dye industry in the late 19th century. Aspirin is a good example of how a small chemical modification of the active substance of an ancient folk remedy can hit the pharmaceutical market and renew itself with novel indications.

8.2.1.2   The many-sided aspirin

Aspirin belongs to the class of nonsteroidal anti-inflammatory drugs (NSAIDs), which include other common over-the-counter drugs such as ibuprofen and naproxen. Aspirin decreases inflammation and concomitant pain, as well as fever. Like all of the NSAIDs, it relieves pain independently of its anti-inflammatory effect.

This drug is not simply a nonprescription painkiller that we use for common aches and pains. Although now replaced by other NSAIDs, aspirin had been widely used regularly by patients struck with inflammatory joint diseases such as rheumatoid arthritis and systemic lupus erythematous. Millions of middle-aged people still take daily small doses of this medication to prevent myocardial infarction and stroke.

However, chronic use of aspirin causes gastrointestinal irritation, which can result in stomach and duodenal ulceration with consequent bleeding. It should also be noted that aspirin should not be given to children due to the increased risk of Reye's syndrome, a potentially fatal disease affecting the brain and the liver.

This discovery and the advent of painkillers with a better side effect profile, such as *ibuprofen* and *paracetamol*, led to a decline in the sales of aspirin in the second half of the 20th century. The renaissance of the drug came when clinical trials in the 1980s found that daily low doses of aspirin prevent stroke and myocardial infarction. It is currently a mainstay drug in cardiovascular prevention. Surprisingly, growing evidence suggests that aspirin also lowers the risk of developing bowel cancer and slows its progression [5, 6]! Even though a consensus on its role in cancer prevention still needs to be made, aspirin may well become one of the oldest drugs to be used as a 21st century target therapy [7].

8.2.1.3 History

The history of aspirin is quite a remarkable one spanning thousands of years. Aspirin, also known as *acetylsalicylic acid*, is actually a synthetic derivative of the naturally occurring salicylic acid. Natural sources of salicylic acid include the bark and leaves of the white willow (*Salix alba*). The first recorded use of its extract is mentioned on 4000-year-old Sumerian clay tablets. Around 400 BC, Hippocrates, the father of medicine, recommended chewing on the bark of the willow tree in fever and pain, as well as drinking extract brewed from willow bark to lessen the pain of childbirth. *Salicin*, a partially purified extract of willow bark, was prepared in 1828. Now that plant-derived extracts were available in purified form, chemists could modify them to form new molecules that might prove more effective, or perhaps safer to use. Salicylic acid was derived from this yellowish substance. Synthetic, industrial-scale production of salicylic acid began in 1878. However, this was far from ideal. Besides its unpleasant bitter taste, it caused vomiting and life-threatening stomach ulcers.

Figure 8.2   White willow (*Salix alba*). *Source*: Wikimedia Commons.

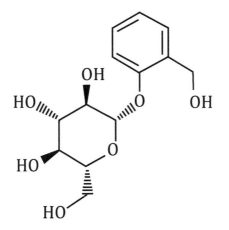

Figure 8.3   Salicin.

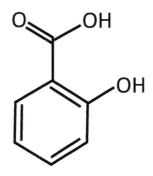

Figure 8.4   Felix Hoffmann (1868–1946). *Source*: Wikimedia Commons.

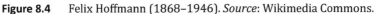

Figure 8.5   Salicylic acid.

Felix Hoffmann, a German chemist working at Bayer, was given the task of finding a more tolerable substitute for salicylic acid. He synthesized acetylsalicylic acid in 1897, the drug that made the Bayer company a pharmaceutical giant. Even though this had already been done by other scientists, it was Hoffman's method that yielded the pure stable form of the drug.

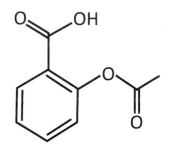

Figure 8.6   Acetylsalicylic acid.

At that time, the manufacture of dyestuffs for the textile industry was Bayer's main profile, and its pharmaceutical research department had only recently been established. Hoffmann is said to have been motivated not only by his superiors, but also by his own rheumatic father's intolerance to salicylic acid. The trade name for acetylsalicylic acid, Aspirin, was coined from "a" for "acetyl" and "spirin" for the German *Spirsäure*, which refers to the genus *Spiraea*, another botanical source of salicylic acid [8].

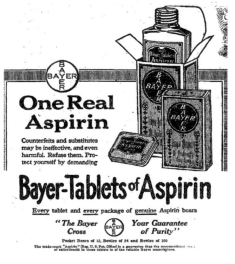

Figure 8.7   Bayer's advertisement of aspirin from 1917. *Source*: Wikimedia Commons.

One would not think that the history of one of the most widely used medicines and one of the most harmful illegal drugs are intertwined. Felix Hoffmann acetylated a number of compounds, including morphine, which resulted in the synthesis of diacetylmorphine. When Bayer employees tried the substance, they found that it made them feel heroic, so they called it *heroin*. Although not an original Bayer discovery, the company was the first mass producer of heroin, which was originally used in pain and cough medications until it was made illegal [9].

8.2.1.4   Closing remarks

We conclude our account by emphasizing that small chemical changes to naturally occurring substances can result in dramatic changes both in their therapeutic efficacy and their side effect profile.

It is also interesting that what starts out as side activity besides dye manufacturing can turn into a billion dollar business.

Furthermore, aspirin is a good example that certain drugs can be used in a number of different conditions, even in the prevention of disease.

Last, but not least, we point out that basic research and the follow-up of thousands of patients in RCTs should complement each other in modern drug development, as illustrated by story of the "coxibs."

Aspirin's antithrombotic properties remained unexploited up until the 1980s, when large clinical trials confirmed it. However, Lawrence Craven, a Californian family doctor, already realized by 1950 that patients who took chewable aspirin preparations for pain relief after tonsillectomy were more likely to bleed. His results went unnoticed since they did not appear in well-known journals. It is a strange twist of fate that Dr. Craven himself died of a heart attack [9]. Another conclusion is that findings that can bring therapeutic consequences for severe conditions like myocardial infarction and stroke may remain hidden for decades if not published in the appropriate medium.

8.2.2   Morphine

People often talk of morphine with a degree of apprehension, since it is associated with severe pain and abuse potential. In terms of pharmaceutical chemistry, it is a compound of great significance. Morphine was the first active agent extracted from a plant, and the first to have analogues prepared from it based on the elucidation of its chemical structure [10]. Since the beginning of recorded human history, humankind has used *poppy*, namely *Papaver somniferum*, to produce opium and use it as pain medication and a narcotic.

Morphine is the archetypal opioid painkiller (analgesic). *Opioids* are a diverse group of compounds that include *alkaloids* in the opium poppy (namely *morphine, codeine,* and *thebaine*), their semisynthetic derivatives, opioid peptides that are normally present in our brain such as *β-endorphin*, and synthetic drugs such as *fentanyl*. Thus, the term "opioid" does not refer to a structural similarity between these compounds but is used to describe drugs that show activity at opioid receptors.

Figure 8.8   An unripe poppy plant (*Papaver somniferum*) exuding latex from the incised pod. *Source*: Wikimedia Commons.

Opioids found their use in anesthesia and surgery as well due to their sedative, anxiety, and pain-relieving (analgesic) properties [11].

Please allow us to present you these compounds in a greater detail, because they adequately exemplify the versatility chemistry can bring into medicine not only with drugs but also with all sorts of formulations.

8.2.2.1 Opium through the ages

Basically, this chapter is not about the history of drugs. However, the social aspects of opioid, especially opium, are so intriguing that we found it worthwhile to include a more detailed historical overview.

Figure 8.9   Raw opium. *Source*: Wikimedia Commons.

It all started with **opium** thousands of years ago, but what is this substance exactly? Opium is produced by making repeated incisions on unripe, green poppy pods. The white sap that oozes out is called latex and is left on the pods overnight. It turns into brown raw opium, which is scraped off in the morning. The most important alkaloids in opium are morphine, codeine, thebaine, and papaverine. Morphine make up 10% of latex [13].

8.2.2.2 Opium in ancient times

Clay tablets from around 2100 BC show that the Sumerians cultivated poppy in Mesopotamia, in what is today Iraq. They referred to the opium poppy as the "joy plant." [14]

According to the Ebers Papyrus from 1550 BC, ancient Egyptians used the unripe poppy pod to prepare medicine that lulls crying children to sleep and to alleviate the pain associated with abdominal tumors.

Given its narcotic effect, it is no wonder that opium was also used for religious purposes in ancient Greece. The ancient Greeks had a number of gods associated with opium, including the god of sleep (Hypnos), the god of dreams (Morpheus), and the god of the night (Nyx). Poppy was believed to be a plant of the underworld [15].

Figure 8.10 The Ebers Papyrus, one of the most important medical records from ancient Egypt. *Source*: Wikimedia Commons.

8.2.2.3 Opium, the aspirin of the Victorian Age

There was a rapid increase in opium use in Western Europe, especially in Britain in the 18th century. Opium was cheap, widely available without restrictions, and sold in various forms such as lozenges, pills, enemas, vinegar, and wine of opium. They were used for all kinds of aches and pains, sleeping problems, cough suppression, to "calm the nerves," and as "refreshment."

Teething and unruly children were given concoctions such as Mrs. Winslow's Soothing Syrup. At that time, the children of working-class people laboring 16–18 h a day in factories were left with "child-minders," who would often give them such syrups to keep them calm. These products were also referred to as the "poor child's nurse." Ladies of the Victorian age drank laudanum, an alcoholic tincture of opium for "female troubles," such as menstrual cramps and menopausal discomfort [16]. Paregoric, a camphorated tincture of opium, was equally popular and is still in use for the treatment of diarrhea in the United States.

Figure 8.11   An advertisement for Mrs. Winslow's Soothing Syrup for teething children from 1885. *Source*: US National Library of Medicine.

Figure 8.12   Opium den in the East End of London. An engraving from an 1874 edition of *Illustrated London News*. *Source*: Wellcome Images.

8.2.2.4   The start of alkaloid research

Until the early 19th century, it was not known what compounds are present in opium. Thus, exact doses of the substance could not be applied. In 1804, a young German apothecary, Friedrich Wilhelm Sertürner, isolated a compound that had not been reported before in scientific literature. He accomplished this by boiling the poppy plant and precipitating crystals from the water using ammonia, yielding a water-insoluble crystal [17]. It evoked drowsiness when he gave the substance to his dog. Thus, he called this compound *principium somniferum*. His findings laid largely unnoticed for a decade because he published them in an insignificant scientific journal.

Sertürner renewed his investigations in 1815 and was not squeamish about testing what he now called *morphium* (after Morpheus, the Greek god of dreams) on himself and three of his young friends. They suffered from severe poisoning symptoms for several days because they had swallowed about 10 times the currently recommended dose of morphine [18]. However, it was worth the trouble. This time, his publication aroused the interest of the famous French physicist and chemist Joseph Gay-Lussac who republished it in the prominent scientific journal *Annales de Chimie*. It was he who altered the term "morphium" to "morphine" in an effort to standardize nomenclature of plant bases [10].

Figure 8.13   Friedrich Wilhelm Sertürner (1783–1841). *Source*: Wikimedia Commons.

Morphine started to make a real impact on medicine after the arrival of the hypodermic needle in the 1850s, and its use in minor surgical procedures for postoperative and chronic pain management, and as an adjunct to general anesthesia was established [19].

Gay-Lussac found Sertürner's findings important not only in that the active agent of opium was discovered. He had foreseen that many other organic alkali compounds would be found in plants [26]. Such naturally occurring nitrogen-containing organic bases are called alkaloids. Gay-Lussac proved to be right: More than 3000 different alkaloids have been isolated from more than 4000 plant species [20]. Among these are the antimalarial *quinine*, derived from cinchona bark and *ergotamine* found in ergot fungi (*Claviceps*), which is used in migraine. Sertürner can justly be regarded as the *father of alkaloid chemistry*.

The isolation of alkaloids from natural sources enabled physicians and scientists to study their biological action using precise quantities. This eventually paved the way for pharmacology to become a separate scientific discipline [10].

8.2.2.5 Flexibility of treatment with various formulations

Morphine is widely used for the relief of acute pain, e.g., in myocardial infraction and traumatic injuries. However, the effect of a conventional tablet or an intravenous dose of morphine sulfate disappears within 3–4 h [11]. Think of how inconvenient it would be for cancer patients that need around-the-clock pain control to take morphine tablets 6–8 times a day. They could not even get a good night's sleep.

Extended (controlled)-release capsules enable the management of chronic pain in conditions such as cancer, shingles, and peripheral artery disease. They release morphine sulfate over 8–12 h, so it is enough to take them twice every day, and there is less fluctuation in opioid levels in the blood [21]. These long-acting opioids are used when treatment with NSAIDs have failed.

Cancer patients frequently experience a sudden onset of transitory but intense "breakthrough pain" superimposed on continuous chronic pain. A combination of immediate- and extended-release morphine capsules is an adequate treatment option in these cases [22].

Opioids are also used for the relief of postsurgical pain. An interesting form of treatment is not formulation but device related. This is called *patient-controlled analgesia* (*PCA*), in which the patients themselves set the dose of morphine given intravenously. A small battery-operated pump is connected to a syringe containing pain medication and to the intravenous line. The pump delivers a constant amount of the drug, and the patient can get more if needed by pressing a button. PCA pumps are safe to use because a time interval has to pass between subsequent doses and the total amount of the drug that can be self-administered is within safe limits [23].

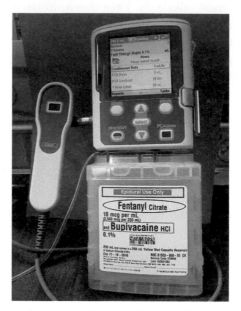

Figure 8.14 A patient-controlled analgesia infusion device. *Source*: Wikimedia Commons.

There are forms of administration of opioids other than the oral or intravenous route. An extended-release *liposomal form* of morphine sulfate can be injected into the outermost part of the spinal canal called the epidural space. This technique is frequently used to provide pain relief after cesarean section.

Morphine sulfate enters the bloodstream via the lower rectal veins when administered as a *suppository*. This is useful when the patient has nausea, vomiting, difficulty in swallowing, or bowel obstruction.

Cancer patients that have developed tolerance to other opioids can be prescribed *skin patches* that release a constant dose of *fentanyl*, a synthetic opioid over 3 days. This allows better pain control and obviates the need for repeated injections or tablets. The fentanyl skin patch and other forms of the drug will be discussed later in this section.

Like fentanyl, *butorphanol* is a synthetic opioid derivative. When used in tablet form, the drug undergoes extensive metabolism by the liver before it reaches the bloodstream. A nasal spray offers the advantage of bypassing the liver; thus, a greater fraction of the dose administered will be present in the systemic circulation by bypassing the liver. Another added benefit is that it has prompter onset of action than conventional tablets [11].

8.2.2.6   Opium fading into background, or not?

It took a hundred years after several unsuccessful attempts to figure out a method of morphine extraction that skips the opium stage. This was desirable because opium production is an extremely laborious process that cannot be mechanized. Just think of incising each and every poppy pod on acres of land and only collecting raw opium the next day!

In 1931, a young Hungarian pharmacist, János Kabay (1896–1936), patented a method that allowed the extraction of morphine and other alkaloids from *poppy straw*, a waste product of poppy cultivation. Poppy straw is the harvested fully mature and dry poppy plant excepting its seeds. This process has revolutionized morphine production throughout the world. It became economical by way of mechanization and the use of raw material that had hitherto been considered waste. Poppy straw has no abuse potential unlike opium, and the whole alkaloid extraction procedure can be performed under controlled circumstances [24].

Figure 8.15   A US marine greets local children working in a poppy field in Helmand Province, Afghanistan. *Source*: Public domain image from the International Security Assistance Force.

8.2.2.7   Morphine's impact on drug development

As you can already see, a number of events inseparable from chemistry contributed to the development of medicine as we know it today, one being the elucidation of the chemical structure of morphine in 1923 by Sir Robert Robinson [10]. This ushered in the era of semisynthetic derivatives of opium alkaloids and synthetic opioids with antagonist and partial antagonist actions. Some of these drugs are stronger than morphine. Some can be used for milder pain, while some are less likely to cause respiratory depression. Some others have less abuse potential, and there are compounds that can be used in the treatment of withdrawal symptoms.

Morphine is a phenanthrene derivative with two planar rings and two aliphatic ring structures. The free hydroxyl group linked to a benzene ring and attached by two carbon atoms to a nitrogen atom is the structural determinant of opioid activity. Relatively simple modifications of morphine molecules lead to the formation of a number of compounds that differ in their analgesic activity. By substitutions at the hydroxyl groups of morphine, compounds with varying properties can be produced. These include codeine, *oxycodone, diacetylmorphine* (better known as *heroin*), *hydrocodone*, and *hydromorphone* [11]. These are called semisynthetic opioids. Some synthetic opioids retain elements of the original morphine structure (e.g., *pethidine*).

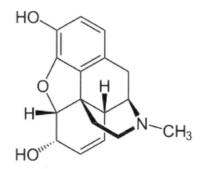

Figure 8.16   Morphine.

Thebaine is another naturally occurring compound in opium that can be converted into a number of other medically important opioids such as buprenorphine and oxymorphone.

In 1898, a cough medicine launched onto the market appeared to work as a painkiller, which is more potent than morphine. It might be surprising for you that it was none other than *diacetylmorphine*, also known as *heroin*, one of the most potent habit-forming drugs ever made. Heroin was used for only a short time during a respiratory infection, and patients did not become addicts, or else it was continuously given to patients with chronic tuberculosis, thus no withdrawal symptoms occurred [27].

8.2.2.8 More opioids in medicine

Next, let us present an array of opioid compounds that found use in medicine. They provide added benefit to patients not only in that their mechanism of action is slightly different from that of morphine. Chemists and drug developers found ingenious ways of augmenting the effect of some of these compounds by creating forms of administration other than conventional tablets or injections.

8.2.2.8.1 Codeine

Codeine is the second major opium alkaloid and was isolated by Robiquet in 1832. Most codeine is produced from morphine by methylation instead of extracting it from opium. When used as a painkiller, it can ease mild to moderate pain (e.g., backache, headache) and often comes in combination with paracetamol. However, it has only one-fifth of the pain-relieving activity of morphine.

Codeine can be found in various cough mixtures, its ability to suppress cough being equal to that of morphine without negative effects on breathing [11]. Although codeine carries less abuse risk, it is on the list of controlled drugs in a number of countries.

8.2.2.8.2 Apomorphine

In 1869, a disabled German chemist and a young English chemist added concentrated hydrochloric acid to morphine and heated it in a sealed glass tube. The result was a degradation product, later named apomorphine, that turned out to be a potent emetic (medicine that makes you vomit), but also brought relief to patients with Parkinson's disease and erectile dysfunction [10].

Levodopa is a staple drug in the treatment of Parkinson's disease. However, response to the drug wanes after a few years of treatment. It is frequent that patients "freeze in place" before the next dose of the drug and cannot initiate movement. Apomorphine quickly reverses this debilitating symptom and is referred to as "rescue therapy." The drug can be self-administered or injected by a family member up to five times a day as needed [28].

Apomorphine as a sublingual tablet is used in the treatment of erectile dysfunction. It has a pivotal role in the detective novel *Sad Cypress* by Agatha Christie.

8.2.2.8.3 Pethidine (meperidine)

Pethidine (known as meperidine in the United States) was the first fully synthetic drug resembling the morphine structure. Synthesized in 1930, the drug was mistakenly thought to have atropine-like properties and was initially used as a drug to treat smooth muscle spasm in renal and biliary colic. When revisited almost a decade later, pethidine was found to demonstrate one-tenth of the pain-relieving potency of morphine with a rapid onset and short duration of action. Just like heroin, it was also marketed as medication for moderate pain without an addiction potential [10].

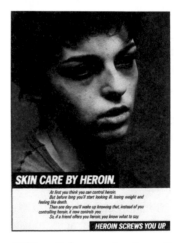

Figure 8.17 A striking 1986 anti-drug poster from the United Kingdom. *Source*: US National Library of Medicine.

Pethidine is of great value in pain control during childbirth since it does not decrease the force of contractions of the womb. However, it can repress the baby's breathing.

8.2.2.8.4   Naloxone

Sometimes when chemists look for a potent agonist, they find a potent antagonist. This is what happened to Jack Fishman, who saved many lives from overdose by patenting naloxone in 1961 [10]. Pure opioid antagonists such as naloxone and naltrexone displace opioids from their receptors.

An opioid overdose results in suppressed breathing and eventually death and is, thus, a medical emergency. The antidote is naloxone given intravenously. It rapidly reverses the effects of opioids and restores breathing. Naltrexone is used in the treatment of opioid dependence. One tablet of naltrexone prevents the action of strong agonists such as heroin for up to 2 days. Naltrexone decreases the craving for alcohol as well and can help patients recover from alcohol addiction [11].

8.2.2.8.5   Methadone

Methadone is structurally unrelated to morphine but has similar pharmacological properties. The drug is used in the treatment of opioid (especially heroin) addiction.

Withdrawal symptoms may occur within 4 h of the last dose of the drug abused and make it difficult for patients to overcome their addiction. Patients also have to cope with constant craving for the drug.

In opioid replacement therapy (ORT), the drug abused is replaced with a similar but less potent compound, namely *methadone* or *buprenorphine*. Methadone is a long-acting drug, and one tablet a day can prevent withdrawal symptoms and craving. The drug is continued for at least a year; then the dose is gradually decreased and eventually stopped. The onset of the abstinence syndrome is slower with methadone, as is the development of tolerance and physical dependence. Although withdrawal symptoms last longer, they are less intense than those seen, e.g., in heroin withdrawal [11].

8.2.2.8.6   Tramadol

Tramadol, a synthetic analogue of codeine was first synthesized in 1962 in an attempt to produce an opioid with a better side effect profile. The drug has weak activity on the μ-opioid receptor, and its mechanism of action is explained by blockade of serotonin and noradrenaline reuptake. This means that tramadol has less addiction potential than most other opioids. It is widely used for the control of pain after surgery [11].

8.2.2.8.7 Buprenorphine

Buprenorphine is more potent than morphine as a pain reliever, but there is a ceiling to its suppressant effect on breathing. Skin patches are available that release the drug over one week.

It can also be used as an alternative to methadone in ORT for heroin addiction. This can be explained by the fact that buprenorphine is a *mixed agonist–antagonist* drug. It acts on μ-receptors as a partial agonist and as a weak antagonist at κ- and δ-receptors. It binds very strongly to μ-receptors, and this means that heroine or any other opioid abused produce less euphoria when injected "on top" of treatment for withdrawal.

When used as treatment for heroin addiction in the tablet form, buprenorphine does carry an abuse potential, because the tablets can be crushed and injected intravenously. Drug developers found an ingenious way of discouraging this practice by including *naloxone* in the tablets [11].

8.2.2.9 Opioids as cough medicine

You can find opioids such as *codeine* and *dextrometorphan* even in cold and cough medicines. The most effective drugs for cough suppression are opioid compounds. They can achieve this effect at doses below those needed to produce pain relief. Their mechanism of action as a cough suppressant is unclear and seems to be unrelated to their effect on pain and breathing.

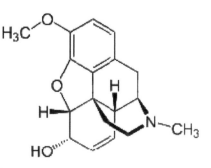

Figure 8.18   Codeine.

Interestingly, there are stereoisomers of opioid molecules that lack painkilling activity and are not addictive, but can work as cough medicine. Stereoisomers have the same sequence of atoms and molecular formula, but they differ from each other in the spatial arrangement of their atoms. One such compound is *dextrometorphan*, the "twin" molecule of a methylated derivative of levorphanol [11].

8.2.2.10 Opioids for diarrhea

The next two compounds provide further examples that opoids can have other functions besides being painkillers.

Loperamide is an opioid that does not enter the brain, so it does not cause euphoria and is not abused as a recreational drug. However, it remains active on μ-receptors in the intestines and causes constipation. Loperamide is available over the counter for the treatment of diarrhea [11].

Paul Janssen, one of the most prolific drug developers of the 20th century, synthesized a number of pethidine derivatives in the 1950s in an attempt to find potent painkillers. One of the compounds he produced was *diphenoxylate*. It did not work as a pain reliever, but it demonstrated a constipating effect like morphine [10]. It is another nonprescription drug for diarrhea.

8.2.2.11 Fentanyl in its many forms

As mentioned earlier, medicine as we know it today would not exist without one of the most productive drug developers of all time, Paul Janssen (1926–2003). The Belgian scientist and his team discovered more than 80 drugs that made a huge impact on several areas of medicine, including psychiatry, anesthesia, pain management, and infectious diseases [29].

Dr. Janssen knew that even small chemical modifications to a molecule can result in significant clinical differences. More than 4000 phenylpropylamine derivatives were synthesized and tested in his laboratories for opioid activity, of which antidiarrheals and antipsychotics were launched onto the market [10]. One of these derivatives was *fentanyl*, a phenylpiperidine whose pain-reliving activity is 50–100 times greater than that of morphine due to being more lipophilic.

Fentanyl injections are widely used in *surgical anesthesia* as injectable painkillers. It has a rapid onset but brief duration of action (30–60 min when administered intravenously) because it soon disappears from the brain and enters the muscle and fatty tissue [11].

When fentanyl became available as a skin patch in the 1990s, it allowed greater flexibility in the management of chronic (especially cancer) pain. This formulation is of huge benefit for patients with persistent moderate-to-severe pain that no longer responds to any other medication (e.g., NSAIDs and other opioids). Fentanyl patches are also useful when the patient is unable to take tablets because of vomiting or difficulty in swallowing. Patches are very convenient to use because they only need to be changed every three days.

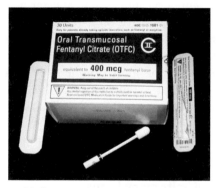

Figure 8.19   Fentanyl lollipop. *Source*: Wikimedia Commons.

How the skin patch delivers fentanyl to the patient's body is very interesting. The patch is a so-called "drug-in-adhesive" formulation. An almost constant amount of fentanyl per unit time is released from the matrix. A concentration gradient between the matrix and the skin drives the release of the drug. Fentanyl accumulates in the skin to form a depot, from where it gradually enters the bloodstream. The patch remains in place for 3 days, and a steady level of fentanyl is reached in the blood by the end of another 3-day application of the patch [30].

There are **transmucosal formulations** of fentanyl too. These include lozenges, lollipops, mouth and nasal sprays, tablets that patients put under their tongues, and films that stick to the cheek from the inside. They produce quick pain relief because the drug is absorbed directly to the bloodstream through the mucous membranes [11].

8.2.2.12   Closing remarks

We have come a long way from the opium poppy to modern pain management. Finally, let us sum up some of chemistry's medical contributions that can be traced back to this plant.

Extraction of the active agent from an ancient herbal remedy and deciphering its chemical structure made it possible to synthesize related compounds that treat symptoms of various kinds and severity.

This gave impetus to the isolation of further compounds from plants and even fungi. Pharmacology was born along the way, because precise quantities of these active agents could now be used for the study of their biological actions.

Morphine production became possible at an industrial level as well as cost effective thanks to a resourceful young Hungarian chemist. János Kabay proved that medicine can be made even out of waste.

The development of innovative new formulations brought better pain relief action and decreased abuse potential of certain opioids. The importance of the former is underscored by the fact that cancer is one of the leading causes of morbidity and mortality in the developed world.

8.3   Inflammation Management

The kidneys have two tiny organs called adrenal glands on their top. The outer layers of the glands produce two kinds of corticosteroid hormones that influence virtually each and every cell in our body. Glucocorticoids have important effects on the immune system and on metabolic processes, whereas mineralocorticoids regulate salt balance in the kidney.

In this section, we discuss the glucocorticoids, also known as the anti-inflammatory corticosteroids, a group of drugs with one of the most versatile disease indications.

8.3.1   Glucocorticoids

In a number of conditions, the immune system mistakenly destroys healthy cells because it recognizes them as foreign bodies. These so-called autoimmune disorders can affect almost any organ, the most frequent parts of the body affected being the skin, joints, the gastrointestinal tract, muscles, and blood vessels. A common autoimmune disease is rheumatoid arthritis, which can impair movement so much that the patient becomes permanently bedridden, if left untreated. Steroids can bring significant improvement in the quality of life for such patients by reducing inflammatory processes.

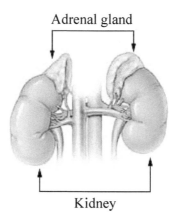

Figure 8.20   The adrenal glands are the source of glucocorticoids in our body. *Source*: cancer.gov.

Organ transplantation is another situation when suppression of the patient's immune system is very important. Without corticosteroids and other immunosuppressants, the body would reject (get rid of) the donor organ since it is not an exact match of the patient's own tissues.

But where is the role of chemistry in all this? Corticosteroids exemplify that isolation of compounds naturally present in our body, and description of their chemical structure and biological function can result in the development of medications that provide relief in serious conditions where the immune system is involved.

8.3.1.1   Glucocorticoids: the magic bullet for autoimmune diseases?

Allergic, inflammatory, and autoimmune conditions represent the most frequent indications of these glucocorticoids. Since they have an effect on the function of most cells in the body, glucocorticoids can influence a number of diseases in almost any organ system.

These include allergic reactions (e.g., to drugs, bee stings), asthma, eczema,* rheumatoid arthritis,* systemic lupus erythematosus,* connective tissue diseases such as polymyosytis, inflammatory bowel diseases (Crohn's disease,* ulcerative colitis*), and multiple sclerosis. Symptoms of these diseases can range from mild to life-threatening. Brief descriptions of the diseases marked by asterisk are given at the end of this part.

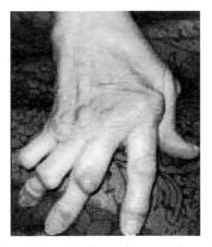

Figure 8.21   Hand   deformed   by   rheumatoid   arthritis.   *Source*: nihseniorhealth.gov.

The course of autoimmune disorders has many ups and downs, with occasional, often unpredictable, flare-ups of the disease and periods of moderate or no symptoms at all (periods of remission). Steroids are used for a short time when symptoms of the disease suddenly reappear or worsen.

However, they are not preferred as drugs for long-term maintenance therapy. Their dose is gradually decreased, and the patient gets another immunosuppressant drug to maintain remission. These precautions are taken in order to minimize the possible side effects of corticosteroid treatment that we will discuss later. When physicians need to resort to steroids as long-term maintenance therapy despite every effort, their aim is to find the minimum dose of the drug that keeps the patient's symptoms under control.

In organ transplant recipients, glucocorticoids are used alongside other immunosuppressants in the prevention and treatment of rejection. Certain kidney diseases and leukemia can also be treated with steroids.

Glucocorticoids can be beneficial even in certain *non-immune-mediated conditions*. One example is increased pressure within the skull due to a brain tumor. *Bethamethasone*, a glucocorticoid with a low degree of protein binding, is given to pregnant women to stimulate adequate maturation of the fetal lung if there is a risk of premature delivery.

Since glucocorticoids are produced by the adrenal gland, it is only logical that they are also used when function of this tiny organ is impaired, as in Addison's disease,* congenital adrenal hyperplasia,* trauma, infection, and shock.

It should be noted that administration of these substances interferes with the hormonal regulation of corticosteroid production by the pituitary gland. This is why large doses of glucocorticoid drugs can suppress corticosteroid production by the adrenal gland itself. Thus, patients can even die if these drugs are suddenly withdrawn [31]!

Patients are often afraid of being prescribed steroids because of several common misconceptions related to their use, such as women ending up looking like men by taking these drugs. These partly stem from the media spotlight on athletes and bodybuilders who illegally used anabolic steroids and testosterone to build muscle and boost their endurance. Many people wrongly identify these substances with steroid medications prescribed by doctors.

8.3.1.2 Versatility of use explained

The mechanism of action of glucocorticoids all comes down to steroid receptors in the cytoplasm of cells that transport the hormone into the nucleus. Gene expression is altered by the steroid–receptor complex interacting with short DNA sequences called glucocorticoid response elements. This interaction is regulated by specific proteins in each tissue, explaining the different effects of steroids in certain parts of the body.

Steroids affect the immune system by inhibiting the functions dependent on white blood cells, with a decrease in the number of lymphocytes [32]. This is why steroids are used in autoimmune conditions and to prevent rejection-transplanted organs.

8.3.1.3 Corticosteroid treatment: A double-edged sword rather than a magic bullet

Even though glucocorticoids are the mainstay drugs in the treatment of dozens of diseases, it is important to keep in mind that steroids are a double-edged sword in the medical armamentarium. These drugs can cause a number of side effects because they influence the function of virtually each and every cell in the body. The side effects of glucocorticoid treatment can all be traced back to the significant influence of glucocorticoids on carbohydrate, protein, and fat metabolism. The group of symptoms seen in prolonged exposure to high levels of glucocorticoids is referred to as *Cushing's syndrome*.

Glucocorticoids increase blood glucose levels and cause muscle wasting by enabling the breakdown of protein. Fat is broken down in the extremities while it is deposited on the trunk, the face (giving rise to a "moon face"), shoulders, and back (often referred to as a "buffalo hump"). Corticosteroid treatment can also result in the thinning of skin and bones (osteoporosis), poor wound healing, acne, purple stretch marks (striae) on the abdomen, increased blood pressure, sleep disturbances, and a significant risk of stomach and duodenal ulcers. Changes in mood and behavior might occur and can manifest themselves as depression, confusion, irritability, and even suicidal thoughts. Patients on steroids are more prone to infections, and dormant tuberculosis may flare up again [33].

It is no wonder that "steroid phobia" lingers in public consciousness. However, these are only possible side effects that do not affect every patient taking steroids. One has to take at least 100 mg of *hydrocortisone* or more (or the equivalent dose of another steroid) for more than 2 weeks for these changes to develop [31]. A compromise has to be made by examining if the benefits of steroid treatment outweigh the side effects and the damage caused by the disease if left untreated.

8.3.1.4 Some clever ways to bypass side effects

Physicians have devised a number of ways to avoid the aforementioned unwanted effects of glucocorticoid treatment. One approach is tapering the dose of the drug soon after the patient's symptoms improve, and giving the drug every other day in order to avoid disruption of the body's own production of steroids.

Chemistry also came up with some smart solutions to minimize side effects. Local therapy minimizes the amount of steroid that enters the bloodstream and reaches other others. Corticosteroid dosage forms other than conventional tablets have been developed that allow the delivery of large amounts of the drug to diseased tissues with reduced effect in other parts of the body.

There are topical preparations for skin conditions, eye drops, inhaled steroids for asthma, injections that can be administered directly into the joint for joint disease, and *hydrocortisone* enemas for inflammatory bowel disease (namely ulcerative colitis).

Nasal sprays containing *beclomethasone, budesonide,* and *mometasone* are available for the treatment of allergic rhinitis. One or two puffs up to three times a day result in so little steroid absorbed into the bloodstream that no systemic effects are produced.

Inflammatory bowel disease involving the rectum and sigmoid colon (the last parts of the large bowel) can be treated with hydrocortisone enemas, foam, or suppositories, although 15–30% of the dose is still absorbed. A controlled-release tablet containing budesonide, a potent synthetic analogue of prednisolone, is also available. It releases the drug only at the site of action, i.e., at the final part of the small bowel and the beginning of the large bowel, with only 10% of the drug reaching the bloodstream [31].

8.3.1.5 Skin-deep treatment with steroids

Soon after the introduction of *hydrocortisone* in the 1950s, researchers found a place for glucocorticoids in the treatment of inflammatory conditions of the skin, too. Steroids can be used topically, i.e., in the form of ointments and creams applied on the skin. Thus, diseases such as *eczema* and *psoriasis* became locally treatable with steroids without the risk of aforementioned side effects elsewhere. However, the skin may become as thin as cigarette paper with prolonged use of these preparations!

From a chemical point of view, topical steroids exemplify that adding all kinds of side chains to existing molecules can result in medications of varying potencies. *Betamethasone*, a 9α-fluorinated steroid, does not present any added benefit as compared to hydrocortisone when used on the skin. However, a compound over 300 times as potent as topical hydrocortisone is produced by adding a valerate chain to betamethasone's 17-hydroxyl position. Acetonide derivatives of fluorinated steroids (e.g., *triamcinolone acetonide*) also present good surface activity. Besides the actual active steroid ingredient, what also matters in the efficacy of topically used creams and ointments is the kind of vehicle or ointment base being used.

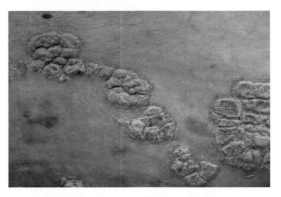

Figure 8.22   Psoriatic plaques. *Source*: Wikimedia Commons.

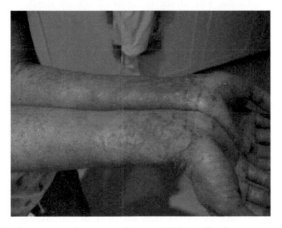

Figure 8.23   Eczema on the arms. *Source*: Wikimedia Commons.

There might be certain skin lesions that are unresponsive to steroid creams and ointments, such as psoriatic plaques. In this case, one mode of treatment is injecting triamcinolone acetonide directly into the lesion in question. This relatively insoluble steroid is released over 3–4 weeks at the site of administration [34].

8.3.1.6   Corticosteroids and Asthma

Asthma is a common disease driven by inflammatory processes in the airways and characterized by recurrent episodes of breathing difficulty.

With aerosol treatment, corticosteroids can be delivered directly to the inflamed airway mucosa with only minimal absorption into the bloodstream. Steroids used for this purpose include *beclomethasone* and *budesonide*. Small amounts of these drugs do enter the circulation, but they do not produce unwanted systemic effects because they have very short half-lives.

The same therapeutic effect can be achieved with fewer side effects with four puffs twice a day (400 µg in total) of the beclomethasone spray as with 10–15 mg of tablet-form prednisolone.

Drug developers presented another interesting approach to further reduce the risk of systemic side effects that result from inhaled corticosteroids being absorbed into the bloodstream. Ciclesonide was approved in 2008 in the United States for the treatment of asthma. The drug is inhaled as an inactive prodrug, and esterases in airway epithelial cells turn it into an active compound by cleaving it. When this active product enters the circulation, it is tightly bound to proteins and, consequently, has limited access to glucocorticoid receptors [35].

8.3.2   The Steroid Story

We provide a short history of steroid drugs in order to demonstrate what impact the elucidation of the composition and function of substances normally present in vertebrates can make on medicine.

The adrenal gland weighs only a couple of grams. However, it is essential for maintaining life, as first demonstrated on dogs, cats, and guinea pigs by the French physiologist Brown-Séquard in 1856.

Extracts prepared from the adrenal glands of cattle that were potent enough to be clinically used were first prepared in the 1930s. These preparations were referred to as *cortin* and were used in the treatment of Addison's disease.

In 1936, Tadeusz Rechstein was the first to isolate a pure, crystalline compound from this extract, which later turned out to be a steroid. This discovery opened a new chapter in hormone research, and several other steroids were later isolated from the adrenal gland.

Another breakthrough came in 1948, when one of these compounds, namely *cortisone*, was found to have remarkable effects on the symptoms of rheumatoid arthritis, a debilitating inflammatory joint condition. The importance of these findings is underscored by the fact that the 1950 Nobel Prize in Physiology or Medicine was jointly awarded to the scientists who discovered the structure and function of hormones of the adrenal cortex.

Soon after the introduction of cortisone and hydrocortisone into clinical practice, side effects became apparent. Besides the issues we already detailed, another problem with these drugs is the retention of salt and water in the body, resulting in edema.

Efforts were made to separate the anti-inflammatory properties of corticosteroids from their salt and water retaining activity. It was found that added potency can be achieved by halogenation with a fluorine atom at position 9α of hydrocortisone, and an extra double bond in ring A results in increased glucocorticoid activity.

This encouraged the synthesis of numerous anti-inflammatory corticosteroids in the 1950s and 1960s, including *methylprednisolone, dexamethasone*, and *triamcinolone*. With the fermentation of cortisone, *prednisone* and *prednisolone* were obtained. These compounds are five times as potent as hydrocortisone in reducing inflammation without further propensity to cause edema [36].

8.3.2.1 Steroid synthesis made easy

The role of chemistry in medicine lies not only in the discovery of drugs for certain diseases, but also in devising ways of synthesizing these drugs in commercial quantities in a cost-effective way.

As already mentioned, it was discovered way back in the 19th century that the secretions of the adrenal glands are essential for life. However, extracts of the adrenal cortex potent enough to enter clinical practice were not available until the 1930s.

At first, cholic acid and deoxycholic acid obtained from ox bile, a cheap by-product from abattoirs, were used as starting material for the commercial production of cortisone. The process involved 37 steps, high pressures and temperatures, and expensive solvents. The price of 1 g of cortisone produced in this way was a staggering $260!

Steroid synthesis requires hydroxylations at certain carbon atoms that are very difficult to bring about with conventional chemical methods. The process was reduced to only 11 steps and made cost effective when it was discovered that *Rhizopus* fungi can perform these specific hydroxylations. Thus, the introduction of microbes brought down the price of cortisone to about $9.70 [37]!

8.3.3 Closing Remarks

Overall, corticosteroids provide relief in a wide variety of diseases, but doctors are walking a thin line between benefits and side effects when dealing with these drugs.

The versatile use of corticosteroids is also explained by the fact that they come in all shapes and sizes. This makes treatment safer by enabling local application of these drugs. Corticosteroids are another example that chemistry can make a difference to medicine not only by way of discovering and synthesizing drugs but also by coming up with new formulations of existing medications.

In the history of medicine, there are countless examples of drugs derived from herbs. However, you do not necessarily have to find a plant in order to find something of potential medicinal use. It is sometimes enough simply to examine how the mammalian body works to draw conclusions that can later result in new medicines.

Steroid synthesis at an industrial scale illustrates that chemistry can be combined with biotechnology. When chemistry struggles, microbes succeed.

At the time when cortisone introduced, there was a deep-seated belief that drugs of natural origin are superior to synthetic drugs [36]. The development of synthetic corticosteroids that entered clinical use refuted this long-held view and encouraged medicinal chemists to produce synthetic analogues of other naturally occurring compounds.

Brief description of some of the diseases in which corticosteroids are used **[33]** (marked by an asterisk in the text)

Addison's disease: A rare, chronic disorder in which the adrenal glands produce insufficient amounts of glucocorticoid and mineralocorticoid hormones. It presents with various symptoms, including weakness, low blood pressure, and hyperpigmentation, and can lead to life-threatening adrenal crisis.

Congenital adrenal hyperplasia: A group of genetic disorders in which the production of cortisol, aldosterone, or both by the adrenal glands is impaired. There is either deficient or excessive production of these hormones, resulting in the alteration of primary or secondary sex characteristics.

Crohn's disease: An inflammatory bowel disease (IBD) that can affect the gastrointestinal tract from mouth to anus. Symptoms include abdominal pain and diarrhea, fever, and weight loss. Bowel obstruction and inflammation of the eye, joints, and skin can also occur.

Cushing's syndrome: A disease caused by chronically high blood levels of corticosteroids. This can be the result of tumor in the pituitary or adrenal gland, or the use of glucocorticoid drugs.

Eczema: Also known as atopic dermatitis, it is an inflammation of the skin. It causes red, weeping, crusting patches that itch intensely.

Rheumatoid arthritis: A chronic autoimmune disease as a result of which joints may become painful and deformed. The disease can also cause damage to the lungs and the membrane around the heart and the lungs.

Systemic lupus erythematosus: An autoimmune disease that can affect virtually any part of the body. It occurs most often in young women and can affect joints, skin, kidneys, blood vessels, and the nervous system.

Ulcerative colitis: Another IBD that causes sores on the wall of the intestines. Its main symptoms are frequent bloody diarrhea mixed with pus.

8.4 Psychotherapeutic Agents

Chemistry has revolutionized the way we think about psychiatric disorders by the development of drugs for the treatment of mental diseases. Deep-rooted prejudices that had been prevailing over debilitating conditions such as depression and schizophrenia have been lifted.

The fact that these diseases turned out to be amenable to drug treatment with medications that modify impaired biochemical processes in the brain helped psychiatry establish itself as a full-fledged medical discipline. In this part, we discuss chemistry's role in the development of the antidepressant drugs.

8.4.1 Antidepressants

Depression is the second most common mental illness in the general population after anxiety. It is estimated that 5–8% of people have depression at some point in their lives in most countries [39]. The disease can occur at any age, but it typically affects adolescents and young adults in their 20s and 30s. By definition, depression is an intense feeling of sadness that interferes with a person's normal day-to day life. Patients are no longer interested in activities they used to derive pleasure from. An untreated episode of depression usually lasts about half a year, sometimes even more, and some people have recurring bouts of the disease.

The exact cause of depression is unknown, but a sad event, biochemical processes, psychosocial factors, and a genetic predisposition certainly play a role in its development. It should be noted that sad mood resulting from the death of a loved one, a serious illness, and other emotionally distressing events is not the same as depression. However, it can turn into actual depression if these feelings do not subside after a certain period of time [40].

Besides the person afflicted with it, depression places a high burden on families as well as on the economy, given its prevalence. Since the advent of the antidepressant drugs as adjuncts to psychotherapy, 80% of patients can now be successfully treated [41]. Without chemistry, these medications would be nonexistent, and the paradigm shift in the way medicine and society regarded mental illness would not have occurred.

In this part, we aim to highlight how chemistry changed the social and scientific view of a disease by way of the development of the antidepressant drugs.

Figure 8.24 *Melancholia* by Albrecht Dürer, 1514.

8.4.1.1 From black bile to neurotransmitters

Up until the 20th century, depression was referred to as "melancholia," reflecting the ancient Greek belief that the presence of too much black bile causes the persistent feeling of sadness in this condition (with *melan* meaning "black" and *kholé* meaning "bile").

According to other long-held ideas, mental illness can be related to evil spirits and the moon's lunar cycles (hence the term lunacy) [39].

Later in the 20th century, the psychoanalytical approach represented by Sigmund Freud was prevalent. Mental illness came to be regarded as the result of a conflict between the conscious and the unconscious mind, with depression being regarded as anger turned into self-hatred [42].

Before the introduction of *iproniazide* and *imipramine*, the first antidepressants, there was only a limited choice of treatment options, with electroconvulsive (electroshock) therapy being the only one with a significant success rate. We would now shiver at the thought of some of these treatments. In the early 20th century, depressed patients were often prescribed amphetamines, opiates, and barbiturate sleep cures. Insulin-shock treatment was another spine-chilling therapeutic option for schizophrenia and depression. An hour of coma was induced with increasingly large doses of insulin, which often resulted in the death of patients [43].

In the first decades of the 20th century, scientists managed get a better understanding of how chemicals that occur naturally in the nervous system transmit messages by traveling from one nerve ending to the other. These molecules are referred to as the monoamine neurotransmitters. In the 1940s and 1950s, serotonin, norepinephrine, and dopamine were identified. Improved technology enabled scientists to more accurately measure their levels and found them to be related to behavior. However, the first antidepressants were found not as a result of these advances, but by mere serendipity. We give you a short account of their discovery later in this part.

When the mechanism of action of the antidepressant drugs was elucidated, a new hypothesis on the development of depression was proposed that pointed to the presence of a biological basis of the disease. Overall, antidepressant drugs act by enhancing the actions of the neurotransmitter amines serotonin and norepinephrine. Consequently, the monoamine hypothesis of depression suggests that the disease results from a decrease in the activity of these amines in the central nervous system.

This recognition radically changed the way depression had been previously viewed. At first, this hypothesis was such a brave new idea that part of the psychiatric community of the 1950s claimed that treating depression with drugs is a real error, since it would not allow patients to discover the internal personality conflicts that underlie their disease [39].

The fact that a scientific explanation from a neurobiochemical point of view was found for a mood disorder lifted some of the stigma associated with it. Many individuals afflicted with this disease felt ashamed because of the stigma on mental illness. Depression came to be regarded as a medical problem amenable to drug treatment, just like diabetes or hypertension [44]. Body and mind were no longer regarded as separate entities.

8.4.1.2   Mechanism of action and side effects explained

The first antidepressants were the *monoamine oxidase inhibitors* (MAOIs) and *tricyclic antidepressants* (TCAs).

MAOIs (e.g., *phenelzine, tranylcypromine, selegiline*) prevent the enzymatic degradation of serotonin, norepinephrine, and dopamine, which means that more neurotransmitters will be released on the stimulation of a nerve cell.

Neurotransmission is not only about certain chemicals crossing the synapse between two adjacent neurons. Once in the synapse, neurotransmitters can be taken back into the presynaptic nerve cell that released it and be reused when needed. TCAs (e.g., *imipramine, amitriptyline*) inhibit this reuptake process.

There are a number of important issues with these drugs limiting their application. The problem with MAOIs is the so-called "cheese effect," which may result in an episode of extremely high blood pressure that may even be fatal. That is, they interact with foods that contain high amounts of tyramine, e.g., cheese, chocolate, and pickled herrings. Depressed people often care little about their lives and find comfort in food; therefore, many of them are unlikely to make dietary restrictions.

Besides serotonin, TCAs influence a number of other (i.e., adrenergic, cholinergic, and histaminergic) neurotransmitter systems in the central nervous system and elsewhere. This results in unpleasant side effects such as dry mouth, difficulty urinating, heart palpitations, blurred vision, and constipations. Only low doses of these drugs can be given at the start of treatment so that patients can develop tolerance toward these disturbing side effects. TCAs can easily be overdosed (either by accident or on purpose), because the lethal dose of these drugs is as little as five times that of the therapeutic blood level. If you consider the fact that many depressed patients have suicidal thoughts, you see that prescription of these drugs raises concerns.

Drugs that interfere only with serotonin and leave alone other neurotransmitters have improved safety and tolerability. These are the *selective serotonin reuptake inhibitors* (SSRIs, e.g., *fluoxetine, sertraline, citalopram*). Patients can start taking therapeutic doses of these drugs from day one, and their overdose rarely results in death [41]. These antidepressants not only relieved patients from many bothersome side effects, but they also represent a milestone in drug development, which we will discuss later.

8.4.1.3 Largely replaced but not forgotten

Although SSRIs and subsequent agents are currently the most widely prescribed drugs for depression, none of these exceeds the effectiveness of tricyclics. TCAs and MAOIs still have a place in the treatment of depression as alternative agents. Depression is not a heterogeneous entity; therefore, it cannot be treated with a "one size fits all" approach. TCAs are particularly useful in patients presenting with weight loss and sleep problems, whereas MAOIs are preferred in patients with significant anxiety and hypochondriasis [45].

Medicine is currently moving toward personalized care rather than giving the same treatment to each and every one with a certain diagnosis. This means that patients with different subtypes of the disease as well as those who are more prone to develop side effects are identified, and the choice and dosage of drugs are tailored to the individual. Therefore, the availability of different kinds of drugs for one particular disease is an especially important achievement.

An interesting aspect of medicinal chemistry is that a single compound can provide benefit in multiple conditions, not only in one particular disease.

Interestingly, TCAs affect not only mood but they have painkilling properties, too. They can be used in the management of neuropathic pain related to diabetes. Patients with chronic joint and muscle pain can be prescribed *duloxetine*, a serotonin–norepinephrine reuptake inhibitor (SNRI).

The MAO-inhibitor *selegiline* is unique in that the drug is used in two different doses for two separate conditions that target two subtypes of an enzyme. At low doses, it blocks only monoamine oxidase B, the enzyme responsible for the breakdown of dopamine. Therefore, this drug is useful in Parkinson's disease, which is caused by an insufficient amount of dopamine in the brain. Higher doses of the drug inhibit monoamine oxidase A as well (the enzyme that degrades serotonin). This can be achieved with a selegiline transdermal patch that enables better absorption of the drug.

Besides depression, SSRIs and SNRIs can also be used in the treatment of anxiety disorders such as posttraumatic stress disorder, obsessive compulsive disorder, social anxiety disorder, panic disorder, and generalized anxiety disorder. Eating disorders such as bulimia and premature ejaculation are further indications of the SSRIs [41].

8.4.1.4   Drugs from rocket fuel

Interestingly, some molecules make quite a convoluted journey until a place is found for them in medicine.

It was Emil Fischer, one of the greatest chemists in history, who discovered a compound called phenylhydrazine back in 1874. When the German army was running out of ethanol and liquid oxygen around the end of World War II, hydrazine was used as rocket fuel. Large stocks of this compound were given to chemical companies to use it as they wish. Experimentations with hydrazine resulted in the synthesis of *isonicotinyl hydrazine* (also known as *isoniazid*) in the 1950s. This compound had already been produced back in 1912, but somehow it had been forgotten for almost 40 years [39]. It was found to have powerful action against tuberculosis. Surprisingly, it was noted that the drug elevated the mood of tuberculosis patients. They became more sociable, slept better, and gained weight [46]. Thus, a side effect in one treatment indication of the drug turned out to be useful as the main positive effect in another disease.

Iproniazid, a compound chemically related to isoniazid, soon entered clinical practice as a drug for depression and was followed by similar drugs. They are referred to as monoamine oxidase inhibitors because they act by blocking the enzyme that destroys the monoamines serotonin, norepinephrine, and dopamine in the brain.

Even though the MAOIs fell out of favor, with only a few that currently remains in use, their discovery paved the way for new treatment options for a couple of non-psychiatric illnesses. Iproniazid was withdrawn from the market soon after its introduction because of potential liver toxicity. Thus, all structurally related compounds were thoroughly examined in order to reduce side effects to as little as possible. This effort resulted in the discovery that some of these compounds have significant antitumor properties! One of these, *procarbazine*, went on to become a widely used chemotherapeutic drug in the treatment of Hodgkin's disease, a cancer that starts in the white blood cells [47].

A beautiful thing about medicinal chemistry is that even compounds that do not show the effect expected by their developers can sooner or later be applied in drug therapy. This was the case with *benserazide*, a compound that comes in a combined formula with levodopa for the treatment of Parkinson's disease. It was synthesized as a potential MAOI, but it could not be used as an antidepressant because it failed to enter the brain. However, it did block an enzyme called DOPA-decarboxylase that degraded the drug levodopa in tissues outside of the brain. Benserazide is currently available in combination with levodopa [47].

8.4.1.5 Serendipity succeeded by rational drug design

The SSRIs ushered in a new era in the development of psychotropic drugs, which was no longer based on the serendipitous discovery of compounds that proved useful in certain diseases [48]. *Rational drug design* emerged as a new strategy in drug discovery.

Overall, the concept behind this process is identifying the enzyme or receptor involved in a given disease, which is to be targeted by the drug to be designed. Then, the three-dimensional structure of this enzyme or receptor needs to be elucidated. Using this knowledge, a molecule is designed and synthesized that binds tightly to this specific biological target and none else. The aim is to create more effective drugs with a better side effect profile as compared to older medications.

In the 1960s, American biochemist Julius Axelrod and his colleagues figured out how different types of antidepressants work. In 1970, Julius Axelrod was awarded the Nobel Prize in Physiology or Medicine along with two other scientists "for their discoveries concerning the humoral transmitters in the nerve terminals and the mechanism for their storage, release and inactivation." [49] They realized that it would be possible to design a drug that influences only serotonin but has no effect on other neurotransmitters. This way, a number of problems related to the older nonselective antidepressants could be avoided.

Emerging techniques such as X-ray crystallographic analyses and nuclear magnetic resonance enabled the study of the three-dimensional structure of compounds. Scientists could find out which part of the molecule examined is responsible for a certain biological activity. Based on this knowledge, they started synthesizing new compounds to enhance the desired effect.

Figure 8.25 Julius Axelrod (1912–2004). *Source*: US National Library of Medicine.

In 1970, scientists at the American pharmaceutical company Eli Lilly started tinkering with a molecule structurally related to diphenhydramine, an antihistamine. This compound showed some antidepressant effect. Analogues of this molecule were synthesized in an effort to find a potent SSRI with little effect on other neurotransmitters. Of the more than 250 molecules examined, *fluoxetine* became a commercial success. The compound is widely known under its trademark name Prozac® and was approved by the FDA in 1987 [39].

The list of rationally designed drugs is increasing. It includes a number of anti-HIV medications, sleep medications such as *zolpidem*, many of the atypical antipsychotics used in schizophrenia as well as *sildefanil* (better known as Viagra®) for male erectile dysfunction.

8.4.2 Closing Remarks

Chemistry can not only make an improvement in people's lives with new drugs, but also provoke a radical change in society's attitude toward certain diseases.

We briefly mention that a similar revolution took place in the attitude toward schizophrenia as well, another debilitating psychiatric condition. A number of parallels can be drawn between the history of antidepressants and antipsychotics. The first effective antipsychotics were also discovered by chance in the 1950s and had multiple sites of action, causing unpleasant side effects. Subsequent drugs were rationally designed in an effort to minimize these. The antipsychotics also had a significant social impact. For millennia, schizophrenia used to be associated with evil spirits and witchcraft, and patients were often submitted to "treatment" that violated their human dignity. With the advent of antipsychotics, it was recognized that impaired balance of dopamine in the brain is responsible for the disease. These drugs enabled the discharge of many patients from mental institutions who could now go on with their lives within the community.

The antidepressants demonstrate that drug discovery can encourage basic scientific research into complex biochemical processes. Just think of the work of Julius Axelrod and his colleagues. As exemplified by rational drug design, this can equally work the other way around, when understanding a biological process results in the creation of new drugs.

In chemistry, it is very common to "recycle" compounds. This was how a compound once used as rocket fuel turned out to be an appropriate starting material for drug synthesis.

8.5 Hormonal Contraceptives

In the long course of history, women and men have gone to great lengths to make sex just for fun without having a baby possible. The arrival of a highly effective form of contraception, known as the Pill, in the 1950s fundamentally changed the way we humans exist. The pill reinvented sex and launched a revolution. No longer in fear of having a child, sexual interaction between men and women took a totally different course. The ability to control the way how families expand deeply affected how women see themselves at home and at work.

8.5.1 Pearls on the Pill

With almost half of pregnancies being unintended in the United States today, it seems that the issue of contraception is just as relevant as it has been thousands of years ago [50]. Out of the 60 million American women in their child-bearing age of 15 to 44 years, 40 million are currently using some form of birth control. Today, the majority of them are on the pill, which translates into a staggering number of 11 million users in the United State [51].

Besides the pill, hormonal contraceptives are available in a dazzling variety of preparations, including oral pills, long-acting injections, transdermal patches, vaginal rings, subcutaneous rod implants, and intrauterine devices. Hormonal contraceptives contain either a combination of an estrogen and a progestin or a progestin alone. These are among the most reliable birth control methods. With typical use, these methods are 91% effective. If all else fail, emergency hormonal contraception, notoriously known as the "morning-after pill," is also available to avoid getting pregnant, effective if taken within 72 h [52].

Most women who use the pill do so to prevent pregnancy; however, 60% of women take the pill for other reasons than contraception. Combination hormonal contraceptives are used in a bewildering variety of clinical setting, like managing painful, heavy, or extended menstrual periods, alleviating the symptoms of menopause or keeping socially disabling acne and excessive hairiness in females under control. Interestingly, females on the pill have reduced risks of ovarian cysts and cancer [53].

These health benefits come with a price though. Nausea, dizziness, and migraines are well-known side effects. Estrogens enhance the coagulability of blood, causing rare but serious thromboembolic side effects, such as myocardial infarction, stroke, deep vein thrombosis, and pulmonary embolism. Depression of sufficient degree might require patients to quit therapy. Since the introduction of the low-dose combined oral contraceptives, the toxicity of these drugs has fallen, but cannot be ruled out [54].

8.5.2 Prelude to the Pill

The straight-up bizarre contraceptive methods females resorted to throughout the ages caused much needless suffering, pain, and of course a great number of unwanted pregnancies. The long history of pregnancy prevention is full of desperate measures that would make any woman cringe today.

Preceding intercourse, ancient Egyptian women would insert a paste made of honey, sodium carbonate, and crocodile dung into their vaginas. The ancient Chinese concubines topped that by drinking hot mercury to rule out pregnancy. In ancient Greece, women drank a very similar toxic cocktail containing lead and mercury, used by blacksmiths to cool their materials with.

Actually, the poisonous sludge did not get out of style throughout the centuries. During World War I, women were volunteering to work in metal factories just to get hold of waste fluid full of lead and mercury, pushing themselves into sterility, severe brain damage, and kidney failure.

There were some less sophisticated contraceptive solutions around as well. Coitus interruptus, or the withdrawal method, is mentioned even by the Old Testament itself. No matter which culture or age, some silly and weird ideas were deeply rooted in public conscience regarding pregnancy control. Greeks held the firm belief that women could prevent sperm entering the womb by holding their breath during intercourse and sneezing afterward. As those were believed to be magical numbers, agile Persian women jumped backward seven or nine times after having sex to dislodge any sperm. Funnily enough, not too long ago in the United States, during the 1960s, women took notorious Coca Cola douches the morning after, rinsing their lady parts with fizzing cola.

In the 1700s, females would cut a lemon in half and use it as a diaphragm, trusting in the spermicidal effects of the acid in the citrus.

Condoms have been around from ancient times. Initially made of animal intestines, they were later fashioned from linen as of the 1500s, when a syphilis epidemic spread across Europe. The discovery of rubber vulcanization in 1839 by Charles Goodyear soon led to the birth of the US contraceptive industry manufacturing rubber condoms and diaphragms, making the aforementioned terrifying methods a thing of the past.

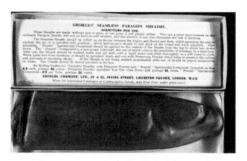

Figure 8.26 A reusable condom, formerly known as a rubber sheath from the 1940s, in its original packaging. *Source*: Science Museum/SSPL.

In the 1950s, four people—the founder of the birth control movement, a controversial scientist, a Catholic obstetrician, and a wealthy feminist—joined forces to make a little pill that reinvented sex. Finally, the secret was revealed how to separate sex for recreation and pleasure from sex for making babies!

8.5.3 Conceiving the Pill

A fabulous group of four renegades is credited with the discovery of a safe and effective birth control method. One of them was suffragette Margaret Sanger, a strong advocate of the right of women to decide for themselves when they want to get pregnant. She was the founder of the first birth control clinic in the United States. Her friend Katharine McCormick was a wealthy feminist who paid the bills for the research effort into the pill, led by a brilliant Jewish doctor named Gregory Pincus. Last but not the least, it was a Catholic obstetrician and gynecologist, John Rock, who took care of the clinical trials of the pill on women. The story of the pill is the biography of these rebel crusaders rolled into one.

It was one of the world's leading experts on mammalian reproduction, Harvard scientist Gregory Pincus (1903–67), who really got this started in the 1930s. People thought of him as a kind of Frankenstein because of his research into in vitro fertilization. His ideas that one day babies will be born in laboratories really scared people. He was ousted from Harvard and never got a job at any other university. He was devastated, angry, and determined to prove the world he was right and able to do big things. He started from scratch, built his own laboratory in a garage, going literally from door to door in his community to collect funds for his hormonal research. He became a real fringe character when Margaret Sanger enlisted his help in 1950s in making a birth control pill.

Margaret Sanger (1879–1966), a suffragette and the founder of the first birth control clinic in the United States, was a firm supporter of pregnancy prevention in the early 20th century and onward. She was the soul behind the search for the pill. Having seen her mother die after having given birth to too many children because of her father's sexual drive, she took the cause very personally. As a young woman, she went to work in the slums of

New York City, where women were either having ten children or putting their lives at risk by getting poorly performed and unsafe abortions. Doctors could only tell these women to go and sleep on the roof if they do not want to get pregnant again. Still, Sanger was convinced that sex was great power and joy as well as an important part of developing the psyche. She really wanted to find a way for women to be able to enjoy sex without fearing that it was going to lead it to pregnancy. In the 1950s, there were all forms of contraception available, like condoms, diaphragms, and intrauterine devices of dubious efficacy. These were all controlled by men and doctors. In a climate of prejudice against premarital sex, it was inconceivable to ask your doctor for those things if you were not married. It was Margaret Sanger who coined the word birth control, putting a particular stress on the word control. Her idea was that if women could control when they want to get pregnant, they could control their lives and control the whole world around them. She was very explicit about her expectations regarding the form of novel birth control she had in mind. She was looking for a pill that women could hide in their purses so that men do not notice. Also it had to be something that would allow women to turn their reproductive systems on and off. During her 40-year-long quest looking for a birth control pill, every scientist she approached turned her down saying she was building castles in the sky. Surprisingly, when Sanger came to Pincus, he already had the answers.

Figure 8.27   Gregory Pincus experimenting on a rabbit. *Source*: Corbis.

Figure 8.28   Margaret Sanger on the steps of a courthouse in New York in 1917, taken during a trial accusing her for opening a birth control clinic. She was found guilty. *Source*: Wikimedia Commons.

Pincus figured out that the female body produces its own contraceptive called progesterone when pregnant, so she cannot get pregnant again in order to protect the fetus. This meant that by giving a woman progesterone, it is possible to trick her body into thinking it is pregnant, and there is your birth control!

All they needed was the money. With birth control and sex education illegal in most US states, government and university funding or support from the pharmaceutical industry was out of the question. That is the point where one of the wealthiest women in America, Katherine McCormick (1875–1967), came into the picture. Interestingly, McCormick and Sanger were old friends, two determined feminist who got to know each other in the suffrage movement. They were both in their seventies when they teamed up to work on the pill. With their reproductive lives past, it was too late in many ways for them. Still they did not give up on their long-cherished dream: empowering women with a birth control pill and the freedom to choose their lives.

McCormick's seemingly storybook marriage to the richest man in the world turned into a nightmare when her spouse went mad on their honeymoon. This handsome, affluent man, diagnosed with schizophrenia, would spend the rest of his life in his own personal insane asylum in his California mansion.

McCormick hired the best doctors and nurses in the world to oversee her husband's care. As one of the first women to graduate from the Massachusetts Institute of Technology, she was no slouch when it came to research. She devoted her time, expertise, and money toward finding a cure for her husband's schizophrenia. She never had any children for the fear that schizophrenia was inherited and became excited about the search for woman-controlled contraception. With the death of her husband, she inherited an immense fortune of $750 million. Funding research into the elusive birth control pill was an obvious choice for her. With her degree in science, she not only wrote the checks but oversaw the whole research process. In the course of her investigations into schizophrenia, she learned about hormones and the work of Gregory Pincus.

Figure 8.29 Katharine McCormick with her husband. Their marital happiness was cut short by her spouse's mental illness.

Pincus, having no experience in human trials, needed a doctor onboard to bridge the gap between animal experiments and clinical studies. His name was John Rock, a well-respected, charismatic Catholic gynecologist. In spite of being really observant, going to mass every day, Rock was willing to fly in the face of the Catholic Church on seeing the suffering of his patients. He felt great empathy for women having more children than they can handle. Many of his patients were overwhelmed by 10 children or children who were all born with horrible birth defects. Besides, he also thought that sex is not just for procreation, but an important part of love, relationships, and marriage. Rock was a great candidate to bring onboard because he was willing to confront the church and urge it to rethink its take on this issue.

Figure 8.30 Katharine McCormick in a laboratory. *Source*: MIT Museum Collection.

Rock, one of the world's leading fertility specialists, was eager to get involved in the pursuit for the pill alongside Pincus. Bear in mind, anything that had to do with birth control was considered illegal, so they could have easily ended up in jail if they were caught red handed. They had to be very clandestine about their business. Pincus tested progesterone first on Rock's patients seeking help with their infertility. They thought these patients would not get pregnant while on the drug anyway, but it would be a great opportunity to see the effects it produces. Then they went into insane asylums, treating women they did not even ask.

However, they were not an ideal group of clinical subjects. For their studies, Pincus and Rock needed healthy young women who were having sex. What is more, contraception was outlawed in the mainland United States. Accidentally, Pincus was invited to give a lecture in Puerto Rico, which turned out to be the most ideal ground to test the pill. The availability of American doctors and nurses, good hospitals, combined with enormous poverty, a large population of young women, and no birth control laws proved to be an excellent playing ground for Pincus and his team. At the start, if you were a female medical or nursing student in Puerto Rico, where the clinical trials for the pill took place in the 1950s, and you were unwilling to participate because of feeling sick or dizzy on the pill, you failed the semester. Pretty dramatic! The experiments were not going well with the medical and nursing students. So Pincus and his team ended up in the poorest communities, where women were in desperate need of birth control. These women were blissfully ignorant of the fact that this was an experimental drug. All they knew was that the medication came from America, and if it is American, then it must be effective. So they put with all the side effects to avoid getting pregnant.

Figure 8.31   Dr. John Rock. *Source*: AP photo.

Given the massive doses in which progesterone was administered, the side effects were terrible and hard to tolerate. These most commonly involved nausea, dizziness, and migraines so severe that some participants dropped out of trials. But the ultimate goal was to make sure nobody gets pregnant on the drug to make it ready for approval. That is why the doses were much higher than they needed to be, and the team was willing to overlook the side effects. After thalidomide, any drug with widespread potential use in women of child-bearing age would be tested in a far more cautious regulatory environment. The team took great risks in enrolling women to experiment on. They were taking healthy young women, giving them a new drug that was not meant to cure anything, and they did not know exactly what they were doing. By today's standards, they used outright sneaky methods to investigate the effects of a potential birth control pill. Women in Rocks' fertility clinics, insane asylums, and the slums of Puerto Rico were used as lab animals. The irony of that is that the goal was to give women freedom. It sounds horrible by today's standards, but the guidelines at the day were pretty well within those parameters.

Figure 8.32 Women making brassieres at the Jem Manufacturing Corp. in Puerto Rico, 1950. *Source*: International Ladies Garment Workers Union Photographs.

Feeling he is racing the clock against the opposition from the Catholic Church and the federal government, in 1956 Pincus decided to take on an aggressive attitude and get the drug approved by the FDA. It was a bold act as the drug designed for healthy young females for daily use had only been tested on 60 or so women, who used it for a mere 6 months only!

In Mexico City, a young chemist in his 20s, Carl Djerassi, worked out how to produce cheap synthetic progesterone from plants. At the time, the production of progesterone depended on slaughterhouse animals, so it was almost prohibitively expensive to get hold of progesterone. Djerassi's great contribution was to figure out how to synthesize progesterone, how to make it more powerful at small doses, and how to make it into a pill. They realized that they could give a much lower dose of progesterone than they were originally giving out, and this made the pill a safer drug.

Having burned himself once, Pincus wanted to avoid controversies, so he decided he will not get the progesterone preparation approved by the FDA as a birth control pill. Instead, he got it licensed for menstrual disorders in 1957 under the name Enovid, with the warning label saying "this pill will likely prevent pregnancy." Once the word was out, women began lining up at their doctors' offices by the thousands with "irregular cycles," getting the pill prescribed off-label for birth control. Shortly after, Enovid was approved for birth control in 1960, and then it took off like a house on fire. Just five years after the pill's FDA approval, more than 6.5 million US women were taking the birth control pill.

Why was the formula revised and estrogen added to the pill? Actually, a batch of progesterone pills seemed to have fewer side effects. A closer look at the laboratory later revealed that the pills were accidentally contaminated with a small amount of estrogen. Luckily, the contamination did not turn out to be a disaster but worked magic!

Pincus then developed the idea for the morning-after pill, but then he developed leukemia in the early 1960s, which may have been connected to all the chemicals he had been working with in the makeshift laboratory that he built.

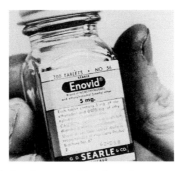

Figure 8.33   Enovid, the first combined oral contraceptive pill. Approved in 1957 by the FDA for the treatment of menstrual disorders, it soon became officially approved for birth control in 1960. Enovid was discontinued in the United States in 1988. *Source*: Library of Congress.

Neither of the pioneers ended up getting rich by making the pill. They made an unselfish decision that they would not keep this thing to themselves, so that it could reach the markets as quickly as possible. McCormick was in her eighties when the pill became approved. For her, the biggest accomplishment was to be able to buy it. She must have been quite a sight when she walked into a drugstore with a prescription and asked her pharmacist for the birth control pill!

8.5.4 Closing Remarks: The Power of the Pill

The benefits of contraceptive use are far reaching. Couples can have happier marriages. The pill takes out the fear of pregnancy from spontaneous intercourse, making a blissful sex life possible. Couples also have the power to control the number of children they desire. They can take control of their lives and delay becoming parents. They can go to college instead and start careers. Women are free to explore more opportunities, finding fulfillment in their studies and jobs. The benefits are clearly reflected by income, mental health, and relationships as well as child well-being. Contraceptives also help women achieve healthier pregnancies by allowing them to time and space births. This all translates into maternal well-being, and a well-documented decrease in the risk of prematurity, low birth weight, and infant mortality. On the flip side, some argue it has contributed to the destruction of the family unit and encourages premarital and casual sex. People are also debating whether it increases the number of teen pregnancies and divorces. One thing is for sure: the pill has changed the way we view women forever.

8.6 Insulin

One of the most compelling human dramas and greatest achievements in modern medicine is the discovery of insulin. Out of the 285 million patients worldwide who suffer from diabetes, a shocking number of 50 million depend on insulin [55]. Their lives would be in the balance without the medication.

8.6.1   Insulin and Diabetes

Carbohydrates and sugars are the most important fuels in our food. The digestive enzymes break these food components down into glucose. Insulin is a protein hormone our pancreas produces to convert glucose into energy. Insulin allows glucose to move from the blood into the cells, thereby lowering blood sugar.

Diabetes mellitus is a chronic disorder in which blood sugar levels are abnormally high because of a lack of insulin production or effect. We call it type 1 diabetes when the pancreas is unable to make enough insulin, and type 2 diabetes when the body cannot effectively use the insulin it produces.

The body gains a significant proportion of its energy from glucose. A lack of insulin production or effect means the body cannot get enough glucose from the blood into cells. Without enough insulin, your body begins to break down itself by using fat and muscle as an alternative fuel.

No wonder weight loss and tiredness can be one of the earliest signs of diabetes. The symptoms of diabetes mellitus are easy to recognize and involve a triad of increased thirst, urination, and hunger.

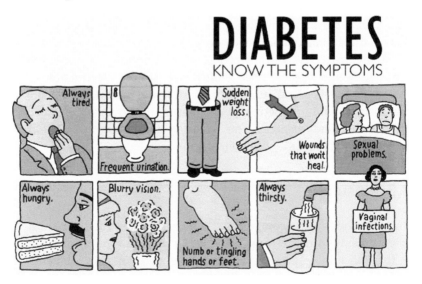

Figure 8.34   The symptoms of diabetes come in many forms. *Source*: www. thediabetesprotocol.co.

Diabetes patients may experience a number of serious, long-term complications. The main concern is that high blood glucose causes blood vessels to narrow, leading to heart attacks, strokes, peripheral, arterial, and nerve disease, kidney failure, poor vision, skin breakdown, and even erectile dysfunction.

As already mentioned, diabetes can come in two forms. One of them is type 1 and the other is type 2 diabetes. Type 1 diabetes is marked by sudden onset and marked symptoms. It usually affects children in whom the pancreatic insulin-producing cells are destroyed as part of an autoimmune process. These patients require daily replacement of insulin in order to stay alive. Interruption of the insulin therapy can lead to a life-threatening complication, known as diabetic ketoacidosis. It is characterized by a buildup of toxic ketones that are released from fatty tissue that the body uses its fat reserves as an alternative fuel instead of glucose. Warning signs of diabetic ketoacidosis in type 1 diabetes include excessive hunger or thirst, vomiting, and a characteristic nail-polish like breath odor.

In contrast to type 1 diabetes, type 2 diabetes is a real chameleon. It has subtle signs of increased urination, thirst, and tiredness appearing gradually, sometimes over a period of years. Type 2 diabetes is much more common than type 1 diabetes. It accounts for 90% of people with diabetes around the world. It is largely the result of obesity and lack of exercise. It has a more insidious onset and usually starts in adulthood. It is usually well managed by noninsulin anti-diabetic drugs, but some patients require the addition of insulin to their drug regimen to keep their glucose under control. When the blood glucose levels get very high, type 2 diabetics may develop severe dehydration, which may lead to confusion, drowsiness, and coma, a condition called nonketotic hyperglycemic-hyperosmolar syndrome.

The diagnosis of diabetes is pretty straightforward, as suggested by tell-tale clinical signs and high blood sugar levels. People with diabetes have to make sure to keep their blood glucose levels in check in order to make complications less likely to develop. This means a diet low in carbohydrates and fat, getting the right amount of exercise, rigorous daily blood sugar monitoring, a broad array of drugs, and, last but not least, insulin therapy.

In order to achieve proper diabetes control, insulin comes in rapid, intermediate, and long-acting basic forms, divided by speed of onset and duration of action. It is injected into the skin either several times a day with a pen, or given continuously from an insulin pump. Science is not at the final stages of developing insulin. New forms of insulin, such as nasal, transdermal, oral, and inhaled insulin, are in the pipeline.

Figure 8.35   First inhaled insulin has been available since February 2015. *Source*: Sanofi S.A.

8.6.2   Life with Diabetes in the Past

Diabetes has been a pretty frightening and clearly recognizable disease. It has been known for at least 3500 years. In diabetes, excess sugar is not only found in the blood but in the urine as well. This sweet taste had been described in urine by the ancient Chinese, Egyptians, Indians, and Persians, who noted that the sugary urine would attract ants. They used to call it the "sugar sickness." The term diabetes mellitus actually comes from a Greek expression and stands for "siphoning" or "passing through honey." Funnily enough, it was known as the "pissing evil" back in the 17th century referring to the excess urination the disease involves. Diabetes was equivalent to a swift, brutal, and painful death sentence, with no real treatment, and a life expectancy of as little as three weeks upon diagnosis. Patients wasted away. They were little more than living human skeletons, unable to drink, eat, or walk. No surprise, diabetes used to be described as "the melting down of flesh and limb into urine."

The cause of diabetes was a mystery until the early 20th century, and doctors could not do anything about it. The only thing they realized was the fact that sugar made things worse. At that time, the only way to "control" diabetes was through a restricted-calorie diet. It comprised minimum amount of carbohydrates and sugar needed to sustain life. This regimen, with food intake sometimes as low as 500 calories a day, required an inordinate amount of will power. Besides, it drained patients of their strength and energy. Doctors talked about it as an under-nutrition approach, but patients talked about it as starvation. Instead of dying shortly after diagnosis, this diet could buy diabetes patients some extra time to live, but only for about a year. Actually it was not uncommon to see patients starve to death on these harsh diets!

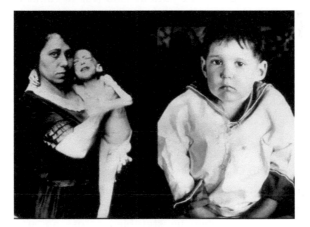

Figure 8.36 Photographs of a 3-year-old child treated for diabetes: before and after insulin. The left one was taken on December 15, 1922, before insulin was available. He was put on the notorious "starvation regime" and weighed 15 pounds. He was one of the first patients being treated for insulin. The right photograph was taken on February 15, 1923, and the child weighed 29 pounds. *Source*: Eli Lilly & Co Archives.

8.6.3 Prelude to the Discovery of Insulin

In fact, it was not until the first decades of the 20th century that some weird experimental efforts set the stage for learning what diabetes is and how we treat it.

Interestingly, it was a young medical student in Berlin in 1869, Paul Langerhans, who discovered the cells in the pancreas that produce a hormonal secretion called insulin. These cells would later be called the islets of Langerhans.

Later on in 1899, another German, a pathologist called Oskar Minkowski (1858–1931) was the first to show the connection between pancreatic secretions and diabetes. He discovered that if he took the pancreas out of a dog, it instantly developed a severe disease indistinguishable from diabetes and quickly died. Sugar was found in the dogs' urine after flies were noticed to be feeding off the urine.

In 1906, George Zeulzer (1870–1949), another native of Germany, went so far as to inject an extract made from calf pancreas into a dying diabetic patient who was in coma. Although the patient later passed away, it is amazing nonetheless that he could achieve temporary improvement.

8.6.4 Discovery of Insulin

An unknown former frontline Canadian surgeon named Frederick Banting (1891–1941) forever changed the course of modern medicine. At the age of 30, totally broke after returning from the First World War, he opened a small floundering surgery practice in London, Ontario, struggling to make ends meet. His meagre earnings forced him to take a position as a demonstrator in the local medical school. Banting was preparing a lecture for the medical students when he stumbled across an article in a medical journal about the pancreas and the insulin-producing islets of Langerhans. This tiny event triggered a monumental discovery. In October 1920, the inquisitive-minded Banting jotted down in his notebook a little idea for research on diabetes. It seemed like such a crazy idea that even Banting doubted its potential success. He wanted to remove the pancreas of hundreds of laboratory dogs to isolate the cells that produce insulin. From these extracts, he would then produce pure insulin to regulate the blood sugar of human patients.

To make his idea work, he returned to his alma mater, the University of Toronto, where he met with world-renowned physiologist and expert in carbohydrate metabolism, Richard Macleod (1876–1935). They did not hit it off well. Macleod was a senior professor, very articulate and learned. As a scientist well acquainted with the literature on the subject, he was unimpressed with Banting's range of knowledge about diabetes and the pancreas. Besides he had his qualms about the soundness of Banting's idea. Banting was a young doctor who knew almost nothing about what he was trying to talk about, and had never even treated anyone with diabetes. But Banting was a dreamer, with just enough salesmanship to win over the academic.

Figure 8.37   Captain Banting. *Source*: Canadian Diabetes Association.

Figure 8.38   Richard Macleod (circa 1928). *Source*: Wikimedia Commons.

So Macleod eventually decided to give him a laboratory with minimum equipment, ten dogs, and an assistant. Macleod reluctantly asked two of his graduate students to help Banting out. The two graduate students tossed the coin to see who would work with Banting. Charles Best (1899–1978) won the coin toss, and nobody knows the name of the one who lost. Best was a 21-year-old science graduate and a medical student. He was bright and ambitious and looked forward to working with Banting on this strange idea in the summer of 1921.

8.6.5 Making the Experiments Work

With a single lab and only a handful of dogs to work with, the conditions for research were inadequate. Still Banting kept pushing on. He was a stubborn and determined person who had risked everything on one idea. He had given up his practice; he was out of money and, in a way, had just gambled everything.

Figure 8.39 Charles Best and Frederick Banting on the roof of the University of Toronto's Medical Building, with one of the first diabetic dogs to be treated with insulin in 1922. *Source*: University of Toronto.

The experiments involved removing the pancreases of dogs, thereby making them diabetic. Their pancreases were ground up into an injectable extract. Their idea was that the pancreatic extract made this way, if injected a few times a day into the dog, will help the dog regain health. Amazingly, after a summer of many setbacks and failures, the team was keeping a severely diabetic dog alive with injections of this extract. The team quickly ran out of dogs though. Banting and Best were so desperate for more animals that they resorted to catching stray dogs on the street. They quickly realized, however, that they required a larger supply of organs than their dogs could provide. So they just went to the slaughterhouse; it turned out that fresh beef or pork pancreases worked just as well.

8.6.6 Getting the Elixir Right

The trouble was that the results they got in terms of glucose control were maddeningly inconsistent. The challenge was to get the alcohol level in the extract just right, so that both pure enough and nontoxic for human consumption. Macleod invited James Bertram Collip (1892–1965), a biochemist in the department of physiology at the University of Toronto, to help Banting and Best with purifying their extract for clinical testing in humans to quicken the experimental pace.

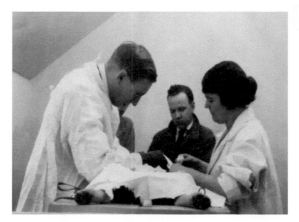

Figure 8.40 Banting performing surgery on a dog. *Source*: University of Toronto.

Macleod set all the researches up in separate laboratories. So a race was up to see which lab would be the first to refine the extract enough for use in human trials. As a result of an uneasy agreement of four people who did not like each other and who occasionally literally fought each other, nonetheless within 18 months, this team gave the world an effective treatment for diabetes. By the end of 1921, they already figured out exactly the right amount of alcohol to purify Banting's elixir.

8.6.7 First Diabetes Patients on Insulin

The team was eager to start testing on humans. But on whom should they test? Banting and Best clearly had confidence in the insulin and began by injecting themselves with the extract. As part of a phenomenon called hypoglycemia, their blood glucose dropped significantly and they felt weak and dizzy, but they were not harmed. They figured out that they can get better by eating sugary food such as honey or drinking orange juice. After the group had experimented enough to gain an understanding of the required doses and how best to treat hypoglycemia, their insulin was deemed ready to tried on patients.

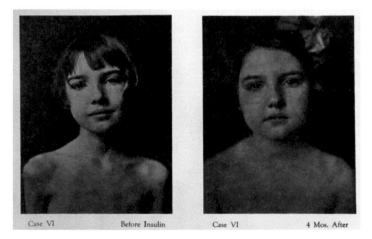

Case VI   Before Insulin   Case VI   4 Mos. After

Figure 8.41 Before and four months after insulin treatment. *Source*: Wellcome Images.

In January 1922, a severely diabetic 14-year-old boy from Toronto, Canada, Leonard Thompson, was chosen as the first person with diabetes to receive insulin. The child was reduced to 65 pounds and was within a few weeks of death. The extract they gave him was still so impure that as a result of the first insulin injection, Leonard had a 7.5 cm callus at the injection site on his left buttock. The first test, in fact, failed. Two weeks later, they resumed treatment, and this time, they got spectacular success. Leonard flourished with the extract and showed signs of marked improvement after the insulin injection. Leonard lived another 13 years before dying of pneumonia.

8.6.8 Getting the Nobel Prize: Resentments Resolved

In February 1922, the team published a paper on the successful development of a purified pancreatic extract made of insulin that saved the life of a young diabetes patient. The news of the successful treatment of diabetes with insulin rapidly spread outside Toronto. Not surprisingly, Banting and Macleod were jointly awarded the 1923 Nobel Prize in Physiology or Medicine for the discovery of insulin for the treatment of diabetes. The fact that the Nobel committee chose only Banting and Macleod for the award caused much animosity. Eventually, the researchers divided the prize between the four of them. Banting shared his part of the prize money with Best, and Macleod shared his with Collip.

8.6.9 Romantic Stories behind the Mass Production of Insulin

The early experimental patients were a lucky few. After surviving nearly 3 years on a starvation diet that nearly killed them, they came back to life on getting Banting's elixir in the human trials. The discovery was soon proclaimed as a miracle cure for diabetes, but that turned out to be not quite right. The remaining masses still waited, with their lives hanging in the balance, for someone who learned how to mass produce the drug.

By the winter of 1922, Banting and his team were hand making small batches of the miracle hormone and administering it to their patients. In the early days, they got all the pork and beef pancreases that they could. They grounded these up and then began the extraction process. However, the Toronto team soon became overwhelmed and could barely keep up even with a small number of patients in clinical trials. The results were astounding, however, and this made a man named George Clowes really enthusiastic about the challenge of mass production of insulin. He happened to be the research director for a large US drug company called Eli Lilly & Co of Indianapolis. The University of Toronto immediately gave the pharmaceutical company license to produce insulin free of royalties. As soon as 1923, about 1 year after the first test injection, insulin became widely available. The medical firm Eli Lilly was producing enough insulin to supply the entire North American continent.

Figure 8.42 Labeling boxes of insulin at Eli Lilly in 1924. *Source*: Eli Lilly & Co.

Figure 8.43 August and Marie Krogh. *Source*: NovoNordisk A/S.

Meanwhile, a Nobel Prize winning Danish doctor August Krogh (1874–1949) traveled to Toronto for a personal interest. During a lecture tour of the United States, Dr. Krogh heard of the discovery of insulin in Canada. Dr. Krogh's wife, Marie, also a physician, was diagnosed with diabetes. Dr. Krogh knew that the only hope for survival was the availability of insulin in their home country. So he raced to Toronto to ask permission to manufacture insulin in Denmark. After obtaining the license, he returned to Denmark to start what eventually became NovoNordisk Insulin Laboratory. So the two big insulin producers, Eli Lilly and NovoNordisk, have their roots back in 1922: Eli Lilly because of an enterprising research director, and NovoNordisk because Dr. Krogh's wife had become diabetic and he wanted to know how to make insulin so that he could save her life. Marie Krogh survived, and when insulin hit the commercial market in late 1923, millions of patients around the world saw renewed chance of life.

The history of insulin is studded with some remarkable stories. One of them has to do with Eva Saxl, a diabetes patient, and her husband, Viktor. Fleeing from the Nazis in Europe, Eva and Viktor took refuge in a Jewish ghetto in Shanghai, China. When Shanghai was cut off from medical supplies and drugs, Viktor made his own homemade insulin to save Eva's life. Following the methods of Banting's team described in a medical textbook, he managed to extract insulin from the pancreases of water buffaloes. Viktor's homemade insulin was the lifeline to the survival of not only his wife, but also to another 420 patients with diabetes. Actually, they supplied all of the diabetes patients in the Shanghai ghetto for four years, between 1941 and 1945. Amazing couple in diabetes history!

For many years, drug companies derived the hormone using pancreases from stockyard, taken from slaughtered cows and pigs. The insulin they produced was known as "animal insulin." In 1955, insulin became the first protein to be fully sequenced. That work resulted in the 1959 Nobel Prize for Frederick Sanger. The early diabetes researchers did not know how lucky they were. It turned out that the basic structure and makeup of animal insulin were almost exactly the same in different animal species, so insulin from cows and pigs also worked in humans.

Figure 8.44   Eva and Viktor Saxl. *Source*: http://www.diabetes.co.uk/.

Figure 8.45   Acids, alcohol, and pancreatic tissue were separated, bathed, and mixed in this laboratory of a 1946 insulin factory in Germany. *Source*: Getty Images.

Once the protein's sequence is known, it is possible, in theory, to recreate it synthetically. In fact, insulin was the first protein to be chemically synthesized in a laboratory in 1963. But researchers were unable to produce much of it. For 60 years after Banting's group isolated insulin, diabetes patients relied on hormone purified from animals. Up until the 1980s, insulin was extracted from the pancreas glands of swine and cattle slaughtered for food. It took about 8000 pounds of animal pancreas glands to produce 1 pound of insulin. Animal insulin worked well on the whole, but it was not an exact match with the human hormone and sometimes caused adverse reactions. More commonly, the injections either caused fat deposits, making the skin look lumpy, or destroyed fat in the underlying tissues, causing indentation of the skin. Sometimes the injections involved severe local or systemic reactions. So the search was on to produce a form of insulin chemically identical to human insulin to make production more efficient as well as side effects less marked.

8.6.10 Lab-Grown Human Insulin

In 1978, insulin became the first human protein to be manufactured through biotechnology. A San Francisco-based team of researchers from the City of Hope National Medical Center and the fledgling biotechnology company managed to synthesize human insulin in the laboratory. The team inserted the gene for human insulin into bacterial DNA using a method called DNA recombinant technology. Large amounts of the so-called recombinant human insulin can be produced this way. Humulin, as the commercial product was called, revolutionized diabetes treatment when it became widely available in the early 1980s. Today, almost every patient with diabetes uses recombinant human insulin instead of animal insulin. This product is identical to actual human insulin, has the advantage of being less likely to allergic reactions than animal insulin [56]. Therefore, it is the safest form of insulin any patient can get.

8.6.11 Closing Remarks: The Impact of Insulin

Insulin is one of the biggest discoveries in medicine. Although insulin does not cure diabetes, it has changed the brutal realities of sufferers of diabetes. When it came, it was like a miracle. People with severe diabetes and only days left to live were saved. For the first time ever, people were living on insulin with diabetes. Diabetes patients who received insulin shots recovered from comas, resumed eating carbohydrates (in moderation), and realized they had been given a new lease on life. And as long as they kept getting their insulin, they could live an almost normal life. With the advent of insulin, life expectancies are not measured in weeks anymore but in tens of years.

Figuring out how to inject insulin gene into self-replicating cells through DNA recombinant therapy means an endless supply of pure human insulin is available, without depending on animal extraction. DNA technology has unleashed a whole new wave of developments. Long-lasting basal insulin and rapid-acting more efficient insulin are available today.

Figure 8.46 Dr. Arnold Kadish of Los Angeles, California, devised the first insulin pump in the early 1960s. *Source*: Medscape.

Administering insulin used to require a brutal routine every morning, noon, and dinnertime. Needles had to be sharpened and boiled up after every use to avoid injection-site reactions. Testing blood sugar also required an extreme effort by today's standards. There were no quick blood tests or easy home kits. Readings required boiling urine samples in a test tube with chemicals multiple times each day. Today, insulin is coupled to easy-to-use devices to improve the way patients use insulin. Home-testing devices have gone a long way to give patients more control over their own management. Delivery systems have dramatically improved and include thin needles, disposable syringes, pens, pumps, and more.

But the power of the insulin story is most evident in the children whose lives insulin saved. Many of them lived into their eighties and nineties, who survived 70 years with diabetes, thanks to the miracle of insulin. These elderly people stand as a shiny example of the power of this discovery and the millions of lives it saved.

References

1. Centers for Disease Control and Prevention (CDC). Achievements in public health, 1900–1999: Control of infectious diseases. *Morbidity and Mortality Weekly Report*, 1990; **48**(29); 621–629.

2. Mavromoustakos, T., Durdagi, S., Koukoulitsa, C., Simcic, M., Papadopoulos, M. G., Hodoscek, M., and Grdadolnik, S. G. Strategies in the rational drug design. *Current Medicinal Chemistry*, 2011; **18**(17): 2517–2523.

3. Arzimanoglou, A., Ben-Menachem, E., Cramer, J., Glauser, T., Seeruthun, R., and Harrison, M. The evolution of antiepileptic drug development and regulation. *Epileptic Disorders*, 2010; **12**(1): 3–15.

4. Landau, R., Achilladelis, B., and Scriabine, A. (eds.). *Pharmaceutical Innovation: Revolutionizing Human Health*. Philadelphia: Chemical Heritage, 1999, pp. 247.

5. Rothwell, P. M., Wilson, M., Elwin, C. E., Norrving, B., Algra, A., Warlow, C. P., and Meade, T. W. Long-term effect of aspirin on colorectal cancer incidence and mortality: 20-year follow-up of five randomised trials. *Lancet*, 2010; **376**(9754): 1741–1750.

6. Rothwell, P. M. Aspirin in prevention of sporadic colorectal cancer: Current clinical evidence and overall balance of risks and benefits. Recent Results. *Cancer Research*, 2013; **191**: 121–142.

7. Chen, J., Tang, H., Wu, Z., Zhou, C., Jiang, T., Xue, Y., Huang, G., Yan, D., and Peng, Z. Overexpression of RBBP6, alone or combined with mutant TP53, is predictive of poor prognosis in colon cancer. *PLoS One*, 2013; **8**(6): e66524.

8. Goldberg, D. R. Aspirin: Turn of the century miracle drug. *Chemical Heritage Magazine*, 2009; **27**(2).

9. Sneader, W. *Drug Discovery: A history*. West Sussex, UK: Wiley, 2005, pp. 116–117.

10. Sneader, W. *Drug Discovery: A history*. West Sussex, UK: Wiley, 2005, pp. 90–91.

11. Katzung, B. G., Masters, S. B., and Trevor, A. J. (eds.). *Basic and Clinical Pharmacology*, 12th ed. New York: McGraw-Hill Medical, 2012, pp. 543–561.

12. Katzung, B. G., Masters, S. B., Trevor, A. J., and Kuidering-Hall, M. *Katzung & Trevor's Pharmacology Examination & Board Review*, 12th ed. New York: McGraw-Hill Medical, 2012, pp. 271–278.

13. Shevelev, V. A., Bamlpvsky, A. I., and Mravyova, V. I. Mechanical drying of raw opium. *Bulletin on Narcotics*, 1958; (2): 6–7.

14. Norn, S., Kruse, P. R., and Kruse, E. History of opium poppy and morphine. *Dan Medicinhist Arbog*, 2005; **33**: 171–184.

15. Rosso, A. M. Poppy and opium in ancient times: Remedy or narcotic? *Biomedicine International*, 2010; **1**: 81–87.

16. Center for Substance Abuse Treatment. *History of Medication-Assisted Treatment for Opioid Addiction. In Medication-Assisted Treatment for Opioid Addiction in Opioid Treatment Programs.* Maryland: Substance Abuse and Mental Health Services Administration, 2005.

17. Schwarz, S., and Huxtable, R. The isolation of morphine. *Molecular Interventions*, 2001; **1**: 89–91.

18. Altman, L. K. *Who Goes First? The Story of Self-experimentation in Medicine.* California: University of California Press, 1987, pp. 89–92.

19. Bagetta, G., Cosentino, M., Corasaniti, M. T., and Sakurada, S. *Herbal Medicines: Development and Validation of Plant-Derived Medicines for Human Health.* Florida: CRC Press, 2011, pp. 396.

20. Alkaloid. In *Encyclopedia Britannica Online.* Encyclopedia Britannica Inc. 2014.

21. Nicholson, B. Morphine sulfate extended-release capsules for the treatment of chronic, moderate-to-severe pain. *Expert Opinion on Pharmacotherapy*, 2008; **9**(9): 1585–1594.

22. Stratton Hill, C. *Guidelines for Treatment of Cancer Pain. The Pocket Edition of the Final Report of the Texas Cancer Council's Workgroup on Pain Control in Cancer Patients.* Collingdale: DIANE Publishing, 1991, pp. 12–13.

23. Momeni, M., Crucitti, M., and De Kock, M. Patient-controlled analgesia in the management of postoperative pain. *Drugs*, 2006; **66**(18): 2321–2337.

24. Bayer, I. Manufacture of alkaloids from the poppy plant in Hungary. *Bulletin on Narcotics*, 1958; (2): 21–28.

25. United Nations. *A Century of International Drug Control*. New York: United Nations Publications, 2010, pp. 85–86.

26. United Nations Office on Drugs and Crime. *Afghanistan Opium Survey 2007*. New York: United Nations Publications, 2007.

27. Sneader, W. *Drug Discovery: A history*. West Sussex, UK: Wiley, 2005, pp. 116–117.

28. Chen, J. J., and Obering, C. A review of intermittent subcutaneous apomorphine injections for the rescue management of motor fluctuations associated with advanced Parkinson's disease. *Clinical Therapeutics*, 2005; **27**(11): 1710–1724.

29. Scientific Achievements. *Dr. Paul Janssen Award for Biomedical Research*. http://www.pauljanssenaward.com/scientific.html.

30. RADAR (Rational Assessment of Drugs and Research). *Fentanyl patches (Durogesic) for chronic pain*. https://www.nps.org.au/_data/assets/pdf_file/0016/23731/fentanyl.pdf

31. Katzung, B. G., Masters, S. B., and Trevor, A. J. (eds.). *Basic and Clinical Pharmacology*, 12th ed. New York: McGraw-Hill Medical, 2012, pp. 697–708.

32. Katzung, B. G., Masters, S. B., Trevor, A. J., and Kuidering-Hall, M. *Katzung & Trevor's Pharmacology Examination & Board Review*, 12th ed. New York: McGraw-Hill Medical, 2012, pp. 343–346.

33. Porter, R. S., and Kaplan, J. L. *The Merck Manual of Diagnosis and Therapy*, 19th ed. New Jersey: Whitehouse Station, 2011, pp. 169, 172, 305, 332, 663, 792–799.

34. Katzung, B. G., Masters, S. B., and Trevor, A. J. (eds.). *Basic and Clinical Pharmacology*, 12th ed. New York: McGraw-Hill Medical, 2012, pp. 1072–1074.

35. Goldenberg, M. M. Pharmaceutical Approval Update: Ciclesonide (Alvesco) Inhalation Aerosol. *Pharmacy and Therapeutics*, 2008; **33**(3): 150–151.

36. Sneader, W. *Drug Discovery: A history*. West Sussex, UK: Wiley, 2005, pp. 180–183, 207–210.

37. Renneberg, R. *Biotechnology for Beginners*. Massachusetts: Academic Press, 2008, pp. 135.

38. Porter, R. S., and Kaplan, J. L. *The Merck Manual of Diagnosis and Therapy*, 19th ed. New Jersey: Whitehouse Station, 2011, pp. 169, 172, 305, 332, 663, 792, 796.

39. Wrobel, S. Science, serotonin, and sadness: The biology of antidepressants—a series for the public. *The FASEB Journal*, 2007; **21**(13): 3404–3417.

40. Porter, R. S., and Kaplan, J. L. *The Merck Manual of Diagnosis and Therapy*, 19th ed. New Jersey: Whitehouse Station, 2011, pp. 1538–1542.

41. Katzung, B. G., Masters, S. B., and Trevor, A. J. (eds.). *Basic and Clinical Pharmacology*, 12th ed. New York: McGraw-Hill Medical, 2012, pp. 697–708.

42. Haddad, S. K., Reiss, D., Spotts, E. L., Ganiban, J., Lichtenstein, P., and Neiderhiser, J. M. Depression and internally directed aggression: Genetic and environmental contributions. *Journal of the American Psychoanalytic Association*, 2008; **56**(2): 515–550.

43. Shock therapy. In *Encyclopedia Britannica Online.* Encyclopedia Britannica Inc. 2014.

44. López-Muñoz, F., and Alamo, C. Monoaminergic neurotransmission: The history of the discovery of antidepressants from 1950s until today. *Current Pharmaceutical Design*, 2009; **15**(14): 1563–1586.

45. Katzung, B. G., Masters, S. B., Trevor, A. J., and Kuidering-Hall, M. *Katzung & Trevor's Pharmacology Examination & Board Review*, 12th ed. New York: McGraw-Hill Medical, 2012, pp. 521–538.

46. Ramachandraih, C. T., Subramanyam, N., Bar, K. J., Baker, G., and Yeragani, V. K. Antidepressants: From MAOIs to SSRIs and more. *Indian Journal of Psychiatry*, 2011; **53**(2): 180–182.

47. Sneader, W. *Drug Discovery: A history*. West Sussex, UK: Wiley, 2005, pp. 397.

48. Vaswani, M., Linda, F. K., and Ramesh, S. Role of selective serotonin reuptake inhibitors in psychiatric disorders: A comprehensive review. *Progress in Neuropsychopharmacology and Biological Psychiatry*, 2003; **27**(1): 85–102.

49. The Nobel Prize in Physiology or Medicine, 1970. http://www.nobelprize.org/nobel_prizes/medicine/laureates/1970.

50. Finer, L. B., and Zolna, M. R. Shifts in intended and unintended pregnancies in the United States, 2001–2008. *American Journal of Public Health*, 2014; **104**(S1): S44–S48.

51. Jones, J., Mosher, W. D., and Daniels, K. Current contraceptive use in the United States, 2006–2010, and changes in patterns of use since 1995. *National Health Statistics Reports*, 2012; No. 60. http://www.cdc.gov/nchs/data/nhsr/nhsr060.pdf, accessed March 20, 2013.

52. Fortner, K. B., Szymanski, L. M., Fox, H. E., and Wallach, E. E. (eds.). *The Johns Hopkins Manual of Gynecology and Obstetrics*, 3rd ed. Philadelphia: Lippincott Williams & Wilkins, 2007, pp. 344.

53. Jones, R. K. *Beyond Birth Control: The Overlooked Benefits of Oral Contraceptive Pills*. New York: Guttmacher Institute, 2011.

54. Katzung, B. G., Masters, S. B., and Trevor, A. J. (eds.). *Basic and Clinical Pharmacology*. New York: McGraw-Hill Medical, 2012, pp. 728–730.

55. *Diabetes in the UK 2010: Key Statistics on Diabetes*. www.diabetes.org.uk, accessed on April 15, 2015.

56. Ho, L. T., Lam, H. C., Wu, M. S., Kwok, C. F., Jap, T. S., Tang, K. T., Wang, L. M., and Liu, Y. F. A twelve month double-blind randomized study of the efficacy and immunogenicity of human and porcine insulins in non-insulin-dependent diabetics. *Zhonghua Yi Xue Za Zhi (Taipei)*, 1991; **47**: 313–319.

Chapter 9

Regenerative Medicine: Repairing Body

Young Ha Kim

Korea Institute of Science and Technology,
5, Hwarang-ro 14-gil, Sungbuk-gu, Seoul 02792, Republic of Korea

yhakim@kist.re.kr

Nowadays, our life has extended to nearly 100 years as the economy has improved and medicine has developed. Nevertheless, no one wants to have an unhealthy but long life. Now various artificial organs and tissues such as dental implants, artificial eye lens, and hip and knee joints are available to enjoy longevity better than before. In addition, many patients waiting for organ transplantation can survive with the help of artificial hearts or kidneys. Figure 9.1 shows various artificial organs currently available.

Materials used to treat diseases or implant things, including artificial organs, called biomaterials, must not be toxic and should not cause side reactions when applied to the body. Our body has a complete protection system that can kill bacteria and destroy all foreign bodies. We should choose or develop materials that trigger less foreign-body reactions, termed biocompatible materials. Biomaterials are metals, ceramics, and polymers (a scientific name for plastics and rubbers).

Chemistry: Our Past, Present, and Future
Edited by Choon Ho Do and Attila E. Pavlath
Copyright © 2017 Pan Stanford Publishing Pte. Ltd.
ISBN 978-981-4774-08-6 (Hardcover), 978-1-315-22932-4 (eBook)
www.panstanford.com

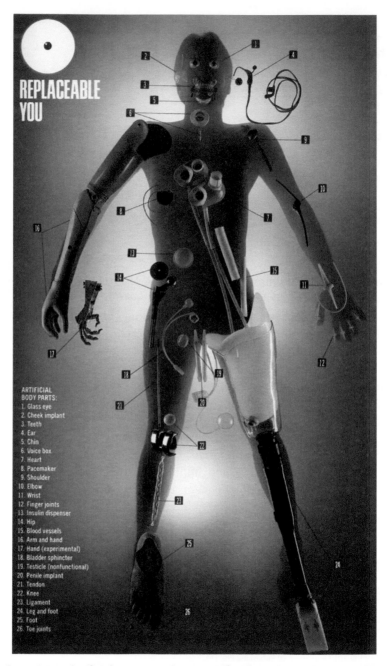

Figure 9.1   Artificial organs. *Source*: *Life Magazine*, February 1989, page 56.

The biological function of a liver or pancreas is too complicated to be replaced or mimicked by artificial materials. All the organs and tissues in the body are produced by cells. Therefore, the cell functions are fused with materials to develop the artificial organs, so-called biohybrid organs. The cells are cultured on the material surfaces, inside the pores of spongy materials or mixed with hydrogels (materials containing much water) to produce organs and tissues. An artificial liver or pancreas is a typical example. Furthermore, the cells are seeded in the biodegradable polymeric scaffolds and then cultured to produce organs/tissues, and the materials are gradually degraded for a period. Finally, the scaffolds are replaced by the organs or tissues. This technology is called tissue engineering and is now the hottest topic in the world. Recently, artificial skin, cartilage, blood vessel, and cornea have been developed by this technology, which demonstrate functions similar to those of real organs.

Our body is completely capable of recovering wounds and regenerating damaged tissues by cells and growth factors controlling cell functions. Therefore, this function is utilized to heal diseases and regenerate tissues/organs, which is called regenerative medicine. Regenerative medicine and tissue engineering utilize the functions of cells to create real organs/tissues and, therefore, will be one of the key technologies in the future.

9.1 Artificial Organs

An ABC TV drama aired in 1974, *The Six Million Dollar Man*, was very popular. The hero was an astronaut, who was almost killed in an air crash. However, he got artificial limbs, arms, and eyes through modern biomedical engineering and became a bionic human. He could run faster than a car and had superpower to beat bad guys. The sister drama, *The Bionic Woman*, was also aired in 1976. These dramas showed the age of artificial organs in advance.

9.1.1 Artificial Kidney

The first artificial organs to be developed were artificial kidneys. Dr. Kolff developed them in 1943 in the Netherlands. Kidneys remove urea from the blood. If kidneys lose their function, patients would have too much urea in the blood and ultimately die. Dr. Kolff succeeded in treating patients by using artificial kidneys made of polymeric membrane (very thin plastic films).

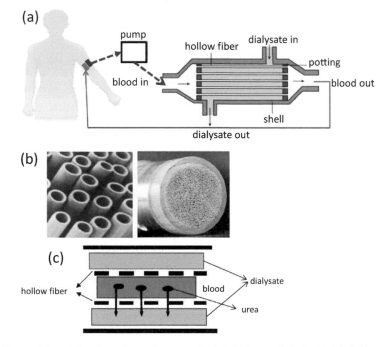

Figure 9.2 The function of an artificial kidney: (a) Artificial kidney treatment: blood (dotted blue line) containing urea is cleaned (urea is removed) during the passage through the module and returned to the body (red solid line). (b) Hollow fibers potted in a module. (c) Mechanism of urea dialysis (hemodialysis): urea in the blood moves to the dialysate by the difference in concentration.

Figure 9.2 shows the concept of artificial kidneys. The blood taken from a patient flows through hollow fibers. The hollow fibers are extremely thin, having a diameter of less than 0.2 mm, and thousands of such fibers are potted together in a container (Fig. 9.2b). The urea in the blood is passed through the tiny holes in the membrane wall and washed out with liquid (called as dialysate) flowing outside the fibers. So urea moves from the blood (high concentration of urea) to the dialysate (low concentration of urea) by the difference of concentration (such a behavior is called dialysis, Fig. 9.2c). But the blood cannot pass through the membrane; therefore, the rest of the cleaned blood is returned to the body. Dr. Kolff applied cellophane to the membrane. Cellophane is a kind of regenerated cellulose. Cellulose is dissolved in carbon disulfide (the solution is named Viscose) and coagulated back in the acidic water to get fibers (named Rayon) or films (named cellophane). Rayon is shiny and strong (therefore, called artificial silk) and was very popular before Nylon was invented. Cellophane passes liquid but not solid particles; therefore, Kolff applied it to artificial kidneys.

Now various polymers are used as artificial kidneys. Another kind of regenerated cellulose, Cuprophan, produced from the cupric-ammonium solution of cellulose is most common. In addition, other polymers such as cellulose acetate or polysulfone (a special, strong plastic) are also included.

However, an artificial kidney just removes urea from the blood but cannot restore the malfunctioning kidney. Therefore, patients are usually worsened slowly. Practically, patients survive on artificial kidneys until they get a transplantation of kidney for a complete recovery.

9.1.2 Artificial Heart

Which organ in the body is the most important? Of course, the brain controls all the functions of our body. But we do not yet understand the structure and function of a brain in detail, and it is still a dream to mimic the function of a brain artificially. The heart is the most important one practically. All the cells in the body need oxygen and nutrients to produce energy by a process called metabolism. The heart circulates blood throughout the whole body to deliver oxygen, nutrients, and metabolism wastes. Even in the case of brain death, the body keeps running if the heart functions normally.

Figure 9.3 shows the structure of a heart. The heart has the size of a fist. It has two upper rooms (atrium) and two lower rooms (ventricle) in both sides. The blood from the body containing much carbon dioxide comes into the right atrium through a big vein (vena cava), goes out to the lungs via a pulmonary artery to get oxygen, comes back to the left atrium via a pulmonary vein, and is finally pumped out from the left ventricle via a big artery (aorta). So the left ventricle is the real pump that circulates blood to the whole body. The heart beats more than 100,000 times a day to deliver about 7000 liters of blood. The blood stream is controlled to flow only in a single direction by four heart valves located between the atrium, ventricle, pulmonary artery, or pulmonary vein. What a precise and robust pump in the world to work for nearly 100 years!

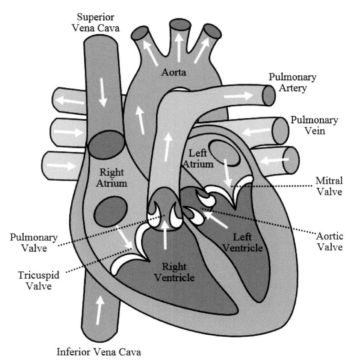

Figure 9.3   The structure of a heart.

There are also diseases in the heart. Many babies are born with holes at the atrium or ventricle innately, and the probability is high as 0.8%, which is a higher value than generally recognized. Hearts with high blood pressure or blood sugar have more load to circulate blood. Heart tissue cells get oxygen and nutrients via blood vessels, called coronary artery, on the surface of the heart. But the coronary artery is very often occluded by fats, and then the heart tissues in the area do not get oxygen and lose the beat function to ultimately result in a heart failure.

An artificial heart is a symbol for artificial organs and has been developed for a long time. Heart transplantation was first introduced in 1967 by Dr. Barnard in South Africa. Since then, artificial hearts have been developed very actively in several universities, including Utah University, University of Texas, University of Cleveland, and University of Pittsburg. The National Institute of Health (NIH) has supported those programs with big funds. The first one, the Liotta model designed by Dr. Liotta and Cooley in Texas Heart Institute, was successfully implanted on a patient for 3 days before transplantation.

Figure 9.4 shows the artificial heart implanted in 1982 in Utah University. This model was named Jarvik-7 after the designer, Dr. Jarvik. The upper oval part was made of Dacron polyester woven and coated by polyurethane and the lower part was of aluminum, also coated by polyurethane. Multilayer sheets also made of polyurethane are located in between. Polyester is poly(ethylene terephthalate), called PET in short, which is used to make the representing synthetic fiber (the trade name is Dacron) and also bottles for soft drinks. Polyurethane is an elastic and tough synthetic rubber applied to wheels of roller skate and to artificial leather and Spandex fabric. The polyurethane multilayer sheets are moved up and down by compressed air to pump the blood. Therefore, the patient was dependent of an air compressor. This artificial heart was placed in the position of the original one instead and, therefore, classified into a total artificial heart. The patient was on television every day to draw interest all over the world and survived for 112 days. The artificial heart itself operated successfully, but he died of an infection. Based on this result, the Food and Drugs Administration (FDA) finally approved a similar artificial heart for clinical application (HeartMate by Thoratec company) in 1994. Such a total artificial heart is not an implantable one. A patient must be hung to an air compressor and, therefore, is not able to have a normal life. It is used while waiting for a heart transplantation.

Thereafter, the development of artificial hearts moved into two directions: one for the left ventricular assist device (LVAD) and the other for the implantable artificial heart inside the body. Both are operated by a rechargeable battery; therefore, the patient is free to enjoy a regular life. An LVAD shown in Fig. 9.5 is additionally attached to the original heart to demonstrate advantages in terms of function and cost. There are already more than 10 kinds of commercial products, including the Novacor LVAD developed by World Heart Corporation.

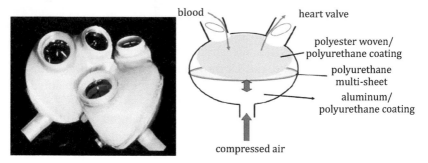

blood — heart valve

polyester woven/ polyurethane coating

polyurethane multi-sheet

aluminum/ polyurethane coating

compressed air

Figure 9.4   Structure of the total artificial heart Jarvik-7. *Source*: Left: http://commons.wikimedia.org/wiki/File:JARVIK_7_ artificial_heart.jpg. Right: drawn.

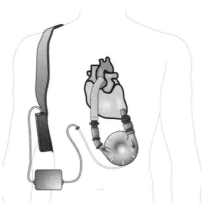

Figure 9.5   LVAD attached to the heart, battery operated. *Source*: http://en.wikipedia.org.

Figure 9.6 shows an implantable artificial heart, AvioCor, developed by Abiomed in 2006. An electric motor in the core moves from right to left to press the blood sacs in both sides to pump the blood. The power is supplied not by a wire through the skin but by a wireless system to avoid possible infection. It is reported to have the longest application for 512 days.

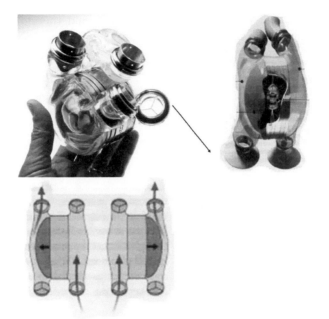

Figure 9.6 Implantable artificial heart AvioCor. A motor in the center pushes the blood sacs in both sides to circulate blood; battery operated. *Source*: http://www.abiomed.com.

9.1.3 Artificial Eye Lens (Intraocular Lens)

Eye lenses are clear but become yellowish turbid in the case of cataract (Fig. 9.7). The cause of cataract is not yet completely understood, but aging is surely one of the reasons. Practically, 85% of old people above 65 years suffer from this disease. The lenses of such patients should be removed and replaced by thin plastic lenses, called intraocular lens (IOL). As shown in Fig. 9.8, an IOL has a transparent center for light passage and a hanging part (called haptic) outside. The diameter is 5~7 mm. There are various designs made of different materials.

The surgical method of inserting IOL has improved with the development of materials. The first lenses were made of hard poly(methyl methacrylate) (PMMA), used for hard contact lenses. PMMA is a clear plastic used for store signs (usually called acryl sheets) and rear lamps of cars. In order to remove the old turbid lens by pressing out and inserting the hard IOL, the cornea must be cut about 14 mm. Such a big wound needs to be sutured after the operation, and it takes time for recovery, and therefore the patient should stay in the hospital overnight.

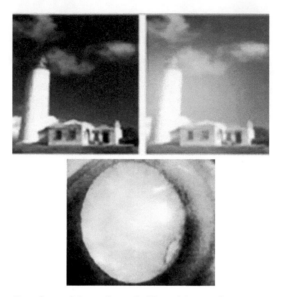

Figure 9.7 Regular vision (top left); vision of a cataract patient (top right); yellow and turbid lens of a cataract patient (bottom). *Source*: Seoul University hospital.

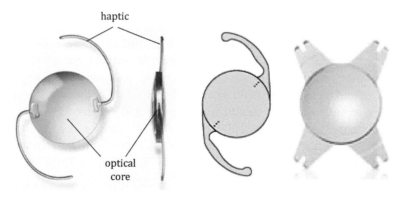

Figure 9.8 Various artificial lenses (IOLs). *Source*: Seoul University hospital.

Then a new technology was developed, which reduced the size of the wound to about 7 mm. The old lens is broken into pieces by an ultrasonic wave and sucked out. Recently, soft IOL has been invented, which can be folded, as shown in Fig. 9.9c. Now, the wound size has been reduced to 3.5 mm, which does not need to be sutured at all. Patients can recover very fast and, therefore, do not need to stay in the hospital overnight. The new soft IOL is made of poly(hydroxyethyl methacrylate) (PHEMA), used for soft contact lenses or silicone rubber (known as Silastic rubber). Silicone rubber is soft and not toxic and is stable in the body and, therefore, also applied to materials for plastic surgery, including artificial noses.

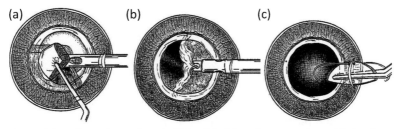

Figure 9.9   Insertion of soft IOL: (a) The front surface of the lens capsule is removed partially. (b) Turbid lens is broken into pieces by an ultrasonic wave and sucked out. (c) Soft IOL is folded and inserted. *Source*: Seoul University hospital.

Figure 9.10 illustrates the eye shape after the insertion of IOL. The IOL is located inside the lens capsule, replacing the original one. The haptic of IOL is fixed at the edges. The front side of the lens capsule removed partially for the operation is not regenerated at all to stay forever.

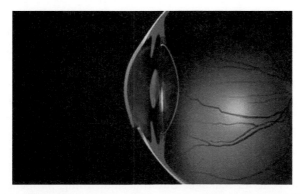

Figure 9.10   IOL inserted inside the lens capsule. *Source*: Seoul University hospital.

The IOL is one of the most successful cases of artificial organs. All patients are happy to get their vision back without any complications. Now millions of peoples implant IOL every year throughout the world. However, the present IOL has a fault. Our eyes can control the thickness of natural lenses automatically to adjust the distance. However, the eyes of patients with IOL cannot do this anymore. All we can do is choose the IOL of the right thickness during the operation. Currently, a multifocal IOL capable of controlling distance is on the market but is not yet popular.

9.1.4 Limitations of Present Artificial Organs

Nowadays, we have various artificial organs, as shown in Fig. 9.1. Are the functions of these artificial organs enough to replace natural ones? Can we change our organs/tissues with artificial ones like car parts? Our body is a really fantastic machine, which is precisely self-controlled in multi-steps. Whenever we investigate the structure and function of the body to develop artificial organs, it is awesome as they are really sophisticated and robust.

We use various metals, ceramics, and polymers (plastics and rubbers) for artificial organs. As they are all different in strength and property, we choose the right materials depending on the implant location and period. However, our body has a complete protection system that kills bacteria and destroys all foreign materials. When the foreign object is too big to be destroyed, such as most implants, the materials are surrounded by new tissues and isolated to protect the body. Such a refuse reaction of the body is a big hurdle for developing artificial organs. For example, when we work on an implantable sensor to measure sugar concentration in the body, the device works well outside the body in the lab. But after implantation into the body, proteins are adsorbed on the surfaces of the devices and they are encapsulated by new tissues and lose their function.

Another example of the refuse reaction of body is blood coagulation and formation of clots. When we are wounded, blood clots to prevent its leakage. But the same thing happens when materials have contact with blood. Figure 9.11 shows blood clots formed on the surface of an artificial heart valve. Blood is also clotted on the surfaces and deposited inside artificial blood vessels, which reduces the inner diameter. Therefore, we do not yet have artificial blood vessels with a diameter below 3 mm. The coronary artery on the heart surface, supplying oxygen and nutrients to heart tissues, has a diameter of less than 3 mm. When the coronary artery is occluded, resulting in a heart failure, we cannot use artificial ones; we have to take out a vein from some other part of the body to replace it. We have the same problem of clotting in the case of artificial kidneys. Blood is coagulated on the surfaces of hollow fibers and modules during treatment.

Figure 9.11   Blood clots coagulated on an artificial heart valve. *Source*: Yonsei Seberance hospital.

So the question is: how can we apply artificial kidneys or heart valves? Fortunately, we have drugs (called anticoagulants) to prevent clotting. The most effective one is heparin, a natural polysaccharide produced in the body. Aspirin, known as a drug for treating fever, is also a good anticoagulant. A patient to be treated with an artificial kidney is injected with heparin in advance to inhibit clotting. A patient with artificial heart valves should take the drug for life time. Doctors give aspirin to people with high blood pressure to reduce the chance of a clot and rupture of blood vessels.

Several artificial organs and materials have been explained here. We have no problems at all with IOL and dental implants. But in the cases of blood vessels, heart valves, or hip joints, we must use them although we know the problems. Furthermore, we do not yet have artificial corneas, although we have several transparent plastics already used for contact lenses or IOL. Clear plastic films are not compatible enough with the natural cornea to be attached together.

In addition, most artificial organs have a simple function; a heart pumps blood or dental implants and hip joints support our body. Only an artificial kidney has a special function: removing urea from the blood.

9.2 Regenerative Medicine and Tissue Engineering

9.2.1 Cell Therapy

The cells in our body are continuously produced, while old ones are discarded automatically. The cells produce organs and tissues by a self-controlled mechanism. Furthermore, the damaged tissues are completely regenerated by the cells. Therefore, such an ability of the body has been utilized to study the recovery and regeneration of damaged tissues and organs, which is called regenerative medicine.

One method is injecting cells into the wound from outside, called cell therapy. Recently, we have several successful results in regenerating soft bones by cell therapy. The knee cartilage is located between the upper femur and lower tibia, helping the movement of the bones and reducing friction (Fig. 9.12a). The cartilage tissue is a three-dimensional network composed of hydrophilic polysaccharides and reinforced with hydrophobic collagen. It is very strong and tough even in a wet condition and lasts for years. Such a unique structure cannot be mimicked or chemically synthesized. Nevertheless, cartilage can be hurt during accidents and is also weakened with age. The cartilage tissue has only a few cells (chondrocytes) that produce the tissue but no blood vessels; therefore, it is very difficult to be restored once damaged. We have artificial knee joints to replace it, but the function is still far from the real one (Fig. 9.12b).

For a small damage to the cartilage, tiny holes are drilled in the bones beneath the cartilage so that the blood coming up from the bone marrow, which contains various growth factors, helps regenerate the cartilage tissue. For a large wound, the healthy part of the cartilage is transplanted to the wounded site. When the wound is bigger than 4 cm^2, the autologous chondrocyte cells (of the patient) are implanted. Figure 9.13 explains the concept of such a cell therapy. The healthy tissues are taken out, and the cells are isolated and cultured to increase the number to several millions. The cultured cells are injected back to the wound site and covered with a bone skin so that they are not washed away. Although cell therapy shows positive results in many cases in clinic, it still needs to be improved. We inject millions of cells but cannot estimate how many cells are really active and contribute to the regeneration. We inject a huge number of cells just hoping for a positive result.

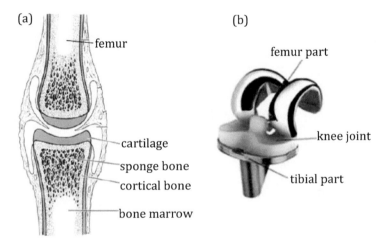

Figure 9.12 Cartilage (a) and knee joint (b). *Source*: http://en.wikipedia. org.

Cell therapy is now intensively studied for the regeneration of other organs. Patients with partially lost heart tissues due to coronary diseases are injected with heart tissue cells for recovery. The terminal nerve system can be regenerated to some extent, but the central spinal one is not. Therefore, the spinal nerve is a target too. Skin cells have been tried to be regenerated in burn wounds. Blood cells have been studied for treating leukemia (a disease that abnormally increases white blood cells in the blood), and islet cells in the pancreas, which produce insulin, have been studied for curing diabetes (a disease that raises blood sugar).

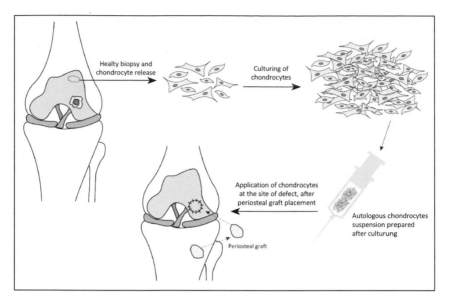

Healty biopsy and chondrocyte release

Culturing of chondrocytes

Application of chondrocytes at the site of defect, after periosteal graft placement

Autologous chondrocytes suspension prepared after culturung

Periosteal graft

Figure 9.13 Autologous chondrocyte implantation. *Source*: Hamlin, M., Draper, N., and Kathiravel, Y. (eds.). 2013. *Current Issues in Sports and Exercise Medicine* (Chapter 2, The Physiology of Sports Injuries and Repair Processes, Figure 15).

Especially stem cells are actively applied in these areas. Stem cells can proliferate themselves and also change (differentiate) into different cells. In the body, an egg cell is fertilized by a sperm to form a single embryonic stem cell, which proliferates and differentiates into a huge number of various cells, producing bones, blood, tissues, and organs (Fig. 9.14). The embryonic stem cell has superior ability, but there is an ethical problem in its use in research. However, we have other adult stem cells with inferior function, such as bone marrow stem cells, which are rather free to be utilized for studies. An outstanding result in the area of stem cells came out in 2007. Prof. Shinya Yamanaka of Kyoto University, Japan, invented another kind of stem cells by reverse differentiation from adult skin cells. These so-called induced pluripotent stem cells (iPS cells) can proliferate and differentiate into other cells as the embryonic stem cells do. However, there is no ethical problem for their use in research; therefore, research activities related with them will be expanded to a great extent in the future. He got the Nobel Prize in 2012 for the contribution.

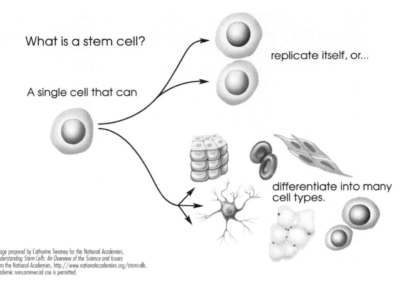

What is a stem cell?

replicate itself, or...

A single cell that can

differentiate into many cell types.

Image prepared by Catherine Twomey for the National Academies,
Understanding Stem Cells: An Overview of the Science and Issues
from the National Academies, http://www.nationalacademies.org/stemcells.
Academic noncommercial use is permitted.

Figure 9.14   Differentiation of stem cells into various cells. *Source*: http://nas-sites.org/stem cells/files.

9.2.2   Biohybrid Organs

The biological function of a liver or pancreas is too complicated to be copied by artificial materials. Therefore, the function of the cells is borrowed and fused with materials to develop biohybrid organs. The cells are grown on the material surfaces or in the tiny holes (pores) of sponge-type materials or mixed with hydrogels (polymers containing much water). This approach utilizes the real function of cells to develop almost natural artificial organs to overcome the limitation of artificial materials.

9.2.2.1   Bioartificial liver

A liver is mostly connected to the blood stream. The amount of blood passing through the liver is as much as 1.5 l/min. All drugs or alcohols entering the body are instantly delivered to the liver to be removed. The weight of an adult liver is 1.5 kg, and among them, liver cells (hepatocytes) are 1.2 kg. So we have a huge number of liver cells. There are two functions of the liver. One is to remove toxic compounds in the blood, and the other is to produce albumin and other proteins. The liver can self-regenerate, but we have many liver diseases associated with virus and cancer. Fortunately, livers can also be transplanted now, but we do not have an enough number of donors.

As it is impossible to mimic the specific function of liver, a bioartificial liver (BAL) has been studied by utilizing liver cells. Liver cells of pigs are generally used as it is easy to be achieved. Figure 9.15 shows a prototype of BAL using a bioreactor. A bioreactor is composed of hollow fibers used in an artificial kidney (see Section 9.1.1). Liver cells are filled inside the fibers with collagen gel and cultured. The blood flows outside the hollow fibers and toxic compounds are removed during the flow. It is difficult to culture liver cells in the laboratory as they live only for 1–2 weeks. However, BAL is still in the initial stage of development. The amount of liver cells used in the device is just about 100 g contrary to 1.2 kg in our body. Also the amount of blood flowing through the bioreactor is only 100 ml/min compared with 1500 ml/min in the body. Nevertheless, it is an important first step toward biohybrid organs.

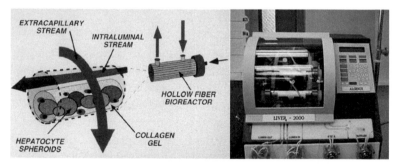

Figure 9.15 Bioartificial liver (BAL) using a bioreactor. *Source*: Hepatix.

9.2.2.2 Artificial pancreas

Diabetes is a disease that raises the content of blood sugar. It is now one of the most serious diseases, and there are more patients in developed countries. There are island-shaped cells, hence called islet cells, in a pancreas producing insulin, a protein controlling blood sugar (Fig. 9.16). There are two types of diabetes. In type I, the function of islet cells is not good enough to produce sufficient insulin and, therefore, more insulin has to be provided from outside. In the more common type II, the blood sugar is not controlled properly even though there is enough insulin produced. Type II diabetes can be treated with the help of exercise, drugs, and diet. There are serious complications in diabetes. As the blood sugar increases, the blood becomes denser, which is more difficult to circulate. In a dangerous case, the terminal tissues do not get enough oxygen, and then the part is dead and must be cut away. Insulin is a protein and digested in the stomach. It cannot be administrated orally but should be injected into the body.

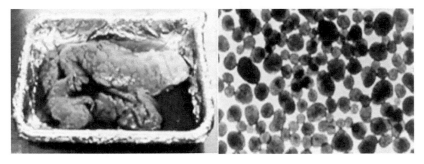

Figure 9.16 Pancreas (left) and islet cells producing insulin (right). *Source*: Seoul University hospital.

It is possible to transplant the pancreas of other persons to produce insulin, but we still have the critical problem of avoiding the refuse reaction of the patient as in the case of other organs. But it is more reasonable to isolate islet cells and transplant them to the patient, which is the concept of an artificial pancreas. However, if the cells from other persons or pigs are just transplanted, they will be destroyed by the defense system of the patient, i.e., by white blood cells, macrophages, or antibodies. We need a system to protect the cells.

If the islet cells are enclosed by a specific membrane that passes insulin and metabolites but expels macrophages or antibodies, the islet cells can be protected inside the membrane to produce insulin. Such a concept is called encapsulation of cells, as shown in Fig. 9.17. The first successful membrane is made of alginate–polylysine–alginate in triple layers. Alginate is a negatively charged natural polysaccharide contained in seaweeds, and polylysine is a positively charged polypeptide. It is recently reported that the blood sugar is successfully controlled by this system. But we still have problems to overcome. The lifetime of the islet cells is just 2 months. We should develop a container to carry the cells and a method to take it out later. We need a more effective membrane to encapsulate the cells too.

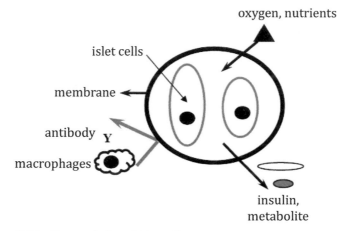

Figure 9.17   Encapsulation of islet cells.

9.2.3   Tissue Engineering

Tissue engineering is the newest technology for producing artificial organs/tissues. Cells are seeded in biodegradable porous scaffolds and grown to produce organs or tissues. During the period, the scaffolds are gradually degraded and replaced by newly formed tissues/organs.

Figure 9.18 explains the concept, which shows a human ear regenerated on a rabbit ear, which is like magic. There are three critical issues when studying tissue engineering: cells, scaffolds, and the culture technology. We should use cells isolated from the patient or stem cells to avoid immunological problem. Many researches are going on how to differentiate stem cells into certain desired cells. But it is still difficult to conduct research on a theoretical basis and is commonly conducted by trial and error.

Second, we need sponge-type scaffolds, which have interconnected pores to deliver oxygen and metabolites effectively to cells. In addition, scaffolds should be strong enough in the beginning but should be degraded properly as the cells produce tissues/organs. Especially, scaffolds should have good affinity for the cells so that the cells adhere and grow well on them. For these scaffolds, we use collagen or other biodegradable synthetic polymers such as poly(glycolic acid) or poly(lactic acid). These polymers are already available as surgical sutures and bone plates. They are gradually degraded; therefore, we do not need to remove them. On the contrary, metallic ones should be removed after healing.

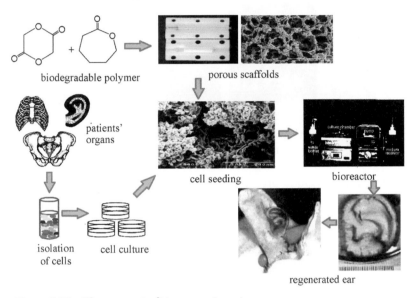

Figure 9.18 The concept of tissue engineering.

The cells in the body sit on a basis called extracellular matrix (ECM) composed of collagen and polysaccharide. They communicate with the ECM and the adjacent cells (Fig. 9.19a). Therefore, it is critical to provide similar surroundings to the cells for effective tissue engineering. There are several methods for making porous scaffolds with same-sized interconnected pores. Most commonly, polymers are dissolved in a solution and added by salts and cast into scaffolds in molds, and then salts are leached out to leave the porous structures (Fig. 9.19b). Nonwoven sheets having pores are also prepared (Fig. 9.19c). More precisely, lattice-type scaffolds are designed and manufactured by a computer (Fig. 9.19d). Recently, live cells are arrayed to a specific shape by a three-dimensional printing method. In spite of these methods, it is not so easy to get ideal scaffolds having homogeneous structures and good affinity for cells. Therefore, real organs are used as scaffolds for ideal tissue engineering. For example, a human heart is taken and the tissues and cells are removed, and then new heart tissue cells are seeded and cultured to regenerate a real heart again.

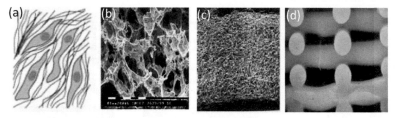

Figure 9.19 Various scaffolds for tissue engineering: (a) cells on the extracellular matrix in the body, (b) porous sponges, (c) nonwoven sheets, (d) lattice-types.

Finally, various proteins, including growth factors, are utilized to activate cell growth and to enhance the generation of organs/tissues. Practically these growth factors control the proliferation, differentiation, and migration of the cells in the body. It is very critical to get the new blood vessel systems well in the early stage to provide oxygen and nutrients to the cells. As scaffolds are used to carry cells in tissue engineering, it has a great advantage to increase the cell seeding efficiency and, therefore, to enhance regeneration compared to cell therapy, which injects only cells.

Figure 9.20 summarizes the recent status of tissue engineering. As it has a great potential to regenerate real organs/tissues, all of them are now targets. Among them, several organs/tissues such as skin, bladder, and cartilage are already applied clinically.

Advanced Tissue Science was the pioneer in commercializing a skin produced by tissue engineering in 1997. The skin was regenerated by the skin cells of a newborn baby. Although it was pretty expensive ($400 per sheet of 2 × 3 inches), it was almost same as real skin. Unfortunately, the company bankrupted in 2002 and discontinued the product.

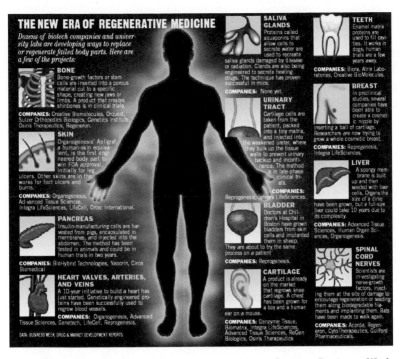

Figure 9.20 Development of tissue engineering. *Source*: *Business Week*, July 27, 1998, pp. 58–59.

Prof. Shin'oka of Tokyo Women's Medical College, Japan, succeeded in regenerating a blood vessel in 2000. A tubular scaffold (10 mm diameter) made of biodegradable synthetic fibers is connected between the heart and the lung, and bone marrow stem cells of the patient are seeded in the scaffold to produce the vessel. The first patient for clinical test was known to be the daughter of Prof. Shin'oka's friend. It was a great decision of the parents to try the new technology by trusting science. Although they had many successful animal tests before to prove safety, it would not have been easy for them to take the risk. If an artificial blood vessel made of polyester fibers was implanted on her, she would have required the same operation repeatedly as she grew since bigger vessels would be needed continuously. She did not have any problem with the regenerated blood vessel as it would grow with her body.

Prof. Okano et al. of the Institute of Advanced Biomedical Engineering and Science, Japan, developed a cell sheet technology that cultivates cells in a sheet to be attached to the wound site in the body. They were successful in healing a patient of coronary artery disease by attaching the heart tissue cell sheet to the wound site. They also demonstrated positive results in partially regenerating cornea and esophagus.

Prof. Atala of Wake Forest University, United States, succeeded in developing an artificial bladder by tissue engineering in 2006. Half of the bladder was grown on scaffolds in the lab and coupled to the other half of the patient.

In addition, several other universities have reported regeneration of heart valves (Fig. 9.21), bones, and cartilages, and they are almost in clinical use. Tissue engineering is the hottest topic in the world now. All countries support the activity. Soon we will have natural organs and tissues through tissue engineering technology to meet the era of real artificial organs.

Figure 9.21 Regenerated heart valves by tissue engineering. *Source*: http://www.bmm-program.nl.

Chapter 10

Transportation

James Wei

School of Engineering and Applied Science, Princeton University,
Princeton, New Jersey 08544-5263, USA

jameswei@princeton.edu

For every plant and animal, transportation is among the greatest necessities for survival and to reproduce and sustain the species. The natural movements of air and water in the environment feed a passive plant with sunshine, carbon dioxide, water, and minerals. A flower plays a more active role in bribing bees and butterflies with nectar, so that they carry its pollen and sperms to the ovaries of other flowers for fertilization. Then the female plants encase the resulting seeds in fruits and nuts for dispersal and germination into new plants. An animal also needs to travel to find mates for courtship and breeding. Some animals travel great distances every year to meet at their breeding grounds, and then lay eggs or give birth to their young. An animal also has the additional need of traveling to find food and water, and the distances to travel may be very large when such feeding grounds are far apart. The great whales circle the oceans of the earth to find food.

Chemistry: Our Past, Present, and Future
Edited by Choon Ho Do and Attila E. Pavlath
Copyright © 2017 Pan Stanford Publishing Pte. Ltd.
ISBN 978-981-4774-08-6 (Hardcover), 978-1-315-22932-4(eBook)
www.panstanford.com

Humans have much greater need for transportation, as we are great consumers of prodigious appetite, from absolute necessities to luxuries and trivia. We go to the supermarket to buy groceries, while some busy families have them delivered to the door. The nomad tribes follow the herd of wild or domesticated animals that move with the warmth and rainfall of the seasons, and thus fertile grazing grounds. We commute from our homes to our offices or factories to work. We visit members of our families and friends; we go to sports events and the theater; and we go on vacation to faraway places. Sometimes we evacuate to escape danger, such as an earthquake or an invading army. Sometimes we invade other lands and remove resources to bring home, and even colonize and settle by evicting or enslaving the original inhabitants.

It is easy to walk and to move a few pounds of goods over a few miles. But if the task is to move thousands of tons of goods over thousands of miles, very powerful methods of transportation are needed. The technology of transportation involves vehicles, power, navigation, and infrastructure. Chemistry contributes mainly through discovering and obtaining sources of raw material, separation and purification of essential material, chemical reactions and transformation from raw material to useful products, and refining the final products through additives, preservatives, protective coatings and compounding.

A new transportation technology not only solves immediate problems, as some of them have profound long-term effects in our lives that the original innovators could not possibly imagine. The Polynesian outrigger made possible long-range ocean voyages and led to the colonization of the entire Pacific Ocean: from Southeast Asia to Tahiti, Hawaii, Easter Island, and New Zealand. The steam engine led to railroad tracks and trains that cross continents and helped the Americans to conquer the West, and the Russians to conquer Central Asia and Siberia. The steamship helped the establishment of the British Empire, which boasted that the sun never sets on their flags called the Union Jack. We can even walk on the moon.

The contributions of chemistry to transportation technology include the following:

- Identifying and obtaining raw materials for technology

- Converting raw materials to intermediate materials by separation, purification, and chemical reaction
- Final production with additives, compounding, and finishing
- Maintenance of technology from deterioration in daily use.

10.1   Vehicles: Frame, Wheel, Ship Planks

The oldest carts and ships have frames that are made of wood, which are later replaced by bronze and steel. In modern times, we have aluminum, which is lighter and suitable for airplanes and spaceships. To reduce friction and wear, cars and airplane landing gears run on wheels, which need rubber tires. The wooden ships are wood planks lashed or nailed together, and the seams between the planks would leak in water and cause the ship to sink. A caulking material must be found to fill in the spaces between the planks, and the best material is tar or asphalt.

10.1.1   Frame Material

Wood is a good material to build frames of vehicles as they are readily found in forests and can be felled, sawed, and joined together with ropes or nails. They are light enough to float on water, but can bear weight and flex, and survive light impacts. Some woods have natural preservatives and do not rot easily, and they include cypress, coastal redwood, and red cedar—rare and expensive, and not available for large-scale use. It was recorded that during the time of Alexander the Great, bridge wood was often soaked in olive oil to prevent the entry of water and insects. Tung oil was used in China, which is a drying oil obtained by pressing the seeds of the nuts of the Tung tree. The modern substitute is linseed oil. The resulting coating is transparent and becomes solid, like plastics. Ship hulls are particularly vulnerable to rot as they are always wet and also attacked by adhering barnacles and boring insects. The protective coating must be waterproof in salt water for many months or years. Some Roman ship hulls were brushed with protective tar. There are many modern wood preservatives made of copper and arsenic, which are brushed on or pressure treated. Chromate copper arsenate is very effective against fungus and insects, gives the wood a greenish tinge, but has toxicity problems.

Figure 10.1 Sumerian chariots with wheels.

Bronze chariots have many advantages over wooden chariots. The Bronze Age is commonly considered to be 3300–1200 BC in the Middle East, which replaced the older Stone Ages. Copper melts at 1084°C and is reasonably easy to handle. However, it is soft and can be easily bent or dented while in use, and needs alloying to make it tough and resilient. There is a form of bronze formed by copper and arsenic, which can be very toxic to manufacturers and users. It has been replaced by the alloy of copper and tin, but copper ores are seldom found in nearby places. The most famous ancient copper ores were found in the island of Cyprus, which became the name of the metal. The ores are called Chalcocite, Cuprite, and Azurite, greenish stones of copper sulfide, oxide, and carbonate. Some ancient inventors before history discovered that these greenish stones hid within them the bright copper, which can be released by smelting. The ores are washed and cleaned, and then roasted in clay ovens with charcoal and not much air, so that the impurities burn away, leaving blistered copper metals. Tin has an even lower melting point of 232°C. The most important tin ore is Casserite, which is SnO_2 or tin dioxide. It was found in Anatolia, which is currently in Turkey, and the most famous ancient mine is in Cornwall in western England. Most of the world production of tin today is in Malaysia and Indonesia. It is a grayish stone that must be smelted with charcoal to remove oxygen to form the metal. These two metals are then melted and alloyed together at the ratio of approximately 12% tin and 88% copper. Bronze does not rust in air, and there are many bronze objects in museums that are thousands of years old, sometimes with a greenish cover of patina.

Figure 10.2   Assyrian war chariots.

The Iron Age followed when it was discovered that iron ores are far more abundant on the surface of the earth and can be found everywhere. Besides being much cheaper, iron is also much harder than bronze. Pure iron melts at 1538°C and requires a much hotter furnace than bronze. Iron ores are usually oxides of iron called hematite or magnetite and contain a great quantity of impurities such as carbon, silica, alumina, phosphorous, and sulfur. In the smelter process with hot charcoal, these harmful impurities must be removed as they tend to weaken the product. The first product of smelting is pig iron or cast iron, which has 2–4% carbon and contains some of the silica as slag, and can be brittle. When this cast iron is hammered out over hot charcoal to remove the slag, the product obtained is wrought iron, which has a carbon content less than 0.3% and is more soft and ductile. Wrought iron is good for making rivets, nails, wires, and chains. Steel has carbon up to 2.1% by weight and gives the material a desirable mixture of hardness, ductility, and tensile strength for numerous applications. The most famous ancient steel may be the Damascus steel, which is a complicated alloy and famous for its durability and ability to hold an edge. It has a distinctive pattern of banding and mottling reminiscent of flowing water and whirls. The most productive method of steelmaking was the Bessemer process in the mid-19th century, which puts molten iron in an oven and depends on a blast of air to burn the excess carbon. The basic oxygen steelmaking (BOS) process has an open hearth and has largely replaced other methods.

Two pieces of iron are often joined together by welding, where the two surfaces are heated to 1500°C to soften and melt together. The two surfaces must be cleaned by the use of a flux material such as rosin, zinc chloride, and borax. A shielding gas such as argon and carbon dioxide may be used to protect the iron surfaces against the formation of iron oxide from contact with air.

Iron rusts in air, especially in wet conditions, and must be protected by coating or alloying. A thin plating of tin on steel makes it safe for handling food storage, which is applied by electroplating or bath in a tub of molten tin. Zinc melts at 420°C and is used in a similar way, which is called galvanizing. A more durable but expensive method is to alloy iron with chromium and nickel to make stainless steel. Nickel has a melting point of 1455°C, which is close to iron, but chromium has a much more demanding melting point of 1907°C.

The density of a piece of wood is approximately 0.7–0.8 g/cc, of stone is approximately 2–3 g/cc, of steel is 7.9 g/cc, and of copper is 8.3 g/cc. When a light-weight material is needed, the most important is aluminum with a density of 2.7 g/cc. The ore of aluminum is bauxite, which is alumina oxide, one of the most abundant materials on earth (ranks third after oxygen and silicon). But the aluminum atom is tied very strongly with oxygen in the ore, and it cannot be smelted with charcoal, like copper and iron. The earliest production method of aluminum was by reaction with metallic potassium or sodium. This was a difficult process, which made pure aluminum more valuable than gold, and was exhibited at the Exposition Universelle of 1855. Napoleon III of France gave a banquet in which, it is said, the most honored guests were given aluminum utensils, while others has to content with gold. The modern aluminum is made through the

Hall–Héroult process, which is electrolytic smelting with electrodes made of carbon. The bauxite has to be in the fluid state, but the melting point is more than 2000°C, which is very difficult to operate. Then it was discovered that bauxite can be dissolved in molten cryolite, a mineral found in Greenland with the formula Na_3AlF_6. The resulting mixture of bauxite and cryolite becomes fluidic at 980°C, which makes it easy for electrolytic separation. The Hall–Héroult process was invented independently and almost simultaneously in 1886 by the American chemist Charles M. Hall and the Frenchman Paul Héroult. Hall opened the first large-scale aluminum production plant, which later became the Alcoa Corporation.

10.1.2   Wheel and Rubber Tire

When we slide an object on the ground, we encounter sliding friction between the surface of the object and the road, which consumes power and creates abrasion and destruction. The rougher the surfaces, the bigger are the power losses and abrasions. Before the invention of wheels, a number of other rolling methods with sleds and rollers of small diameters were used to move heavy weights. The heavy stones of the Great Pyramid and the Stonehenge were moved by supporting them on such rollers. Another method was to slide heavy sleds over frozen surfaces in winter.

The wheel is one of the greatest inventions. The oldest wheels might be pottery wheels, made to facilitate shaping clay vessels by circular motion. The oldest cart wheels were made of disks of solid wood joined together to make a circle with a hole in the center, called the hub. Later, iron bands or hoops were placed around the wooden disks to tie them together and to increase resistance to wear. The load of the vehicle rests on the axles that go through the hubs. When we put load on the axles, we encounter rolling friction between the surface of the wheel and the ground, which is much smaller. We also encounter friction between the axle and the hub. Fortunately, we can cover the hub to keep it smooth and clean and lubricate it with grease or animal fat.

Figure 10.3 Rubber tire.

Before the widespread use of modern smooth roads made of concrete and asphalt, cart wheels had to roll unevenly over gravel and dirt roads and bounce up and down. This can be very jarring on passengers and goods and can damage the frame of the vehicle, especially if it is traveling at high speed. An elastic material that can absorb these shocks and cushion the frame would give a smoother and faster ride. When Christopher Columbus went to the New World, he saw the natives kicking rubber balls in sport and noticed the elastic property of the balls. Natural rubber is harvested by making cuts on rubber trees and collecting the milky and sticky latex to refine into balls. Rubber was imported to Europe and planted in many other tropic places.

Natural rubber is made of molecules of isoprene, which is a small molecule with five carbon atoms; this isoprene is called a monomer, which can be strung together into a chain of polymer. Natural rubber is sticky, deforms easily when warm, and is brittle when cold. In this state, it is a poor material when a high level of elasticity is required. Rubber is composed of long individual polymer chains that can move independently of each other at moderate temperatures, so that rubber can change shape under stress and bounce back after the removal of stress.

However, these movements are very slow at low temperatures and very fast at high temperatures, leading to softening and melting. Crosslinking of chains together would create a giant network that prevents the polymer chains from moving too independently. In 1839, Charles Goodyear invented vulcanized rubber by adding sulfur to rubber latex and heating it in an oven, which caused the crosslinking with a resulting product that stays elastic in all kinds of weather. This led to the widespread use of rubber tires in transportation, and enormous plantations were established in Brazil, Ceylon, Indonesia, and Malaysia. The concept of a pneumatic tire with an inflated inner tube was introduced by John Dunlop in 1887, which led to an even more elastic and shock-resistant tire, suitable for high-speed rides.

Rubber became indispensable to both peace and war, and its availability depended on trade with tropical countries or overseas colonies. By 1925, the price of natural rubber had soared. Synthetic rubber became the goal of researchers in Germany and Russia, countries in Northern Europe without tropical plantations around 1890. During World War II when the Malaysian rubber plantations were under Japanese occupation, the Americans accelerated their research and development. The most successful results are neoprene and styrene-butadiene rubber (SBR).

Rubber tires have another serious defect concerning their rate of wear. They could seldom last for more than 3000 miles, till the discovery that the blending of carbon black can greatly toughen them and make them last up to 40,000 miles. Carbon black is soot produced by burning natural gas in burners with insufficient amounts of air, which create smoke and fine particles of carbon, which are then blended into rubber latex at up to 30% by weight.

10.1.3   Ship Caulking and Tar

A sailor or seaman is often called a tar. A wooden ship is constructed by planks of wood lashed or nailed together. Water can rush or seep through the seams between the planks and can cause the ship or boat to sink. In Genesis chapter 6, God told Noah that to escape the coming flood that will destroy the world, he is to:

> *6:14 Make thee an ark of gopher wood; rooms shalt thou make in the ark, and shalt pitch it within and without with pitch.*

Pitch is a dark resinous substance that is traditionally used to caulk the seams of a boat or ship to make it waterproof. Pitch can be called bitumen or asphalt and can be made from petroleum.

There were many places in the Middle East where petroleum might have seeped from underground sources to the surface and then the more volatile components evaporated, leaving behind the dark and resinous pitch. The modern city of Los Angeles has a place called the La Brea Tar Pit, which preserves the bones of many animals trapped in the sticky tar, such as the saber-toothed tiger.

In the ancient Middle East, the Sumerians used natural asphalt/bitumen deposits for mortar between bricks and stones, ship caulking, and waterproofing. The Greek historian Herodotus said that hot asphalt/bitumen was used as mortar in the walls of Babylon. In some versions of the Book of Genesis in the Bible, the name of the substance used to bind the bricks of the Tower of Babel is translated as bitumen (see Gen 11:3), while other translations use the word *pitch*. Asphalt/bitumen was also used by ancient Egyptians to embalm mummies. The Egyptians' primary source of asphalt/bitumen was the Dead Sea. In approximately 40 AD, Dioscorides described the Dead Sea material as *Judaicum bitumen* and noted other places in the region where it could be found. Natural asphalt/bitumen was slowly boiled to get rid of the higher fractions, leaving a material of higher molecular weight, which is thermoplastic; when layered on objects, it became quite hard upon cooling.

Tar can also be made from coal, wood, or peat. Heating of pine wood causes tar and pitch to drip away from the wood and leave behind charcoal. The state of North Carolina has the nickname of "Tar Heel," which recognizes that in the early history of the state, its economy depended on the production of tar, pitch, and turpentine from its vast pine forests. These products are called "naval store" and were also used to manufacture soap, paint, varnish, and particularly roofing material.

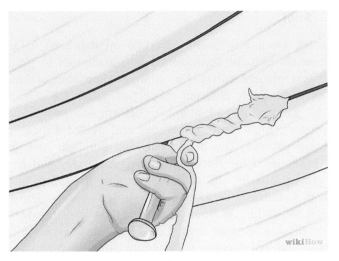

Figure 10.4 Caulking and waterproofing.

10.2 Power, Fuel, Engine

A vehicle moves by the power generated from an engine by the burning of fuel. The earliest vehicles moved by the power of human and animal muscles, generated from burning food, which contains carbohydrates and fat. Ever since the arrival of the steam engine in Scotland in around 1750, the prime power for transportation was the steam engine, which burns wood or coal, and then the internal combustion engines that burn gasoline or kerosene. The ocean-going vessels can burn heavy residual fuel oil, which has a high concentration of sulfur and nitrogen, and spread their fumes to sea plants and animals. In the chargeable electrical vehicles, the electric engine needs batteries, which can be charged at home or at stations from utility electricity generated with a variety of sources, including burning of fossil fuels or biomass, renewables such as wind and solar, hydro, and nuclear power.

The most important transportation engines today are the gasoline engines, diesel engines, and gas turbines. They all use fuels derived from crude oil after distillation and separation, and numerous processes of separation and reaction, and of finishing.

10.2.1 Fuel

The starting point is a barrel of crude oil delivered to an oil refinery, which is to be transformed into a number of products for the marketplace. A barrel of oil contains thousands of different molecules with widely different properties. The price of a barrel of oil (42 gallons or 159 liters) is determined by the values of the products that can be recovered and by the difficulties of the refining process. The best crude oils are light and sweet, such as Brent from the North Sea between Norway and Britain, with a gravity of 0.84 and very little sulfur; the worst oils are heavy and sour, such as Maya from Mexico with a density of 0.93 and 3.4% sulfur. Oil from tar sand can be even heavier than water, or density greater than 1.00. It is easy to refine light sweet crude oil and produce valuable products such as gasoline. The heavy crudes are exceedingly difficult to refine, create a corrosive environment, and produce less valuable products.

A common method of classifying molecules in oil is by their structure and by their boiling points in distillation. Most of the molecules consist of only carbon and hydrogen and are hydrocarbons. Other molecules may contain sulfur, nitrogen, and oxygen, which are considered undesirable as they lower the heating value and can cause noxious emissions. Some heavy molecules may contain metals such as vanadium and nickel and can cause serious problems. Among the desirable hydrocarbons, the common structure types are called PONA:

- Paraffins, with the formula C_nH_{2n+2}, are either straight chains or with one or more side branches. They are considered saturates as the ratio of H/C is greater than 2.0.
- Olefins, with the formula C_nH_{2n} or C_nH_{2n-2}, are either straight or branched. The ratio of H/C is either 2.0 or smaller.
- Naphthene or cycloparaffin has molecules that form one or more ring, in addition to branches. The ratio of H/C is either 2.0 or smaller.
- Aromatics with one or more ring structures and even lower H/C ratio down to 1.0.

In Fig. 10.5, the first row shows a normal paraffin with straight chains and an isoparaffin with branches; the second row shows a cycloparaffin or naphthene and an aromatic with many double bonds.

The first major equipment in a refinery in which crude oil enters is the distillation column, which applies heat to boil off the lighter fractions and leave behind the heavier fractions. This is separation by boiling points, and the usual groupings are as follows:

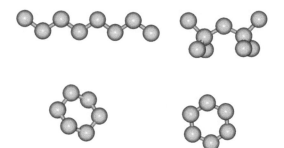

Figure 10.5 Paraffins, cyclo-paraffin, and aromatic.

- Light gases, with four or fewer carbon atoms, used for chemical feedstock, and home barbecue as well as cigarette lighters. They generally boil below 36°C and have to be kept in pressurized containers.
- Gasoline, with 5–12 carbon atoms, is the fuel for motorcars, and generally boil from 36–210°C.
- Kerosene is for lamps, diesel fuel is for trucks and buses, and distillate gas oil is for burning in homes. They generally boil from 210–500°C.
- Residual oil, which includes material that boil above 500°C, contains most of the impurities, including sulfur, nitrogen, and metals. It is further separated and treated to make products such as residual fuel oil used in utility power generation and marine boilers, asphalt for roads and roof paving, lube oil for engines, and wax for candles and cosmetics.

Transportation fuels are the most valuable products of oil refining. They have many requirements, which include:

- Being in the liquid state during operation and with appropriate boiling point ranges
- High heat of combustion, in BTU/pound and in BTU/gallon
- Good performance parameters: octane number in gasoline, cetane number in diesel, etc.

- Acceptable levels for impurities

A transportation fuel should remain liquid at all working environmental conditions, so that it can be easily pumped from the storage tank to the combustion chamber. For a car on road, the temperature range is from the coldest polar conditions to the most blazing tropic. For an airplane flying at a height of up to 50,000 feet, the temperature is considerably colder. A solid fuel like coal would be difficult to shovel into the combustion chamber, and a gaseous fuel like methane would take a very large volume to contain, unless it is at very high pressure requiring compression and thick walls.

A good gasoline must also have appropriate boiling point ranges. This is measured by the distillation curve: We put gasoline in a pot to heat up and note the amount of material boiled off and the corresponding temperature. The most important parameters are the initial boiling point when boiling just begin (such as 100°F), the 10% point when 10% of the gasoline has boiled off (such as 120°F), the 50% point (such as 210°F), the 90% point (such as 340°F), and the final boiling point (such as 410°F). In a car, gasoline is mixed with air and sprayed into the engine cylinder, meets with an electric spark, and must ignite. This requires that enough gasoline evaporates into vapor, so the initial boiling point is specified to be around 36°C. On the other hand, there should not be too much light components so that in summer, there would be too much evaporation to cause a "vapor lock" or a bubble in the fuel line so that liquid fuel would not flow smoothly. If the gasoline has too much heavy components that would not burn in the cylinder, it would slide into the lube oil to dilute it and lower its viscosity. This sets another limit on the maximum end point of 204°C. In winter, we desire gasoline with higher vapor pressure, and in summer, we desire the opposite to compensate.

Fuels are burned to create heat, which powers the vehicle. Other things being equal, the fuel that can create the most heat would be the most desirable. This is measured by either the volumetric method of kcal/ml or the weight method of kcal/g. A kilocalorie is the heat it takes to raise 1000 g of water by 1°C. One can convert to the volumetric heat by multiplying the gravimetric heat and density:

kcal/ml = kcal/g × g/ml

The hydrocarbons are made of only hydrogen and carbon, and their gravimetric heat depends mainly on the ratio of H/C as hydrogen has a much higher heat of 29.2 kcal/g versus carbon at 7.8 kcal/g. Molecules that contain the molecule oxygen, such as ethanol, can be considered to be partially burned already and have a heat value of only 7.1 kcal/g. Let us look at Table 10.1 for properties of gasoline components.

Table 10.1 Properties of light fuels

Molecule	H/C	kcal/g	g/ml	kcal/ml
n-Hexane, C_6H_{14}	2.33	11.5	0.659	7.6
Cyclo-hexane, C_6H_{12}	2.00	11.2	0.779	8.7
Benzene, C_6H_6	1.00	10.9	0.879	9.6
Ethanol, C_2H_5OH	2.00	7.1	0.765	5.5

The first three hydrocarbons are ranked by their ratio of H/C, which explains the differences in their gravimetric heat value or kcal/g. But their volumetric heat value ranking is reversed. A pound of hexane gives more heat than a pound of benzene, but a gallon of benzene gives more heat than a gallon of hexane. The difference is due to the much higher density of benzene, which is a more compact molecule than hexane. By any measure, ethanol gives heat of only 62% of hexane by weight, and 72% of hexane by volume. Thus, ethanol has a low score on the important parameter of miles/gallon, or the cruising range before refueling.

A barrel of crude oil from different sources can have vastly different properties. The more valuable crude oils are light, paraffinic, and with little impurities; the less valuable crude oils are heavy, aromatic, and high with sulfur and metals. An oil refinery tries to convert as much of the crude oil to highly desired products, which are gasoline and kerosene. Before the turn of the century, kerosene for lamps was the prime product, and gasoline was a useless byproduct, sometimes flushed into creeks. After the rise of the automobile, gasoline became the prime product. The usual situation for a refinery is not enough gasoline and kerosene for the market, and too much light gases and residuals.

Many great processes have been invented and improved over the last century that make molecular conversions to boost the volume and quality of gasoline and kerosene to meet market demand, at the expense of the less desirable pools. Since gasoline and kerosene are middle-boiling molecules, a modern refinery creates more middle-boiling molecules by cracking the residuals into smaller molecules, initially by thermal cracking, which is in an empty heated furnace, and more recently in catalytic cracking with the intervention of catalysts that are solid granules with special properties. Eugene Houdry was a French engineer and retired military officer who developed a cracking process in 1922, using a local clay called alumino-silicate. He moved to the United States in 1930 and started a company to supply catalysts to oil refineries. Typical reaction in cracking is as follows:

Figure 10.6 shows an n-paraffin with 12 carbons, which is cracked into an n-paraffin with six carbons and an olefin with six carbons. The original 12-carbon molecule is in the kerosene range, and the subsequent six-carbon molecules are in the gasoline range. Another benefit of this cracking reaction is the much higher octane numbers of the products, which will be explained in the next section. Alumino-silicate catalysts dominated oil refinery for many decades, but they have low activity and poor stability in use. Then came the revolutionary discovery of zeolite catalysts, which are molecular sieves with very regular channels and holes that would let small molecules enter and repulse large molecules. They are far more active and produce more gasoline.

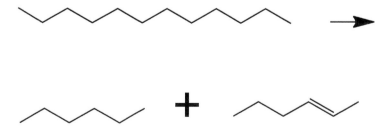

Figure 10.6   Paraffin cracking.

The other approach is by joining smaller molecules from the light gases into middle-boiling molecules by the process of alkylation and polymerization.

Figure 10.7 shows the joining of a four-carbon olefin with a four-carbon paraffin, two molecules in the light gas range, to make an eight-carbon paraffin, which is in the gasoline range.

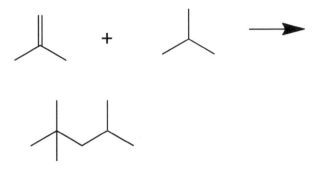

Figure 10.7 Alkylation.

The next most important property of gasoline is the octane number, which is critical to the performance of the gasoline engine. Gasoline is sprayed into the cylinder as the piston rises, to be ignited by the sparkplug, so that the resulting combustion generates heat and power to push the piston back down. Thus, the best time for ignition is when the piston is at the top, or already on the way down. But if the gasoline ignites too early while the piston is still rising, the resulting explosion would oppose the motion of the piston, causing knocking and loss of power. This phenomenon is called "pre-ignition" and must be avoided, especially in demanding situations such as going uphill and accelerating to enter a freeway. It is also worse when the high-performance engine has a high compression ratio, such as 9 to 1, which can generate more power at higher efficiency. The most demanding situation would be in a gasoline-powered fighter airplane, where a small boost in power can mean life and death. The standard used to measure poor knocking is the normal heptane with seven carbons, given an octane number value of 0, and good knocking iso-octane with eight carbons is given an octane number of 100. The family of normal paraffins has the lowest octane numbers, which are even lower when the carbon number increases.

There are two main methods to cope with this low-octane gasoline, by changing the normal paraffins to isoparaffins and aromatics and by additives such as tetraethyl lead. The later solution is the introduction of tetraethyl lead by Thomas Midgley Jr. in 1921 for General Motors. It is the most effective anti-knock agent when added at the rate of 3 cc per gallon of gasoline, which improves performance and efficiency. Unfortunately, the combustion product is lead oxide aerosol, which is discharged into the atmosphere and is a potential pollutant. The eventual ban on lead in gasoline came as a result of the catalytic converter introduced in 1973 to new cars, which can be coated with lead aerosol and loss of activity.

The catalytic reforming process was introduced by Vladimir Haensel of the Universal Oil Products Company (now UOP LLC) in 1949, which is based on a catalyst coated with platinum. It promotes the reactions of naphthenes into aromatics, paraffins into aromatics, and n-paraffins into isoparaffins. The final product has much higher octane numbers. After the ban on ethyl lead, a number of additives were proposed and used in gasoline. The most effective are oxygenates such as methyl-tert-butyl ether (MTBE) and tert-amyl methyl ether (TAME).

10.2.2   Engine, Lubricant

An ideal lubricant should be stable in the environment and should not deteriorate or breakdown. Water is too thin and evaporates. Animal fat and vegetable fat are not very satisfactory as they can turn rancid or breakdown. Mineral oils that are derived from crude oil are a byproduct of oil refining, which are abundant in supply and have the appropriate properties. Lubricating oils are produced from heavy petroleum from the bottom of the refinery distillation, and the best are made from light and sweet crude oils. Two operations are important for improving the properties of lubricating oils: de-waxing and enhancement of viscosity index.

At low temperatures, normal or straight paraffins precipitate and form a crystalline wax, so that the oil may turn into a gel or a solid. The temperature at which this phenomenon takes place is called the "pour point." When an idle engine is cooled to below the pour point, the lubricating oil in the engine becomes gelatinous and it becomes difficult to start the engine. A satisfactory lubricating oil in the temperate zone should have a pour point around –20°C, and even lower in the Arctic. Methods for lowering the pour point include mixing additives and de-waxing before the oil is sold as a lubricant. A number of solvents such as methyl ethyl ketone (MEK) are added to the oil at low temperatures, which forces the wax crystals to precipitate out. Then the wax is separated by centrifugation and filtration, which finds application in candle making and cosmetics.

The purpose of lubrication is to maintain a thin liquid film between moving surfaces to keep them apart without rubbing, which would cause energy loss and erosion. A very viscous fluid such as molasses would be thick enough to keep the surfaces apart but would be difficult to flow and would consume power. A thin fluid such as water would not be able to keep the surfaces apart. An ideal lubricant should have the Goldilock property of not too thick and not too thin, but just right. The viscosity of a fluid reflects on the competition between the random motions of molecules from the temperature, against the strength of the attractive forces between two molecules. A car starts and operates at the temperature of the environment, which may be very hot on a summer afternoon or very cold in a wintry snowstorm. Most fluids have the property of being more viscous when cold and less viscous when hot. The temperature sensitivity of viscosity is measured by the "viscosity index," which compares the viscosities of a given lube at 100°F and 210°F, in reference to a good standard lube and a bad standard lube. A good lube has a viscosity index of 100, and a bad lube has a viscosity index of 0.

It is highly desirable to have a lubricant that is less dependent on the season from hot to cold and on the latitude from sea level to a high mountain. Thus, when the viscosity index of a lubricating oil is too low, it must be improved from a minimum of 40 for a naphthenic lube to an acceptable value of 80–100 for a paraffinic lube. It was found that a class of additives can be used to increase the viscosity index, such as long-chain polymers, which can increase viscosity by entanglement. These long-chain polymers resemble normal paraffins as they are both long straight chains, but the molecular weights of the polymers are much greater than the lube oil molecules.

10.3 Navigation: Compass

How does a traveler, on land or on water, know how to navigate from a starting point to the next destination? The sun generally rises in the east and sets in the west and is in the south during high noon. These readings are precise only two days a year, at the Spring Equinox on March 21 and at the Autumn Equinox on September 21. If you live in the Northern Hemisphere, the sun rises in the southeast and sets in the southwest during October to February; however, the sun rises in the northeast and sets in the northwest from May to August. The other method is to wait till the sun goes down, and look for the Polaris or the North Star at the tail of the Small Bear constellation, which is always at the north. But if you live in the Southern Hemisphere, you will find empty sky near the South Pole, and the nearest thing is the Southern Cross, which is located at 60° south. What do you do when the time is three o'clock in the afternoon, the month is November, the sky is cloudy, and you cannot see the sun or the polar star?

According to legends, the Yellow Emperor of China fought a battle in 2600 BC in fog, in which he invented the south-pointing chariot, with a figure on the roof that always pointed towards the south, no matter which way the chariot turned. More reliable historical records show the earliest reference to magnetism in Chinese literature in the 4th century: "The lodestone makes iron come, or it attracts it." Lodestone is a naturally magnetized iron ore with the formula Fe_3O_4. It took a long time from the discovery of magnetism to the first use of compass for finding directions, probably for divination and building orientation. The device in Fig. 10.8 is a ladle or spoon of lodestone with a smooth bottom, which sits on a flat surface and can rotate to face the north. The first reliable record of its use for military navigation was in a Song Dynasty book dated 1040–44 AD. There is a description of an iron "south-pointing fish" floating in a bowl of water, aligning itself to the south. The device is recommended as a means of orientation "in the obscurity of the night." The first suspended magnetic needle compass was recorded by Shen Kuo in his book Dream Pool Essays in 1088 AD.

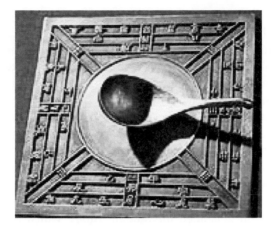

Figure 10.8 Chinese magnetic south-pointing spoon.

Hematite is an iron oxide with the formula Fe_3O_4. Natural hematite is called lodestone, which is found in many parts of the world. Pieces of iron are attracted to the lodestone, and after some rubbing, the iron can become permanently magnetized and can be used as a compass. Some soft iron does not hold magnetism permanently and reverts to the de-magnetized state soon afterward. Some types of iron remain magnetized and become suitable for making permanent magnets and compasses. Modern chemists have synthesized and tested many metals and oxides that have much stronger magnetic properties. There are ceramic and metallic magnets. Alnico is sintered aluminum, cobalt, and iron. The strongest permanent magnets are now made of rare earth elements, especially neodymium and samarium-cobalt magnets, which are critical for the success of high-performance computing and communication equipment, such as the hard drive memory.

A modern car, boat, or airplane communicates with the ground or other vehicles through electronic devices, which must be powered by a portable electrical source: batteries. A battery converts chemical energy into electrical energy and was invented by Alessandro Volta in around 1800. He made the voltaic pile placing a stack of alternating copper and zinc plates, separated by brine-soaked paper disks. When he connected the copper plates to the zinc plates with a wire, he found electricity flowing. He also found that if he wanted a higher voltage, he could increase the number of alternating plates. Now we have many primary batteries, which are single use and disposable, such as the flashlight zinc-and-carbon battery, separated by a packing soaked with ammonium chloride. We also have secondary batteries that are re-chargeable, such as the lithium-ion batteries used in laptop computers and cellular telephones.

Modern automobiles have GPS or global positioning system that tells where the vehicle is and how to navigate. It is based on electromagnetic signals between the car and a string of some 24 satellites that circle overhead at all times. This communication depends on electronic devices made of opto-electronic materials such as silicon and transistors. Before 1947, electrical communication and computation depended on vacuum tubes, which can control and magnify faint signals. Vacuum tubes were large, power hungry, and needed to be replaced after they burned out. The computer ENIAC carried 17,000 vacuum tubes. In 1947, Bardeen, Brattain, and Shockley invented the transistor in Bell Labs, which is based on the element germanium. In 1954, Morris Tanenbaum of Bell Lab demonstrated the first silicon transistor. Unlike germanium, which is rare on earth, silicon is the second most abundant element on the earth crust and available everywhere. The transistor does the work of the vacuum tube, takes up much less space, uses less power, and does not wear out.

The starting point of modern electronics is the manufacture of a cylindrical ingot from a single crystal by the Czochralski process. Molten silicon at around 1500°C is poured at the bottom of a tube, and a silicon crystal is dipped in the molten layer and slowly pulled up to begin the growth of the ingot, which may be 20 cm in diameter and 2 m long. To turn silicon into a semiconductor, it is doped with boron to make the P-type semiconductor and phosphorous to make the N-type semiconductor. The ingot is sawed into many thin wafers and then further processed to make transistors.

The technology escalated rapidly to integrated circuits and very large-scale integrated (VLSI) circuits that package thousands of transistors in a very small space. A wafer of silicon becomes a many-layered structure with tiny features that are getting smaller every year, which makes it practical to package more units into a small space. These steps involve many chemical reactions such as the additive steps of vapor deposition and removal steps such as etching. We have seen an incredible rate of progress for four decades, which is called the Moore's law according to which the number of transistors on a device doubles every 2 years.

10.4   Infrastructure: Roads

A walker can travel through most terrains that are not occupied by trees and big stones. Some obstacles such as mountain passes and swamps need improvement, and rivers require bridges. When wheeled vehicles appeared, roads paved with stone were needed for the passage of heavy carts, and heavy traffic required flatter and wider roads. A good road should be durable and should not wash away by flooding after a rain. The simplest road is made of leveled earth; a more elaborate road has a layer of gravel at the top, which is more durable and can bear more weight. The Roman Appian Way was paved with stones and gravels and could support chariots and other wheeled transports. An ideal chariot road should be smooth so that the wood or iron wheels do not bounce from stone to stone, which is particularly critical if the travel is at a very high speed as in wartime. It should also be domed in the center so that rain water rolls down to the sides and does not accumulate into puddles. Modern roads for automobiles and trucks are mostly made of asphalt and concrete. The runways in airports are also designed to accommodate wheels supporting heavy load and traveling at very high speeds. They have specifications and requirements that exceed those of highways.

Figure 10.9   Gravel paved road.

10.4.1 Tar and Asphalt

The Macadam road is made of 3-inch small stones and topped by 0.8-inch pebbles. When tar is added to bind the Macadam road, it is called tarmac, which makes the road smooth enough for motor vehicles at higher speed. The first tar came from coal-tar, which is produced as a byproduct of coke making for metallurgy, when coal is heated in airless ovens to drive off the water, volatile organic matter, and coal-tar. Coal-tar contains most of the sulfur and other impurities in the coal, so it is smelly and smoky.

Crude oil distillation fraction heavier than 525°C or 977°F is called residual oil, which is usually burned in utility plants and ships and contains many impurities such as sulfur and nitrogen. It is also heavy in asphalt, which can be separated from the fuel oil not by a difference in volatility in distillation, but by a difference in solubility in a liquid solvent. Solvent de-asphalting (SDA) often uses propane as the solvent, and separation takes place in a tower with higher temperatures at the bottom and lower temperatures at the top. The solvent and material with lower weight and more paraffinic would be removed from the top as cleaner fuel oil, as well as wax and lubricating oil. The insoluble, higher weight and more aromatic asphalt separates out at the bottom of the tower as a solid or semi-solid.

The primary use of asphalt/bitumen is in road construction, where it is used as the glue or binder mixed with aggregates to form asphalt concrete. The formula is usually 5% asphalt/bitumen cement and 95% aggregates (sand, stone, gravel). It is very viscous and has to be heated so that it can be mixed. The resulting asphalt roadway is much smoother than a gravel road. The main advantage of an asphalt road is the ease and low cost of preparation, by coating a gravel road with a layer of hot asphalt. An asphalt road also has the tendency to deteriorate after heavy use, and potholes are formed in harsh freezing winters. When water penetrates the soil structure under the asphalt pavement and traffic passes over the affected area, the poorly supported asphalt surface breaks. Continued traffic ejects asphalt and the underlying soil to create a hole in the pavement. Another hazard to asphalt surfaces comes from a vehicle that leaks gasoline or diesel fuel on the roadway, as these fuels tend to penetrate and soften the asphalt.

Figure 10.10 Concrete road.

10.4.2 Concrete

A concrete road is more suitable in major roads with heavy traffic and in harsh conditions. The concrete surface is made of Portland cement and coarse aggregates, sand, and water. The use of concrete became popular from the Roman times, and particular examples are the Coloseum and the dome of the Pantheon. They have lasted 2000 years. Concrete is a composite of coarse granular materials embedded in a hard matrix of cement as binder that glues them together. Roman concrete was made from quicklime, volcanic pozzolana, and pumice. Quicklime is calcium oxide, which is produced by calcination or roasting limestone in a kiln at above 825°C. The reaction converts calcium carbonate into calcium oxide and carbon dioxide. The resulting quicklime is not stable and reacts with carbon dioxide in the air and returns to limestone. It can also react with water and form calcium hydroxide. Cement is mixed with aggregates and sand and poured into slabs in molds, which hardens with time to concrete. The joints between concrete slabs also need to be filled with asphalt. The strongest slabs have steel enforcement rods or mesh to make it more resisting to cracking and capable of bearing more weight. They can also be grooved to provide a durable skid-resistant surface. They are much stronger and durable than asphalt roadways. But they are more expensive and require more time to construct.

Chapter 11

Communication and Entertainment

Attila E. Pavlath

President of the American Chemical Society, 2001

Attila@pavlath.org, AttilaPavlath@yahoo.com

11.1 Introduction

Is chemistry involved in communication and entertainment? The average person, even if took chemistry in high school, rarely thinks about chemistry when using the numerous everyday products. Naturally, if you engage in a conversation about medicine, food, vitamins, or gasoline, even if the person did not take chemistry, he or she will quickly acknowledge that chemistry plays some part in their development, although without realizing its extent in those areas as described in other chapters of this book. The situation is completely different when it comes to communication and entertainment. It is not surprising that when people look at cell phones, televisions, or computers, most of them consider these devises as an assembly of various metallic parts. If pressed hard, they attribute these devices to discoveries by physicists. However, they are unaware that these devises would not have been available for everyday use without dozens of developments through chemistry. Naturally, there is no doubt that physics played an important role, which is undeniable, and that physicists

Chemistry: Our Past, Present, and Future
Edited by Choon Ho Do and Attila E. Pavlath
Copyright © 2017 Pan Stanford Publishing Pte. Ltd.
ISBN 978-981-4774-08-6 (Hardcover), 978-1-315-22932-4 (eBook)
www.panstanford.com

made many discoveries. In fact, chemists and physicists frequently worked together hand in hand. There is no reason to get into a philosophical chicken/egg discussion about which came first. Chemistry will always be there from the point when our brain formulates an idea and converts it into sound or transfers it to paper.

11.2   Process of Communication

Chemistry is in every aspect of communication between two human beings. We have to transmit our thoughts, ideas, and information in everyday life using various means. In science fiction literature, telepathy and mind reading have been proposed, but even if we ever achieve such a way of communication, the role of chemistry cannot be denied. It is supposedly established by the brain through some so far undetermined method creating brainwaves similar to electromagnetic waves used by radio communication. If the brain is capable of radiating such waves, which can be received and understood by another brain, it must involve some types of biochemical processes. The full capability of the brain is not known, but its operation is similar to that of a powerful computer. The simplest way to visualize is comparing it to the case when you have a new computer without the manual and you are trying to find the full scope of what you can do with it. If telepathy actually becomes a reality in the future, it is still hidden in the vast manual of brain to be discovered.

Until we find the secret of its operation, we have to use more "mundane" ways unless we stand within earshot of each other. Strictly, even then we might have to rely on chemistry to create and operate various devices to amplify the sound for overcoming hard hearing or even to filter out background noises. In modern communication, the information we want to transmit is first converted to sound waves, and then the sound waves are converted to electromagnetic waves, which travel at the speed of light. These electromagnetic waves are transmitted to large distances, and the process is reversed at the receiving end. Naturally, we can use the old fashioned way of committing to a piece of paper and sending it through some transportation device, post office truck, stage coach, or pony express; however, the paper and writing devices are already the results various chemical processes. Without chemistry, we probably would be limited to smoke signals, but even in such case, we could debate how the smoke is created.

11.2.1 Advanced Synthetic Materials

Even if you do not realize that a variety of electronic communication devices, such as consumer electronics, cellular phones, and personal computers, rely on components resulting from various chemical processes, the complex collection of these sensitive circuits has to be housed in tough, durable enclosures to protect them from the environment and rough handling. While in some cases, the housing may be made completely from metal, it should be pointed out that to obtain any metal, one has to use chemistry. In nature, very few metals are found as pure element. Perhaps obtaining gold comes the closest to avoiding chemistry, but even emptying Fort Knox would not provide enough material for the millions of such devises used in everyday life. Furthermore, a full metal casing is not only heavy, but also can create short circuits. Therefore, plastics are essential not only because of their lighter weight compared to metals, but also because of their insulating properties. The flow of electrons that make up electrical currents cannot readily penetrate the molecular structure of plastics. By manipulating the structures of molecules and creating new ones, chemists and engineers have produced new materials that are both strong and flexible. These advances have improved impact resistance, reduced the total weight of equipment, and decreased the cost of consumer goods.

However, the major mistake in attributing only physics to the mushrooming use of these devices is not realizing that the components, which make the operation of these instruments possible, are the results of many important chemical developments. Cell phones, radio, television, or movies are all based on inventions made by chemists. You do not have to know how chemists made these possible. Take them away, and our lifestyle would be limited to that of the caveman in the Stone Age.

11.2.2 Communication Devices

Hundred years ago, almost all children experimented by creating a simple device made of two empty metal cans connected with a tight filament, which transmitted our speech through vibration and created resonation at the other end. In most cases, it was probably our imagination that we heard each other through our own homemade device and not through the air, but it was our first experiment using somehow chemical development without realizing that it was chemistry related. Today, our younger generation probably only hears about this from their grandparents.

11.2.2.1 Cell telephone

For them, this is becoming the only one way: cell telephone or mobile in another name. Interestingly, the younger generation is more familiar with it than some of their elders. It is not surprising to see it in the hands of 4–5-year olds. It is nearly always in their hands, almost as they were born with it. The device is used not only for transmitting speech but also for sending pictures, playing games, and connecting to the Internet. In a strange way, the most used application is not talking to each other but sending texts. One can frequently see two children walking next to each other and exchanging text messages while being ignorant about their environment. The more dangerous use is texting while driving. Today, its use is so frequent by everyone, young and old, that a satirical cartoon suggests that nothing has changed since the Stone Age.

Figure 11.1   Nothing has changed! Reproduced with the permission from Chris Madden.

The cell telephone is a device that can make and receive telephone calls over a radio link while moving around a wide geographic area. It does so by connecting to a cellular network provided by a mobile phone operator, allowing access to the public telephone network. The systems went through various steps of improvements. The starting system was labeled 0G, zero generation. This was very primitive and expensive. This first generation was introduced in the mid-1980s. The "1G" systems could support far more simultaneous calls, but still used analog technology. In 1991, the second generation (2G) *digital* cellular technology was launched in Finland by Radiolinja. Ten years later, the third generation (3G) was introduced in Japan, but the ever-increasing use forced additional improvement to first 3.5G, and in 2010 Sprint offered 4G, which has higher data transfer speeds and capacity.

The major components of a cell phone in addition to its casing and the audio setup, i.e., a built-in microphone and receiver, are a battery, which provides the power for the operation of its various functions; a keyboard or touchscreen for inputting; a screen to see the input and view the sent material (text and picture); and a SIM (subscriber identity module) card, which carries the necessary information about the user. The card is of the size of a postal stamp, and it is interchangeable with other communication devices.

The size of the cell phone has rapidly changed during the years. In the 20th century Dick Tracy cartoons, the detective used it as a wristwatch. In 1962, noted science fiction writer Arthur C. Clark actually predicted its availability in the mid-1980s in *Profiles of the Future*.[1]

He was right! While the first real cell phone was demonstrated in 1973, it became commercially available only 10 more years later. It weighed 1 kg, and its size was about five times what we use today. It was not possible to carry the device in pocket. It was only usable for talking, but no texting. Its use slowly increased; especially in the last decade, the growth was exponential, and the cost became quite low, making it available to most of the world. Today, over 6 billion cell phones are in use, covering 87% of the population.

---

[1]Clarke, A. C. *Profiles of the Future* (1962, rev. eds. 1973, 1983, and 1999, Millennium edition with a new preface)

Figure 11.2 Evolution of cell phones. https://en.wikipedia.org/wiki/ Mobile_phone#/media/File:Mobile_phone_evolution.jpg (Public domain).

Even the simplest cell phones fulfill many communication needs. This can be for social or business contact or to contact emergency services from almost everywhere, if needed. Many people have more than one cell phone for business and personal use. Multiple SIM cards may also be used to take advantage of the benefits of different calling plans; a particular plan might provide cheaper local calls, long-distance calls, international calls, or roaming. Today, the capabilities of cell phones include more than just simple communication. These phones are called smartphones. They can take pictures, maintain schedules, and even remind us to do something at a given time. They can provide entertainment by downloading games. Smartphones can be used to scan and transmit checks for deposit to one's account. They can even obey voice command to dial a programmed phone number or recover information.

Where is the chemistry in cell phones?

Chemistry is in every part of a cell phone. First, the power is provided by regular dry batteries. Although recently some cell phones have been introduced that get energy from solar rechargers, this does not change the picture as solar cells are made from special chemicals. Regardless of which one is used, both types require chemistry. The next step is the conversion of sound to electromagnetic waves for transmission. If you have ever seen the design of a radio, you would realize the complexity of this step. In the early stages of wireless communication, radiotelephone was used for this purpose with bulky radio vacuum tubes and electromechanical switches. Chemistry entered already in the simplest vacuum tube, which is like an incandescent light bulb containing a filament and an electrode enclosed in an evacuated glass enclosure. When the filament is heated, electrons are emitted, which head toward the electrode if an appropriate voltage difference is created between them. There are different types of vacuum tubes with more electrodes. These were the basis of electron technology. They started the commercial expansion of any process where sound waves were needed to be amplified, converted to electromagnetic waves for transmission and reconverted. Radio, television, and telephone networks would not have evolved without them.

However, even though better materials were made available for vacuum tubes and the size of the tubes decreased, still they were not only bulky but also had short lifespan and had to be frequently replaced. This happened periodically because sooner or later, depending on the heating, the filament was bound to burn out after an unpredictable time. When used in household devices, such as radio or television, if the operation of the device was faulty, they were all removed and personally checked for efficiency in setups generally available in supermarkets to find out which one did not meet the requirements. If the test results did not measure up to the displayed value, new tubes could be bought immediately. If upon reinsertion into the household device, it still did not work, it was time to take it to the repair shop.

After World War II, intensive research was started to replace vacuum tubes with smaller and lighter devices to make the operation more reliable. Although the research is generally described as solid-state physics experiments, its development depended on chemistry. The main role of the vacuum tube is controlling the flow of the current of electrons, and rectifying and amplifying as needed. For this purpose, appropriate solid-state chemical components, without vacuum, had to be developed, providing the required conductivity only in the required direction as the vacuum tubes. These were labeled as semiconductors.

Semiconductors are defined by their unique electric conductive behavior, somewhere between that of a metal, which conducts electricity, and an insulator, which does not. Various chemical compounds, elements, and their derivatives have such properties, but for practical purposes, only a few, such as germanium and silicon, were found acceptable. A pure semiconductor is not very useful as it is neither a very good insulator, nor a very good conductor. However, one important feature of semiconductors (and some insulators, known as *semi-insulators)* is that their conductivity can be increased and controlled by doping with impurities. For example, the presence of 0.001% of arsenic in germanium increases the electric conductivity by a factor of 10,000. The effect of doping is dependent on what element is being doped with what element and not just on the quantity. However, this can be a problem in their commercial manufacturing since small unintentional changes in purity might considerably alter their conducting characteristics. Chemical processes had to be developed to control purity. The most frequent semiconductor is based on silicon crystals. Devices using semiconductors were at first constructed based on empirical knowledge, before semiconductor theory provided a guide to construct more capable and reliable devices.

The experiments with semiconductors led to the development of the transistor in which components of vacuum tubes were replaced with solid-state circuitry to carry out the same purpose. A transistor is a semiconductor device used to amplify and switch electronic signals and electrical power. It is composed of semiconductor material with at least three terminals for connection to an external circuit. A voltage or current applied to one pair of transistor terminals changes the current through another pair of terminals. Because the controlled (output) power can be higher than the controlling (input) power, a transistor can amplify a signal. The transistor is the fundamental building block of modern electronic devices. It was developed in 1947 by John Bardeen, Walter Brattain, and William Shockley, for which they received the Nobel Prize.[2]

---

[2]"November 17 – December 23, 1947: Invention of the First Transistor" (http://www. aps.org/publications/apsnews/200011/history.cfm). American Physical Society.

It revolutionized the field of electronics and led to smaller and cheaper electronic devices such as radios, calculators, and computers, but its application is evident everywhere.

While the introduction of the transistor created the next step in electronic revolution by replacing the vacuum tubes with smaller and more dependable components in numerous devices used in everyday life, it was still not enough to make lightweight cell phones of conveniently small size. The development of integrated circuit (IC) or microchip created the real revolution in electronics. It is a set of electronic circuits on a small plate (that is where the name of chip came) of semiconductor material, which is silicon in most cases. It is now the basis of every electronic device and supports the structure of modern societies because of the economical way of producing it. The use of the first ICs was demonstrated by Jack Kilby in 1958.[3]

His patent application (filed while at Texas Instruments) defined it as "a body of semiconductor material wherein all the components of the electronic circuit are completely integrated." He received the Nobel Prize for his role in the development. Kilby's ICs made of germanium was the first step; he called them ICs. Even though it did not solve some of the problems and another inventor Robert Noyce modified it by using silicon, which is being used universally today, Kilby's work was designated by the Institute of Electric and Electronic Engineers as a Milestone in 2009.[4]

After this, research continued to place more and more circuits in smaller and smaller places. The greatest driving force behind the research was the need of aerospace projects to have lightweight digital computers onboard for the guidance system of rockets. In the 1960s, an IC contained a few hundreds of transistors; in the 1970s, one chip included tens of thousands; and in the 1980s, the number was hundreds of thousands. In 1989, the number went over a million, and the billion mark was reached in 2005. A 1 cm^2 area can include several billions of transistors.

---

[3]Kilby, J. S. Miniaturized Electronic Circuits, United States Patent Office, US Patent 3,138,743, filed 6 February 1959, issued 23 June 1964.

[4]"Milestones: First Semiconductor Integrated Circuit (IC), 1958" (http://www. ieeeghn.org/wiki/index.php/Milestones:First_Semiconductor_Integrated_Circuit_ %28IC%29,_1958). IEEE Global History Network. IEEE.

However, a chip can be as small as a few square millimeters. The width of the conducting lines in a circuit is now in the 10–100 nanometer range. Today, silicon monocrystals are the main substrate used for ICs, although for specialized applications such as LEDs, lasers, solar cells, and highest-speed ICs, other materials, such as gallium arsenide, are used. It took decades to perfect methods of creating crystals without defects in the crystalline structure of the semiconducting material.

The cost of construction of ICs is low because the components are printed through photolithography in one step instead of making these from individual transistors one at a time. For example in 1962, the price of a chip was $50.00; in within 6 years, it dropped to $2.33. Today we are talking about pennies. Research is still going on to create more advanced ICs. The microprocessors are even more complex ICs, which can control everything in the average household, even a digital microwave oven or blender. While the cost of designing and developing these microprocessors is high, they are used in millions of production units and, therefore, the devices can be manufactured economically.

However, the development of ICs is just one of the contributions chemistry provided for cell telephone. The transmission between cell phones requires satellite stations so that we can reach each other even on the other side of the world. Interestingly, communication satellites were already visualized in 1945 by Arthur C. Clarke, noted science fiction writer, 12 years before Sputnik. At that time, he was an RAF electronics officer and member of the British Interplanetary Society. In a short article in *Wireless World*, he described the use of manned satellites in 24-hour orbits high above the world's land masses to distribute television programs. At that time, this was ignored, but slowly the idea caught on. Today just for the United States, 36 satellites are in orbit and other countries are launching more and more satellites. These stations are even more complex combination of billions of microchips. The outer skin of satellites is made of special materials to protect against meteorites and cosmic radiation. Last but not least, they have to be placed in orbit around the world with rockets powered by high-energy oxidizers and fuel.

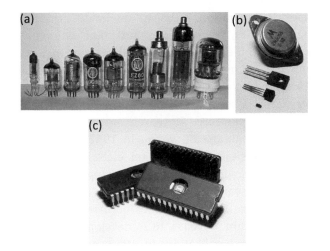

Figure 11.3 Comparison of the sizes of (a) vacuum tube (https://commons. wikimedia.org/wiki/File:Elektronenroehren-auswahl.jpg), (b) transistor (https://upload.wikimedia.org/wikipedia/ commons/2/21/Transistorer_%28cropped%29.jpg), and (c) microchip (https://commons.wikimedia.org/wiki/File: Microchips.jpg).

Finally, to use even slightly more complex wireless communication devices than the one that only transfers sound, cell phones need a screen on which various information or pictures are displayed. This is where again chemistry comes to the rescue. Cell phones have various sizes of screens displaying numbers, writings, and even pictures. This is possible through liquid crystal display (LCD). It takes the place of the bulky cathode ray tubes (CRT) and consumes much less energy. Therefore, it is ideal for equipment that can be powered by batteries, such as a cell phones, but it is the main component of display in video games, clocks, calculators, instrument panels, just to name a few. It is everywhere in our life. An LCD is a flat panel display, electronic visual display, or video display that uses the light-modulating properties of liquid crystals. Liquid crystals do not emit light directly. Its low electrical power consumption enables it to be used in battery-powered electronic equipment. It is an electronically modulated optical device made up of any number of segments filled with liquid crystals. It is more efficient than a CRT, which was the main component of television screen. It actually made

CRT obsolete. Its operation is based on liquid crystals, which change characteristics, appearance, and color on exposure to electric or magnetic file. Liquid crystal is a matter in a state that is between conventional liquid and a solid crystal. Chemists already discovered in 1888 various esters of cholesterol that have different colors at room temperature, which change with the change in temperature. Research on liquid crystals has been under way for almost 80 years. In 1962, RCA laboratories started the development of flat panel electronic displays. The application of electric field resulted in changes in appearance, and this led the research toward replacing the CRT by liquid crystal based flat panel. While changes occurred at 125°C, which was impractical, within 4 years RCA developed liquid crystals for room temperature application. In 1973, new materials were designed for practical applications. It could be applied to small-area LCDs, and this resulted in a wide range of applications for numerous electronic products. This is the full story on how chemistry contributed to worldwide communication through cell phone.

11.2.2.2   Landline communication

While cell phones are owned by almost 90% of the world's population and they are so popular that the landline telephones and telephone booths are rapidly heading toward becoming museum pieces, you can still find landline telephones even in households with multiple cell phones. This is the case especially in certain geographical locations where the wireless reception is occasionally week or unavailable because of the position of the communications satellites. The landline telephone is the original design of communication to conduct conversations between persons at various short or long distances, where they cannot be heard directly. Its only function is to convert sounds to electronic signals, which are transmitted through cables, and replay the sounds at the receiving end. It was invented in 1876 by Alexander Graham Bell.[5]

---

[5]US 174465 (http://worldwide.espacenet.com/textdoc?DB=EPODOC&IDX=US17 4465) Alexander Graham Bell: "Improvement in Telegraphy" filed on February 14, 1876, granted on March 7, 1876.

Although he patented it, there were many others who tried to reap the commercial benefits that he had to fight for his patent rights. From then on, telephone communication became the essential part of business operation household.

The shape of the telephone started out from the simplest design with a handheld microphone and a listening device. This rapidly changed with the addition of various types of dialing devices, first the rotary, and then the pushbutton setup, leading to the now popular portable cell phone-like receiver. This modern design has some of the features of the more complicated cell phones and is wirelessly connected to the base. It can operate at a distance of 50–100 feet from the base but not independently. The major difference is that regardless of its shape, the transmission is not through the air but solid cable connection. In some remote areas, one can still see telephone poles with copper wires. The conversion of sound to electronic signals and their reconversion to sound waves at the receiving end are the same as for wireless communication with the help of chemistry. However, transmission through solid metallic filaments has major problems, such as interaction with electromagnetic radiation, which can affect the quality of the transmitted sound and loss of the intensity. Transmission of multiple conversation is difficult. It also needs a large amount of copper.

Fortunately, chemistry again came to the rescue through the development of optical fibers. What is an optical fiber? It is a flexible, transparent fiber made generally of high-quality extruded glass (silica) or plastic, slightly thicker than a human hair. Actually, even unpigmented human hair was used in some experiments, but obviously this was only to show the wide range of possible materials. It can transmit light between the two ends of the fiber. Fiber optic cables can also deliver an electric current for low-power electronic devices. Transmission by optical fiber has very little (0.2 decibel/km) or no loss. Such remarkably low losses are possible only because ultra-pure silicon is available, which is essential for manufacturing ICs and discrete transistors. In certain cases, however, some impurities, such as germanium and aluminum oxide, are added intentionally to change certain physical properties of the silica fiber crucial for conductivity. Silica can be drawn into fibers at reasonably high temperatures and has a fairly broad glass transformation range. Silica fiber also has high mechanical strength against both pulling and even bending, provided that the fiber is not too thick and that the surfaces have been well prepared during processing. The waves are guided through the fiber by a physical process called total internal reflection. Waves can enter optically dense materials, but depending on the properties of that material, the waves either leave the material or are reflected back to the material when they hit its boundaries. This requires appropriate coating for optical fibers. The core through which the waves proceed is covered by three layers. The complete thickness of the fiber is around 0.8–1.0 mm, but that of the core is only 0.008–0.010 mm. The first layer is called "cladding," which is made of a special material to prevent the wave from leaving. The second layer called "buffer" protects the inside from moisture and physical handling during the manufacturing and installation process. The third one is a jacket, which gives further protection from the environment.

Figure 11.4 Bundle of optical fibers transmitting light (https://upload.wikimedia.org/wikipedia/commons/4/49/Fibreoptic.jpg).

The original inventor of telephone, Alexander Graham Bell, was actually able to transmit voice over an optical beam in 1880, but this was found to be impractical because of atmospheric interferences. However, optical fibers do not pick up any interference from the environment. Intensive research was carried out to find the right material for the fiber. Today, a single fiber can carry millions of voice communication and thousands of television channels. The loss of amplitude is so low that long-distance communication is practical for both voice and television.

11.2.2.3   Other electronic communication devices

Today, those who use cell phones for texting are probably too young to know that there was a different type of "texting" earlier: telex exchange. The telex machine looked like a typewriter of different sizes. At the beginning, telex had its own network independent from telephone network, but in time they were running effectively on phone network. Those attached to the network could deliver text messages to anyone on the network in any part of the world. The message could be typed in, but for faster transmission, it was encoded onto a special tape, which was then read into the line. At the receiving end, the machine typed out the message. On an average, 60–70 words per minute were transmitted. The system incorporated the "answerback" code, which automatically sent back a message acknowledging the receipt. The network was first operated through cable but later through shortwave telex links. While one can still find telex machines outside of museums, they have been replaced by FAX machines. These are used mostly in the maritime industry.

While the first commercial FAX machine was introduced in 1964 by Xerox Corporation, its history goes back more than 100 years. The first FAX machine was patented by Alexander Bain in 1843. This was improved by Giovanni Caselli, who actually established by his Pantelegraph a telefax service between two French cities, Paris and Lyon, in 1865, years before the invention of telephone.[6]

---

[6]"Istituto Tecnico Industriale, Rome, Italy. Italian biography of Giovanni Caselli" (http://www.itisgalileiroma.it/shed/shed0/shed0 /caselli.htm). Itisgalileiroma.it.

However, his machine did not scan the document to be transmitted; it required drawing. It was only in 1881 when Shelford Bidwell's machine was able to scan a sheet. In 1990, Arthur Korn introduced the Bildtelegraph, which was used to transmit photographs of wanted criminals between Paris and London. The idea was continuously improved, and the first wireless transmission was developed by Richard Ranger in 1924, which was used for transatlantic transmission. In the same year, Herbert Ives transmitted the first color picture.

Starting with the FAX machine of 1964, rapid improvements were made to decrease its weight and make it easier to operate. The weight was reduced to 46 pounds, and the transmission took 6 min. Today, the FAX machines are easily portable and faster. They can be connected anywhere to a telephone line. Actually, a FAX machine appears to be a telephone equipment at first glance. It can also be used for voice transmission, but its purpose is to scan and transmit documents with text and/or picture through the telephone line to another FAX machine. The machine can be set for sending and receiving. The receiving machine can serve for both voice transmission and operate as FAX on the same telephone number depending on the position of the setting switch if one does not want to maintain two different exchanges. Generally, the caller will hear a special signal instead of ringing if the machine is set for FAX reception. The sending machine scans the document and digitalizes it, and the resulting data are compressed for faster transmission. The receiving machine reconverts the data and prints it. At the beginning, special thermal fax paper was used, but now standard printing paper is used for printouts. Simple FAX machines only send and receive documents, but today hybrid devices are available, which can copy, fax, print, and scan separately. However, these hybrid machines are generally connected to computers, and the transmission is done more and more through the Internet. Faxing is slowly joining the Telex and the rotary telephone.

One more device is not used by the average person: radiotelephone, which uses radio waves for voice transmission. This can be considered the forerunner of cell phone. However, the operation of a radiotelephone requires radiotelephone operator license and qualified persons. The only exception is when the device makes short range transmission, which is better known as walkie-talkie. This is used in construction sites and even in toys by children. It is bulkier than the modern cell phone. Its main use was in the marine industry for ship-to-ship and ship-to-shore communication through a marine operator to connect to landlines. However, it is gradually replaced by satellite telephones and is kept as a back-up system for safety purposes.

11.2.3   Storing Information

Even before telephone and radio allowed instant communication, there was a need to record ideas, messages and pictures in some form, which could be sent later by some other method or preserved for history. In the modern life with various communication methods this became even more necessary. The historical method, which is slowly becoming obsolete for the younger generation, is writing a letter. Unfortunately, the traditional 3Rs—wRiting, Reading, and aRithmetic—are replaced by 3Ts: Telephone, Television, and compuTer. It is a nightmare for many of them when a blackout occurs.

11.2.3.1   Paper and ink

The role of chemistry in recording something on paper by hand should be evident to everyone, though it is only subconscious, like in many other areas, until we are reminded of what is involved in both paper and ink. Paper is one of the oldest conveniences created by humans, which is still used today. However, even many centuries before paper was developed in 100 AD, humans were looking for various media to record thoughts, words, and pictures. In ancient China, silk was used, while the Egyptians manufactured papyrus for this purpose. Papyrus is made from the inner fibrous pith of the papyrus plant, *Cyperus papyrus*. Thin strips are cut and tightly arranged parallel to each other, and then another layer of strips is glued to the previous layer at right angles. The moist layers are pressed and hammered to form one single sheet. Unfortunately, papyrus is moisture sensitive, and in humid atmosphere, it is subject to decomposition because of mold formation. Some early recordings were lost because of this problem. While the discovery of paper manufacturing made papyrus obsolete, its use was revived in the 19th century, mostly out of curiosity and for attracting tourists. Presently, the papyrus plant is still cultivated in Africa for purposes other than writing, such as making baskets, hats, and mats.

The main component of paper is basically the same as papyrus: cellulose. The history of paper is traced back to China as a necessity to replace silk as a recording material so that expensive silk can be used more for garments. Paper is a thin material produced by pressing together moist fibers and drying them into flexible sheets. Before the invention of the paper machine, the most common fiber source was recycled fibers from used textiles such as cotton, hemp, and linen. In the mid-19th century, the use of wood for the mass production of paper minimized the recycling, but using textiles was still practiced for higher quality paper. In our time, because of increasing environmental concerns, even the recycling of waste paper became an important process, since the worldwide use of paper quadrupled during the past 40–50 years. Using wood as the starting material, a pulp is created through both mechanical and chemical processes. The main two components of wood are cellulose and lignin; the overall quality of the paper is determined by the cellulose. The mechanical process retains most of the lignin, which is slowly oxidized in the presence of light, and the paper will become yellow and brittle in time. Mechanical pulping is cheaper, and the paper is used for newspaper and paperback books. For applications where the long-range whiteness of the paper is crucial (e.g., books, documents), the pulping must be done through the chemical process, which eliminates the cause of these problems by removing the lignin and provides cellulose fibers. It also retains the original length of the fibers in contrast to the mechanical process. In the recycling process, various steps need to be used to separate the fibers not just from dyes but also from various foreign materials before the pulp can be used. The pulp is adjusted to slight basic pH using sodium silicate and/or sodium hydroxide. Bleaching is done by hydrogen peroxide. In addition, other chemicals might be used to obtain usable pulp.

Naturally, fibers are not the only components of the pulp; other chemicals (fillers) are added, such as chalk or china clay, which improve the characteristics of the paper for printing or writing. In the early paper-making process, aluminum sulfate salts (alum) were added to make the paper more water resistant, which was needed to prevent the spread of the ink. Unfortunately, since alum is acidic, it slowly hydrolyzes the cellulose and eventually results in deterioration. Therefore, for sizing purposes, today non-acidic additives are mixed into the pulp and/or applied to the paper web later in the manufacturing process. The purpose of sizing is to establish the correct level of surface absorbency to suit the ink or paint. Paper is used for many purposes other than writing and printing, such as packaging materials, drawing sheets, and toilet tissue, for which additional chemical treatments have been developed.

In the case of ink, the presence of chemistry is much more evident immediately. Ink can range from liquid to paste using dyes or pigments of the desired color. Ink can be a complex mixture of various chemicals: solvents, pigments, dyes, and many other materials. The components of ink affect its flow, thickness, and appearance when dry. The pigment-based inks are used more frequently than those using dyes because they are more color-fast, but they are also more expensive, less consistent in color, and have less of a color range than dyes. The advantage of dye-based inks is that they are stronger and available in a much wider range of color. The disadvantage is that dye may soak into the paper and may cause bleeding at the edges of the images. Both types can be applied by brush, pen, and quill, both for creating a picture or writing a text.

Ink has an even longer history than paper, about 4000 years, starting out throughout Asia. In China, the use of both natural (plant dyes) and mineral (graphite) based inks has been proven in archeological finds. Graphite and soot mixed with fish glue in a fine aqueous suspension were used for brush painting on silk. A wide variety of India inks based on burnt bones, pitch, and tar were identified in various documents created 2500 years ago. It was created in a more scientific way in the Roman Empire, 1600 years ago. An old recipe describes using iron salts mixed with tannin, which gave a bluish-black color and changed to brown after some time. An 800-year-old process for ink making started out from hawthorn branches. The dried bark was mashed and soaked in water, and then boiled with wine to create a thick black paste. After it was dried in the sun, the addition of iron salt and wine finished the process. The development of the printing press by Gutenberg needed further developments for more advanced inks because the ones used for handwriting created blurs. This led to the creation of an oily substance using soot, turpentine, and walnut oil.

Nowadays, the most used black inks are either carbon or iron gall inks. The carbon inks are manufactured from soot and some types of glue as binding agents, which allow the adherence of the carbon particles to the paper. Such inks are chemically stable and not affected by sunlight. Their disadvantage is that humidity affects the appearance of the text and pictures and may be washed off from the surface; therefore, it requires storage in dry atmosphere. The iron gall inks are used since the 12th century, and they were preferred for a long time. However, they are corrosive and can decompose the paper, resulting in brittleness. Many important historical products have deteriorated, and various procedures have to be employed to stop the deterioration.

The discussion of inks would not be complete without mentioning shortly indelible and invisible inks. Indelible inks can create text and drawing that cannot be removed for a certain period. They are used in some underdeveloped countries during election to prevent fraud, repeated voting. Ironically, they can be used for the opposite purpose by marking undesirable eligible voters' finger before election to prevent them from voting. As it frequently happens in our modern society, scientific discoveries can provide important benefits, but unfortunately, unscrupulous individuals can use them for harmful purposes. It is not science but people who commit the crime.

Invisible ink is used for writing, which is invisible either on application or soon thereafter and can later be made visible by some means depending on the type of ink used. Invisible ink can be applied to a writing surface with various writing devices, including fingers. When it dries, the surface appears blank without any changes. The writing and picture can be reconstructed by ultraviolet rays, heat, or chemical reactions. The reconstruction can be made permanent or temporary. The temporary reconstruction may be used in espionage where the person searching for secret messages does not want to alert the recipient that the message has been intercepted. Using either ultraviolet light or an iodine fume, messages can be quickly screened for invisible ink and also read without first permanently developing the invisible ink. The originator will be unaware that the secret message has already been intercepted by someone else. There are various simple and more complicated invisible inks. For example, various acidic juices, such as from lemon and onion, react with the paper and form esters on the application of heat. Writing by cobalt chloride is invisible but turns blue when heated; upon cooling for a while, it becomes invisible again (if not overly heated). Writing by phenolphtalein solution becomes visible in presence of ammonia vapors and may disappear again by certain acidic vapors. Marker pens are commercially available as toy but are also used for security purposes to be applied to protect household items, which can help to identify them in case of burglary.

It should be mentioned that a primitive type of "invisible" writing was used in ancient times. The messenger's head was shaved, and the writing was applied to the scalp. Then the messenger was sent after the hair grew back to a safe length. At the receiving end, the hair was cut to reveal the message. The messenger did not know what message he was carrying; obviously he was protecting it during the journey. The process took time, but in the absence of other methods, this was the best option.

11.2.3.2 Recording sounds and pictures

Interestingly, recording sound was made almost simultaneously with the invention of transmitting through a wire. Phonograph is a device that not only records but also reproduces the recorded sound. The earliest phonograph was developed by Edison in 1877 using a tinfoil sheet cylinder. Alexander Graham Bell, the inventor of telephone, improved it a few years later using a wax-coated cardboard cylinder. Around 1900, Emile Berliner replaced the cylinder with a beeswax coated zinc disc coated on a turntable. After recording, the disc was exposed to chromic acid, which created grooves where the beeswax was removed during the recording. The new device became known as gramophone. The material of the disc changed to vinyl to make easier manufacturing and better endurance. With the advent of transistors, further improvements were made in the quality of tone. However, the use of optical disc and magnetic tape became more and accepted, and today very few gramophones exist outside of the walls of museums.

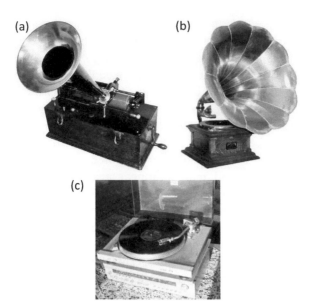

Figure 11.5 Progress in 100 years: (a) Edison's Phonograph (1889) (https://en.wikipedia.org/wiki/Phonograph#/media/File: EdisonPhonograph.jpg), (b) Victor Talking Machine (1907) (https://en.wikipedia.org/wiki/Phonograph#/media/File: VictorVPhonograph.jpg), and (c) Turntable Gramophone (1979) (https://en.wikipedia.org/wiki/Phonograph#/media/ File:Romanian_pickup1.jpg).

The next step in recording was the introduction of magnetic tape introduced by Fritz Pfleumer in 1928 using a ferrioxide-coated paper strip. This developed into the use of tape recorders with magnetizable plastic filmstrips. A tape drive moves the tape from one reel to another through a tape head, which can write, read, and erase as needed. The tape could store sound, picture, and computer data. Its introduction was a new revolutionary step in the development of radio and television broadcasting and later especially in the use of computers. In radio and television, it allowed the recording to be made at any time and played back whenever it was needed. However, separate recording devices were needed for sound and picture. They were not interchangeable. In early computer developments, unparalleled amounts of data could be mechanically created, stored for long periods, and rapidly accessed. For UNIVAC, a thin metal strip of nickel-plated bronze was used. In old movies depicting UNIVAC-like computers, the twirling magnetic reels gave a dramatic effect. Over years, magnetic tape can suffer from deterioration. When the binder of the tape absorbs moisture, it can render the tape unusable. With appropriate care, however, it is very reliable and can be rerecorded many times with a lifetime of 15–20 years. With new developments for faster access, the optical disc drive gained more popularity and use, but because of the lower cost of tape recording, magnetic tapes refuse to retire and are still used in many applications as videocassettes (VHS), though less and less.

The optical disc is a flat, circular disc made of polycarbonate and coated with aluminum on the side on which the encoding is made. The first patent was issued to David Paul Gregg in 1961. It was a very simple form of DVD, but it took almost 25 years until a practical DVD was developed. Rapid research continuously improved the storage capacity and the quality. The capacity increased from 700 MB CDs through 4.7 GB DVD up to 400 GB Blue-ray discs providing the highest quality for movie recording. While all magnetic tapes are rewritable many times, optical disc can be of three types: read-only (CD-ROM), recordable once (CD-R), and re-recordable (CD-RW), where CD stand for Compact Disc. They differ according to the composition of the encoding layer on the disc. The rewritable one is a complex combination of antimony, indium, silver, and tellurium. The recording is made by a laser creating pits and bumps; the smaller these indentation, the higher the capacity of the disc allowing more precise recording. The data can be read back by laser illumination while the disc is spinning up to 4000 RPM. The designation CD actually refers to this characteristic: compact disc. A CD can hold 700 MB of information, but new developments have increased the capacity of optical discs: DVDs can store 4.7 GB and Blue-ray discs can store 25 GB. While one side generally carries a paper label, in contrast to the flappy disc used in early desktop computers, the coated side is not protected and must be treated carefully because scratches or even fingerprint can interfere with recording and recovering information. It is widely used for storing music, video, and data for computer. The latter use is especially the advantage of the optical disc because it can store much larger amount of data and accessed more readily than the magnetic tape, though USB flash drives are gaining more use with their higher capacity, smaller size, and less sensitivity during handling.

11.2.3.3 Photography

There are historical reports that studies on obtaining a picture through a pinhole camera were carried out in China and Greece in as early as 5th–4th BCE. The camera obscura was first reported in the 6th century. Albertus Magnus discovered silver nitrate in the 13th century, and Georg Fabricius found silver halides in the 16th century. Most of the silver compounds are light sensitive, both visible and invisible, and will decompose under appropriate conditions, creating black depositions. In 1694, Wilhelm Homberg reported the darkening effect of light on some chemicals. The next challenge, for centuries, was finding a way to preserve the images obtained by these methods for later viewing, which is where chemistry again provided the solution. This was the start of photography.

The first successful recording was done in 1822 by the French inventor Nicéphore Niépce, but the created image was only visible for a few hours before it disappeared. He worked with Louis Daguerre, who experimented for 3 more years to improve it using a silver-plated copper sheet. He could reduce the 8 h of exposure to 2 h using light-sensitive silver. His continued works reduced the required time to minutes. Since these processes created a negative image, the image had to be photographed again to obtain the positive form.

Others experimented and improved the method. It took 30 years to come up with thin gelatin emulsion coated glass plates. Some experiments were done with paper and plastic films, but the glass plates were not only less expensive but also yielded better quality. The major breakthrough, which started practical photography, occurred in 1884, when George Eastman developed an early type of film to replace photographic plates, leading to the technology used by modern film cameras. In 1891, Gabriel Lippmann developed a process for making natural color photographs. Even though he received the Nobel Prize for his discovery in 1908, it was only of scientific interest and was not practical.

Eastman's first experiments were on coated paper strips. The first plastic film was developed in 1889 using celluloid, which is nitrocellulose (nitrate film) and easily flammable. This was especially a problem in motion pictures where the film was exposed to heat in the projector, and it frequently caught fire especially with silent films, which spent more time in the projector. It took 20 years to introduce a safe film, cellulose acetate, but it was not as strong and cheap as celluloid though it was safer. Slowly, however, celluloid was phased out, and it was discontinued in 1951. For years, pictures were recorded in black and white using the photosensitivity of silver halides, which decomposed to black metallic silver on being exposed to light. The unexposed silver halides were then dissolved during the development process creating a negative image. From the negative, one could get the real picture through the same process.

Monochromatic photography, black and white, became an inexpensive way to record people, objects, and landscape, but intensive research was carried out already from the 19th century to reproduce natural colors. The first color photography experiment was reported by James Clerk Maxwell in 1855 by first taking three black-and-white photographs using red, green, and blue filters, and then superimposing them. Continuing research was hindered because most of the available compounds in the early days of photography were sensitive only to blue light. In 1873, Hermann Vogel reported dye sensitization, which extended the interaction to green, yellow, and red. In 1907, the first commercial color photography, Autochrome, was introduced by the renowned Lumière brothers. The creation of Kodachrome in 1935 can be considered the start of the "color revolution." A wide range of films were manufactured for various sensitivity designated from ISO 25 (slow) to 3200 (fast) to provide for every circumstances. The final step was the instant color film by Polaroid.

While black-and-white photography needs one photosensitive layer, the simplest color photographic films need a minimum of three layers with dyes sensitive to different colors. Red is the first layer on the film, topped by green, and blue is the uppermost layer. In newer films, many more layers are used with various layers to give back the picture more realistically. During development, the exposed silver salts are converted to metallic silver, just as with black-and-white films. But in a color film, the byproducts of the development reaction simultaneously combine with chemicals known as color couplers, which are included either in the film itself or in the developer solution to form colored dyes. Because the byproducts are created in direct proportion to the amount of exposure and development, the dye clouds formed are also in proportion to the exposure and development. Following development, the silver is converted back to silver salts in the *bleach step.* It is removed from the film in the *fix step.* This leaves behind only the formed color dyes, which combine to make up the colored visible image.

The advantage of recording picture on film is that they can be stored in albums and placed on montage for immediate collected viewing. It is estimated that worldwide trillions of pictures are kept commemorating various events in our life. The problem is that these pictures, especially the colored ones, fade. In addition, the pictures made on films have to be developed and cannot be viewed immediately. We want to see our photos shortly after taking them without waiting. This allows us to retake them if we did not get them right. For this purpose, instant cameras were developed using special films, which provided the picture within 15–20 s without the need of development. It found use in taking unalterable instant pictures of crime scenes and also where instant photos were needed, such as in ID cards, drivers' license, or passport.

The development of digital camera solved some of the problems of photography using films. The pictures were recorded as a set of electronic data. This allowed instant viewing. The pictures could be stored indefinitely, and they could be printed out on any paper depending on the required quality. Using a computer, they could be altered, combined, and retouched to change color.

It is evident that any type of photography is based on chemistry, but chemists are not conceited. We have to give equal credit to those who developed the mechanics of various cameras, their components, including lenses. Photography is just another example of cooperation by all branches of science (physical, biological, and medical) to improve our everyday life.

11.2.3.3 Copy machines

A copy machine is used to reproduce on paper any image or writing from another document or flat object. However, this generic name covers various methods. The first was the Thermo-Fax machine, originating from photography. It used an appropriately chemically treated heat-sensitive paper. It was placed on the black-and-white image to be copied and exposed to infrared radiation while heating, which reproduced the image. The method was later developed to create transparencies for presentations. Today, it is used mostly by artists for various artistic creations. It was relatively inexpensive, costing 1.5 cents per sheet. However, the accuracy of copying was not perfect, creating brown instead of black images because of occasional burning.

Various other methods were used in office work, e.g., carbon paper and mimeograph machines, but the development of xerographic copy machine created by Xerox in 1949 made them obsolete. While the name is trademarked, in everyday use, it is called "Xeroxing" even if the machine is not made by Xerox. The name Xerox is from a Greek word meaning "dry writing." The discovery of photocopying belongs to Chester Carlson, who practiced actual "kitchen chemistry" in his home by using a sulfur-covered zinc plate. He named it electrophotography and patented it in 1938. Ironically, he was turned down for commercialization by 20 companies, including some large ones who thought that there was no need for such machine, and that the existing ones (e.g., carbon paper) will never be replaced. Naturally, this was based on the assumption that there would not be any need for a large number of copies. In 1947, Haloid Corporation, which manufactured photographic papers, saw the light and worked with Carlson to develop the first copy machine. Interestingly, they thought that the word electrophotography is too complicated and created the word "Xerox." Not surprisingly, the invention caught fire in public use, and the company expanded rapidly, ultimately changing its name to Xerox. The present process provides a copy instantaneously, but it is a multistep process in the machines. Since then, various modifications have been made, but the basic principle is the same. First it charges a cylindrical drum (positive or negative) with a special semiconductor, photoconductor that becomes conductive by light. Then the document to be copied is reflected by light to the drum on which the charges are eliminated with the exception where black parts of the document did not allow transmission. This creates an electrical image on the drum. In the next step, a toner is applied, which will be absorbed on the area of the drum while it is still charged. Then the particles are transferred to paper with a higher charge than the drum. In the final step, the toner particles are fused to the paper by heat and pressure rollers. It is predictable that the next step of research was to develop color copying, which was commercialized by 3M in 1968 as a Color-in-Color copier, replacing the electrostatic technology with a dye sublimation process. The whole photocopying process created various legal problems such as copying copyrighted materials and even counterfeiting currency.

One final method of copying is scanning. The flatbed scanner is somewhere between a digital camera and the photocopier. It can do more than the photocopier machine. It can provide a greater resolution of the copied object, and also it can make copies of three-dimensional arrangements of limited depth up to 10–12 mm. It can be coupled to a computer to transfer the document in digital form. It is also used by artists, especially color flatbed, to create various artistic designs.

11.3 Entertainment

What types of entertainment are in our everyday life? Music, radio, television, movies, and games. What role does chemistry play in them? Musical instruments are made from various metals, which are the result of various chemical procedures. However, do the structure materials of the instruments make any difference? Are we that desperate to show the influence of chemistry? It is interesting that we find significant differences in the most unexpected places. For example, for an expert violinist and a "gourmet" listener, there is a great difference in the quality of the music provided by an everyday violin and a Stradivari, which is one of the most valuable violins ever made. Why? Jozsef Nagyvary, a chemistry professor at Texas A&M, spent years trying to find out the reason. He analyzed the wood from a Stradivari and was actually able to create violins from specially treated woods to reproduce the same quality of music.

Naturally, chemistry has much more influence, but discussing it in more detail is an anticlimax. It was shown on the previous pages that chemistry provided the basis for communication and photography, which are the components of most entertainment methods. The combination of various developments made the major forms of entertainment in the modern time possible: going to the movies, listening to radio, and watching television. Movies are the results of recording events on photosensitive plastic films and then replaying. Although the first silent movies could be projected simultaneously by trying to apply the sound with an audio, it was not reliable. Special films had to be developed on which both the picture and sound could be recorded simultaneously, synchronizing the voices with the events shown on the film.

The radio is a one way wireless communication device based on the same numerous chemical developments, which made the cell phone possible. While Marconi is credited with the discovery of the radio (wireless telegraphy) for which he received the Nobel Prize in 1909, its role in the catastrophe of the Titanic grabbed the general attention and resulted in continuous improvement. From the primitive crystal radio, vacuum tubes, transistors, silicon chips, and CDs made the high-quality home entertainment system a reality. The FM stations' music through your home entertainment center could not provide you with the relaxation in your home without chemistry.

From there, it is only one step to the television, first to black-and-white and then to color reception. The transmission and reception would not be possible without the most developed ICs and microprocessors. Giant flat screens based on liquid crystals can put you in the midst of a sport event in the other side of the world or even from space. The first system, which is the ancestor of modern television, was created by a 23-year-old young German student, Paul Gottlieb Nipkow, using a spinning disc with holes through which light could pass through to a light-sensitive sensor made of selenium. To be of practical use, the signals had to be amplified and the research was carried out internationally. In 1907, Russian Boris Rosing used a cathode ray tube for amplification. In 1925, Scottish John Logie Baird, using an improved Nipkow disc, achieved 30 lines of resolution. In 1926, Hungarian Kálmán Tihanyi designed a system with fully electronic scanning, quickly followed by Japanese Kenjiro Takayanagi producing 40-line resolution. In 1927, Russian Leon Theremin achieved 100-line resolution. Finally, in 1931, Mexican Guillermo Camarena developed the first "trichromatic field sequential system," which lead to a patent for color television in 1941. World War II slowed further developments. In 1954, the first workable color system was created, but it was slowly commercialized. However, by 1970, very few black-and-white broadcasting existed. Nowadays, the younger generation cannot imagine how their grandparents or even their parents could ever survive without color television.

11.3.1 Computers

Strictly speaking, computers are not communication or entertainment devices, but they are used more and more in these areas in addition to scientific usage. The Skype program is used more and more frequently for instant communication, both with audio and video. Information retrieval is a type of communication. The younger generation uses computers more for entertainment (playing video games) than for calculation. This became possible due to the development of the bulky machines created after World War II to desktop and laptop computers. Laptops and iPads can be easily taken everywhere, including an airplane.

The first computer took up a huge room. The ENIAC and UNIVAC, created in 1946 and 1951, respectively, were heralded as the greatest development for various calculations, both scientific and business. The ENIAC contained 17,468 vacuum tubes, 7200 crystal diodes, 1500 relays, 70,000 resistors, 10,000 capacitors, and around 5 million hand-soldered joints. It weighed more than 30 short tons (27 t), was roughly 8 by 3 by 100 feet (2.4 m Å ~ 0.9 m Å~ 30 m), took up 1800 square feet (167 m^2), and consumed 150 kW of power. It could be programmed to perform complex sequences of operations, but the task of taking a problem and mapping it onto the machine was complex and usually took weeks. The program had to be prepared first on paper, and it took days to enter it into the computer. However, it was still more efficient than the early computers, which were mechanical, not electronic. Five years of planning and building provided the UNIVAC a greatly improved ENIAC, which became commercially available. Naturally, it was still huge. It used 5200 vacuum tubes, weighed 29,000 pounds, consumed 125 kW, and could perform about 1905 operations per second. The complete system occupied more than 35.5 m^2 of floor space.

Both the ENIAC and UNIVAC were milestones, but their inventors could not see 50 years in the future, just as the Wright brothers, after their historic flight, could not visualize that Chuck Yeager's plane will break the sound barrier 40 years later. Interestingly, even science fiction writers drew a blank. One of the greatest sci-fi writers, Isaac Asimov, predicted in 1948 the immense power of computers, but in his short novel "Let there be light," he visualized a super computer with an ever-increasing size ultimately covering the surface of the moon and even outsourcing it to the subspace. He just could not foresee transistor, ICs, and silicon chips, which will miniaturize ENIAC while creating an even faster and more efficient computer.

Figure 11.6 What difference chemistry makes: (a) Programming ENIAC in 1946 https://en.wikipedia.org/wiki/Computer#/media/File:Eniac.jpg. (b) The author working on a laptop in a park in 2011.

The operation of UNIVAC required an expert team to program through punch-cards and to supply data through numerous rolls of magnetic tapes. The results were obtained on dot matrix printers, and the graphs were printed in black and white with X-s and O-s. Compare this to the laptop of today, which true to its name sits on your lap and is powered by a lithium battery for hours. It comes with preinstalled programs for writing and complicated graphing, which can appear in color on the liquid crystal screen. It can be stored on a small stick from where it can be printed out. Programming is simple, and most teenagers can develop desired programs, though some of the older generation might look at it with fear. And if this was not enough, through a simple program called Skype, a user can connect with another user through the Internet and converse face to face in real time without a cell phone. Even modern entertainment enters the picture. Hundreds of games are developed every year for single users or multiple teams. As the modern Wilbur Wright, one cannot even imagine what will be its use in the 22nd century.

11.4   Closing Word

In my frequent talks about the benefits of chemistry in our everyday life, I always close my presentation with a challenge to the audience: Tell me something you use in your life that has nothing to do with chemistry. Generally, this is followed by silence. However, at a recent presentation in a university hall packed with students, one young fellow in the front row, sitting with a laptop computer, raised his hand: "I never use anything related to chemistry." When I silently pointed to his laptop, he was quick to state defensively: This is not chemistry; this is physics.

This is the problem! If the benefits of chemistry are not evident to a scientifically oriented person, we should not be surprised if the average person on the street does not realize that we would be back to the Stone Age without the chemical developments we achieved. In this book, you can read about the vast benefits in the area of energy, transportation, medicine, and food. I hope that adding this short chapter with numerous examples from the areas of communication and entertainment fully convinced you about the importance of chemistry in our life and you will spread it in your friend circles. This should be our challenge to show the non-scientific public what chemistry has provided to their life. Talking to them is the duty of every chemist. No one can leave it to others claiming that it is below their dignity.

PART III
CHEMISTRY AND ACTIVITIES

Chapter 12

Problems and Solutions: Activities of Chemists and Educators for the Public

Attila E. Pavlath

President of the American Chemical Society, 2001

Attila@pavlath.org, AttilaPavlath@yahoo.com

12.1   Public Image of Chemistry

Chemistry is the basis of everything in our life. Without its benefits, we would be back in the Stone Age. Think for a minute and make a list of items you use in 24 h. Include everything: cooking, toiletries, clothing, transportation, medication, and everything else you can think of. Did you find anything on that list that is not the result of chemistry? Even if you just drink rainwater, you have to catch it and store it in some container. Quite difficult to exist without chemistry unless you go back to the Stone Age, is n't it?

12.2   Chemistry Has an Undeserved Public Image

Unfortunately, for the average person, comforts of life such as auto, refrigerator, radio, televisions, or cell phones, which were

Chemistry: Our Past, Present, and Future
Edited by Choon Ho Do and Attila E. Pavlath
Copyright © 2017 Pan Stanford Publishing Pte. Ltd.
ISBN 978-981-4774-08-6 (Hardcover), 978-1-315-22932-4 (eBook)
www.panstanford.com

made possible by chemical developments, are taken for granted. Ironically, they think of the effects of chemistry on our life only when the media reports something that has a possible harmful effect on the environment or health. Is it human nature, or perhaps the faults in human nature, that we are more attentive to catastrophe? A survey conducted at the Indianapolis 500 auto race found that the majority of the attendees came to see some spectacular crash. Movies with spectacular explosions, crashes, and various natural disasters attract many viewers of all ages. Rubbernecking, while driving on the freeway, is frequent, even if it is an accident in the other direction and creates traffic jams. The media play on this human behavior. Occasional problems are reported with big headlines in newspapers and television. Conversely, benefits of chemistry, which makes our daily life more convenient and pleasant, are most often ignored or, at best, reported as short notices. Search your memory and find how many times your hometown or any other newspaper reported in headlines on the front page about some beneficial chemical discovery for our life in contrast to repeated discussions on a possible chemical contamination? Frequently, when the case was a false alarm or an overreaction, the correction is buried in an obscure part of some minor area of interest. The last time when chemistry took front page was after the end of World War II when nylon stockings became available for the general public. While hundreds of women lined up at a department store to buy them on the first day, one attractive young girl decided to try it on immediately while sitting on the sidewalk providing revealing view of her shapely legs. This picture taken by an alert reporter, in the era of today's fashion of miniskirts and bikinis, would not make headlines, but 70 years ago, it was shown on the front page with big letters NYLON IS HERE! How the public perception of chemistry changed since that time is well demonstrated that one of the large chemical companies discontinued its famous slogan "better things for better living through chemistry" because of image problem.

Did you ever think about what would our life be without the many benefits chemistry has provided? Do you think chemistry has received the proper recognition? True, occasionally some chemical developments represent problems. However, nothing is perfect in life and chemistry is no exception! Chemistry, as anything else in life, is not perfect. Can you name any action you have ever made in your life that had zero risk? Every year, 40,000 persons die in automobile accidents, and still millions of people get in their cars every morning, taking the risk as part of their life. Statistically, flying is safer than driving but still involves a risk. However, that does not hold us back from visiting relatives and friends thousands of miles away. Even as a pedestrian crossing the street at a green light, we have a finite chance that a speeding car or a drunken driver hits us. We perform hundreds of other similar routines as parts of our everyday life, which even in our own home can have harmful consequences: There is no zero risk! From the moment we are born, we are exposed to various risks. Yes, chemical developments, as everything else, carry some risks. However, we have to evaluate everything statistically. For any single problem attributed to chemistry, there are hundreds of beneficial developments in our everyday life. Why is it then that the media always focus on problems with large headlines but are mostly silent about the advantages? There is no sinister reason behind it; we have to understand the circumstances in which the media are operating in modern times. In some countries, a dictator government controls them. However, in a free society, the media, both newspaper and television, can survive only if they attract readers, listeners, and advertisements. As mentioned before, the majority of the population is fascinated by spectacular events; the media just cater to their interest. One science editor gave an honest explanation. "In everyday life no news is good news, but the media operate reversely: good news is no news."

Evidently, we must investigate problems caused by accidents, fire, or explosions and take appropriate steps to decrease the frequency of their future occurrence. Some chemicals used in everyday life can have unexpected side effects not known at the time of their discovery. When used in foods or as medicine, the US Food and Drug Administration (or its equivalent in other countries) requires long-range testing to determine possible side effects. However, it must be emphasized over and over that there is no zero risk! The investigations cannot include every possibility. For example, the well-known insecticide DDT eradicated mosquitos in Southeast Asia, savings tens of millions of people from malaria. Its effect on the fragility of bird eggs could not be anticipated.

When you receive a new medicine for your ailment, a long list is provided warning about its possible side effects, even if the possibility might be one out of thousands of uses. The exact human physiology can vary from one person to another; it cannot be predicted with 100% certainty. For example, the harmful effect of Thalomide, which is used for the treatment of multiple myeloma, could not be predicted for pregnant women.

It should be emphasized that when a harmful side effect is discovered, chemistry always swings into action and develops a substitute to minimize or eliminate the side effect. When fluorocarbons, which made modern refrigerators possible and saved countless of persons from food poisoning caused by spoiled foods, were found to interact with the ozone layer, new refrigerants were developed to stop the problem. Similarly, hundreds of medicines or insecticides are created continuously to serve the intended purpose with fewer side effects. Will we ever discover the ideal chemical for any purpose, which has absolutely no possible harmful consequences? Perhaps in the distant future, as predicted in science fiction stories, we will have omnipotent computers, which will be able to design a medication for every individual. Until then, we must inform and convince the public that chemistry is our savior and not our enemy. If we improve the public image of chemistry, we will make great progress toward improving our everyday life.

12.3 Is It Important to Improve the Public Image of Chemistry? Does It Make Any Difference?

The lack of recognition of the many achievements of chemistry improving our life is of major worldwide concern because it has serious effect on the future of chemistry. Any increase or decrease in the public image of chemistry has similar direct effect on the quality of our life. If the public is only informed about the occasional side effects of certain discoveries without giving proper coverage and recognition for the hundreds of benefits, it degrades the image of chemistry. The population will not support chemical developments. Elected members of any legislative body are strongly influenced by public opinion. If the public considers chemistry a menace, politicians will create very restrictive legislations and regulations. Research grants for new discoveries will be cut back. This will delay the discovery of new developments needed for the improvement of our life. Unwarranted rules, regulations, and long bureaucratic processes force the chemical industry to relocate to other less restrictive places. This will result in the export of jobs, or if they stay, the production cost and thus sales price will rise. Negative image also frequently steers young people away from selecting chemistry as profession.

12.4   What Is Needed?

The decline in the image of chemistry created more and more problems for chemical research and developments. Organizations both in the United States and abroad became very concerned about the future of chemistry. It is evident that we must find long-lasting solutions. We might wish for quick Band-Aids to the scars so that we could quickly return to our ivory tower. However, this can only provide temporary, if any, solutions. We cannot behave as ostriches thinking that the problems will go away if we ignore them. Whether we like it or not, we must work on remedying the situation collectively and continuously. The present situation was caused by our own negligence. We ignored the effect of the media: newspapers, radio, television, and social networks such Facebook or Twitter influencing the population. We assumed that the benefits of chemistry are self-evident to everyone. We behaved (and some of us still do) like Archimedes writing in the sand even in his last minutes. He was only concerned with his studies, while the vandals were already breathing down on his neck: *noli turbare circulos meos* (don't disturb my circles). Two thousand years later, the modern Archimedes is depicted by Hollywood movies as an unkempt, absentminded chemist who works day and night in his laboratory frequently even unaware about his own or his family's welfare.

There are many ways to improve the image of chemistry, but none is quick and easy. We are all in this together; we must act now before it is too late. It requires the continuous attention of chemists and chemical societies throughout the world to promote many various activities. Obviously, we have to provide correct and unbiased information to the public. The question is how?

12.5   Science Education

An old philosophical saying states: A child's education starts 20 years before being born. Unfortunately, today's public was not informed adequately about the benefits of chemistry when they attended schools. One of the most important actions must be the improvement of science education in schools, mostly in high schools, but even in lower grades. This can prepare tomorrow's public for the appreciation of chemistry. Chemist parents need to get involved in work of school boards by attending meetings, or hopefully even running for the board. The support of science teachers is very important. We must help them in their work by encouraging them and recognize the effort of outstanding chemistry teachers. Evidently, as any improvement, this also requires financial support, which in turn depends on support from parents to influence politicians and school administrators in this direction.

12.6   Creation of Interest toward Science

Chemistry is very important for the economic growth of a nation. Sometimes we hear statements about not having enough students studying chemistry; therefore, without enough students entering the chemical profession, there will not be sufficient new developments to maintain economic growth. The alarm bells were sounded that sooner or later we will fall behind the developing nations where chemistry is a mandatory subject. Is this true? Is the younger generation less willing to study chemistry today than 50 years ago when I became a chemist? Interestingly, even though we do not teach medicine in high schools, medical schools have no problems for having multiple applicants for their capacity.

There is no question; the present generation is more materialistic than we were 50 years ago. The earning potential of a physician is much higher than that of a chemist. In spite of the higher tuition cost in medical schools, students are more willing to invest in them than in chemistry education. While they are trying to select a challenging intellectual profession, financial benefits are also important for the majority of them! While debate is still going on whether this is true or not, the most important way to address financial and enrollment problems is how we teach chemistry in our schools?

Granted, there will always be some in any generation and in any profession who are first concerned with finances. Very few graduating from medical schools plan to be general practitioners in faraway rural areas. Yet we cannot condemn a whole generation as selfish money-hungry group. Regardless of whether their concern is adequate return for the educational investment, they know what medical science can do. The general public appreciates what physicians can do for their life and respect them, while very few exhibit similar attitude toward physical sciences and their practitioners.

This is where we, chemists, must do important public relation work. Again another philosophical saying states: You can take a horse to the water, but you cannot force it to take a drink. We not only have to improve science education, but we also need to create interest in sciences, especially in chemistry, in the students. For this the first step is getting students' attention, since many consider chemistry as a dull subject. It is not easy to do this with today's generation heavily involved in various social activities. As the first step, 25 years ago, the American Chemical Society started a program to capture their interest: National Chemistry Day.

In the month of October every year, a special day was designated as the National Chemistry Day when, throughout the country, colorful demonstrations were held in public places showing the wonders of chemistry. These were safe and colorful experiments designed to be interesting for both adults and children. Later, this was extended to the National Chemistry Week for the last weeks of October. It created more publicity covered by the media. Seeing the success, other countries have joined this effort.

This can be just the beginning. Colorful demonstrations are entertaining especially for younger children before high school, but many students do not take chemistry in high school, or if it is a mandatory subject, they are glad to be out of the classroom. The first problem is the way we teach chemistry. Unfortunately, we start to instill chemistry in high school students by awing them with atoms and molecules. Not surprisingly, many of them lose interest considering it an esoteric subject. The image of a chemist as stereotyped by Hollywood as an unkempt, sometimes bungling absentminded person comes to their mind. The other image is someone whose work is only causing environmental, food, or health problems.

While spectacular demonstrations will catch the attention of children, teenagers, and parents, they need to be followed with the information that chemistry is more than just experiments for the eyes. They have to show that chemistry can be a rewarding profession because it provides many benefits for our everyday life. This is the most important point for adults but most importantly to students. One of the problems for the reluctance of studying chemistry in high school is that many students are not aware about its benefits and frequently consider it a dull subject. In some countries, schools are experimenting with starting to teach chemistry first by explaining how chemistry benefits our life, before talking about atoms and molecules. This results in more interest. Today's students will be the general public of tomorrow, and this will create a positive attitude toward chemistry and automatic increase in selecting chemistry as a profession. However, regardless of what profession they might choose, they will have a balanced view and will not be taken in by sensationalist newspaper headlines chastising chemistry for occasional problems. Every action we take in our life has the possibility of some undesired effect. We chemists are the ones who when some chemical problems develop immediately swing into action and develop ways to provide the original benefits by eliminating the problems.

Evidently, these demonstrations may convince more students to select chemistry as their profession; however, this is not the purpose. Even if they become lawyers, accountant, auto mechanics, or even politicians, they will have more appreciation for chemistry and they will not be so easily taken by sensationalist media reports blaming chemistry for the problems of our society. The students of today will be the public opinion of tomorrow. If practical results are convincingly shown, the average population will be more activated to take actions. This will improve the public image of chemistry. Politicians will be more willing to support basic and applied research and will be less restrictive with industrial regulations. This, in turn, will encourage industry to increase production, thus creating more jobs. Consequently, the economic status of chemists will improve. While chemists still will not reach the earnings of a plastic surgeon, but even some of the materialistic students will look more favorably to chemistry as profession. At the end, everyone wins and perhaps the old slogan that was discarded because of unfavorable public opinion about chemistry, "Better things for better living through chemistry," will be revived.

12.7   Role of Individual Chemist and Chemical Societies

There is no question that the benefits of chemistry outweigh (by 100 to 1) the problems that might be caused in certain cases. There has to be a worldwide movement to overcome the negative image the media created and attributed to chemistry. Chemical societies can create various events, press releases; however, most of the responsibilities lay on every chemists. We are proud to discuss our discoveries at international, national, and local meetings talking to other chemists. However, this is "preaching to the choir." When the media report some problem, we mope about this and "curse the darkness instead of lighting a candle," so to say. The public does not realize the benefits of our discoveries because we do not talk to non-chemist audiences, small or large, in a layman term. A new analytical method for the quality of asphalt describing the use of a special column in gas–liquid chromatography creates no interest. However, if a press release states that this new way will assure better quality asphalt and will decrease potholes in our roadways, it will capture everyone's interest who use public roads. We must continuously inform/educate the public about the benefits of chemicals and chemical processes, as well as problems that were solved or minimized to sustain and improve our environment (air, water, land, and sea). We must talk about these frequently to friends, neighbors, non-scientific colleagues, talk shows, civic groups, and also induce them to talk to their friends about these.

This is where the responsibility of individual chemists lays regardless of whether they are Nobel Prize winners or entry-level young chemists. We chemists are rightfully proud of our achievements, such as new compounds, structural determination, or a reaction mechanism, and discuss these among us at various conferences. However, we are guilty of not talking about the hundreds of benefits of chemistry in a suitable language to the wider populace whose knowledge of chemistry is rather limited. Their main interest is the effect of chemical discoveries on their everyday life, and not how we achieved it. Reacting hydrogen and nitrogen, which was believed to be inert, at 1200 degrees and 200 atmospheres in the presence of an iron catalyst is a great chemical success, known as the Haber–Bosch synthesis. That the result is ammonia might capture some interest as a household cleaning agent. However, if instead of lecturing about the chemistry involved in this new manufacturing process for ammonia, we simply enlighten our audience that two chemists in some way created for agriculture an artificial fertilizer that resulted in increased food production and decreased cost, they would appreciate chemistry without understanding the process.

While reporting our work in a scientific language at meetings, we forget that the average person is frequently confused, the least to say, by such languages and does not understand the importance of our work. The American Chemical Society created the Chemical Ambassador program, which promotes individual chemists to contact social clubs, schools, and even just non-scientific neighbors to talk about what would be their life without the contributions of chemistry: to describe in layman term what chemists have done, NOT how they have done it.

Naturally, everything depends on whether we can provide the right and most convincing information for these talks. In 2001, the American Chemical Society created the Chemical Technology Milestones Exhibit, which was a major step forward on this road. It summarized what chemistry has done for life for the past 125 years. It was a great success and was presented as an electronic exhibit. Four electronic boards displayed 79 examples in four areas: energy and transportation, information and communication, health and medicine, and food and agriculture. On each of the four boards, the exhibit listed the chronology of the discoveries in the corresponding fields with a pushbutton for each development. Upon activation, it displayed on a TV screen of the corresponding board the details of the discovery's impact on human life. No chemical equations were included. The boards were displayed at the Fall ACS Meeting in Chicago at the 125th anniversary celebration.

The purpose of this exhibit was to increase the public image of chemistry by conveying the importance of the discoveries to the average people without chemical background. The plan was that after the Chicago meeting, it would be moved around in the country to be exhibited in convention centers, museums, and other establishments frequently visited by the general public. However, the bulkiness of the electronic exhibit created transportation difficulties.

This was just the beginning. The Hungarian Chemical Society and the Hungarian ACS International Scientific Chapter came up with the idea of converting the original electronic exhibit to an easily transportable colorful poster exhibit. Prof. Veronika Németh of the University of Szeged (nemethv@chem.u-szeged.hu) volunteered to create it with the help of her graduate student Ms. Nora Rideg, both in English and in Hungarian. This was later converted to a virtual book (www.chemgeneration.com) by the Hungarian and German Chemical Societies. It was made available throughout the world in the form of book, CD, and the Internet. The Korean Chemical Society converted it into a book and distributed 3000 copies in South Korean schools.

This started a worldwide campaign to publicize the importance of chemical developments in our life. The posters made in Power Point presentation could be easily translated to various languages by replacing the original English text, while retaining the colorful artistic background. This was found very useful for the celebration of the International Year of Chemistry in 2011 that many other chemical societies joined the effort. Presently, it has been translated into at least 25 languages and can be freely downloaded from www.chemistryinyourlife.org. In various countries, the exhibits are moved around the country and displayed in schools, museums, convention centers, and other places frequently visited by the population.

Other organizations, even outside the United States, also became more concerned about the problems and are seeking solutions. The problems will not be solved overnight. We have to take many steps continuously to restore chemistry to its deserved place and remove the shadow that was cast on chemistry by undeserved exaggeration of problems.

Nothing is perfect in life and chemistry is no exception. We must be on the alert to minimize or prevent major problems. But it is not enough to be defensive; we must continuously educate the general public about the benefit chemistry did, can, and will provide for our everyday life. The National Chemistry Week brings chemistry in the public spotlight once a year to show the wonders of chemistry. What we have to do is more than showing colorful exhibitions.

This book will be very valuable and can play a highly important role in promoting the public image of chemistry. It could and should also be translated to various languages to provide chemists with an important tool in their talk to the public about the importance of chemistry.

However, this is not the end. While we should be proud of the creation of posters and the books, we cannot sit on our laurel for carrying out a successful, undeniably outstanding job. This is just the beginning! We need a worldwide effort to improve the public image of chemistry. There is a logical next step in this continuous program of educating the public about the benefits of chemistry. Chemistry has been used in everyday life worldwide for centuries, even before we talked about atoms and molecules. There is a vast treasury in the cultural history of many countries, which contains various practical developments in medicine, agriculture, and metallurgy. They were passed on during the time even if by mouth to mouth, though those who used them did not know the scientific explanations of their utility. Nevertheless, these people knew that they were beneficial and working. Whether some superstition and folklore were also involved, it is immaterial. It is evident that chemistry was always an important factor in our life. The medicine men did not know that the bark of willow contained aspirin, but they utilized it as a painkiller. Collecting and organizing the cultural utilization of natural compounds throughout centuries should lay down the basis of another exhibit. This could be more easily viewed and appreciated by everyone regardless of their educational level. This does not mean that those who use only natural (organic?) products do not need chemistry. How would they provide tree barks as a painkiller for 7 billion people? We would have to destroy hundreds of millions of trees. What would be the environmental result? How would they figure to provide food to the ever-increasing world population without artificial fertilizers, insecticides, and pesticides without chemistry? Today less than 5% of the population in the developed countries can have enough organic foods, because the remaining 95% get their food through the benefits of chemistry. Food prices would go up and malnutrition, starvation would occur. Can we exist without chemistry?

The education of the public must be continued by additional new ways. There will always be sensationalist attacks on chemistry for the slightest possible problems whether these are real or not. The best action is counteracting them in a positive way to show what would be our life without chemistry. We should create a Public Image of Chemistry Institute. Its charter should be continuous dissemination of the benefits of chemistry and educating the public. Such an institute can be a worldwide network; we do not have to have an actual building. The Internet gives us the opportunity to disseminate the information provided in these exhibits and book. Such a center must have the following responsibilities:

- Be a watchdog for misrepresentation of chemistry in the media and provide clarification. This would be fortified through a nationwide network where members would report local news stories, good or bad, which needs to be spread or corrected.
- Follow those new discoveries in chemistry that can have direct beneficial effect on our everyday life and explain in layman term how the public will benefit.
- Create similar descriptions on existing inventions for distribution to the media, including radio and television.
- Create a worldwide database of speakers who are ready to talk about the benefits of chemistry to a wide variety of local, national, and worldwide audiences in person and/or through the Internet.
- Be a reliable source both in person and/or for responsible reporters on emerging news stories who want to report facts not sensations.

Will this solve all the problems? Will it eliminate the sensationalist incorrect reports? No! But we cannot wait for the perfect solution. Such philosophy created many of the problems of our profession. Shakespeare wrote 400 years ago: "Our doubts are traitors, and make us lose the good we oft might win by fearing to attempt" and suggested the necessary action: "Our remedies oft in ourselves do lie, which we ascribe to heaven." We have the future in our hands now. Let me quote from the musical I wrote, titled "It's time for a change," the closing lines of a pseudo-Shakespearean monologue:

> "We can make a difference in this world,
> By daring to enter new worlds never dreamed afore,
> The unknown should not make cowards of us all, and
> Lose the opportunity for a better future."

Chapter 13

What Can Chemistry Do for the Future?

Choon Ho Do

Korean Chemical Industry Specialists Association,
30-3 Cheonyeon 1-gil, Geunnam-myeon, Uljin-gun, Gyeongbuk 36332, South Korea

choondo@sunchon.ac.kr

13.1   Introduction

What chemistry has done for us during the past hundreds of years has been described in Part II of this book. It has been the basis of every aspect of our civilization today. Evidently, chemistry can and will do many more things for humanity in the future. It is time to see what chemistry will do in the future.

When we heard words starting with prefixes such as bio-, nano-, smart-, and gene- and words such as stem cells and Hubble telescope, we are already on the doorstep of our future society. When we speak of the future, the following ideas may quickly across our minds quickly: Doris Day's song, "Que Sera Sera (Whatever Will Be, Will Be)," or Annie's song, "Tomorrow," saying, "The sun'll come out tomorrow. Bet your bottom dollar that tomorrow there'll be sun." Or, someone may recall science fiction films such as "Alien," "Star Trek," and "The Martian." Or,

Chemistry: Our Past, Present, and Future
Edited by Choon Ho Do and Attila E. Pavlath
Copyright © 2017 Pan Stanford Publishing Pte. Ltd.
ISBN 978-981-4774-08-6 (Hardcover), 978-1-315-22932-4 (eBook)
www.panstanford.com

some other thoughtful intellectuals may think of The Prophecies by Nostradamus.

Knowing the future in the short and long term is important not only for everyone's own life, job security, or fortune, but also for our society, our nation, and our world; otherwise, we are unsure whether our future is secure, safe, or prosperous or sustainable. Nowadays, we are predicting, imaging, designing, preparing, and simulating the future. We invest our efforts, energy, money and time personally, socially, nationally and globally to make our future better. Therefore, contemplation on the future is not just an activity of curiosity seekers but is a necessary and real occupation of industries, business, and societies. We frequently read and watch feature articles and news items on the future from broadcasting stations, newspapers, magazines and scientific societies, such as BBC, CNN, *Scientific American*, the *New York Times* and numerous other media, such as the International Union of Pure and Applied Chemistry (IUPAC), the American Chemical Society (ACS), the Royal Society of Chemistry (RSC), and the German Chemical Society (GDCh).

There are many changing factors that make difficult for us to predict the future. They come not only from scientific discoveries and inventions but also from social changes and from Mother Nature, including climate change, volcanoes, earthquakes, tsunamis, etc.

In contrast to a bright and rosy future, there are many predictions of a dark future. These warnings about the future are also helpful for us to be able to prevent or avoid disasters. Although there are numerous references to authors writing about the future, we would like to mention a few here. George Orwell's novel *1984*, published in 1949, predicted a dreadful future with the dictatorship and surveillance of the Big Brother. Although the year 1984 is now past "future," it gives us the meaning of "Big Brother" and privacy. In the late eighteenth century, Thomas R. Malthus predicted in his book *An Essay on the Principle of Population* that the population would increase faster than we could provide food on Earth. He warned us of overpopulation, the shortage of food and consequent famine. In 1962, Rachel Carson described the detrimental effect of indiscriminate use of pesticides on the environmental effects in "Silent Spring." There is a flood of warnings from the United Nations and various organizations and many concerned scientists and scientific societies on climate change. Best-selling author Alan Weisman investigated the chances for humanity's future in his 2013 book *Countdown: Our Last, Best Hope for Future on Earth*.

Naomi Oreskes and Erik M. Conway co-authored the book *The Collapse of Western Civilization: A View from the Future* in 2014. On social change, Fareed Zakaria, the host of CNN's *Fareed Zakaria GPS* wrote the book *The Post-American World* in 2008. Klaus Schwab, the founder of the World Economic Forum (WEF), predicted the upcoming technological revolution that will fundamentally alter the way we live and work in his 2016 book *The Fourth Industrial Revolution* [1].

On chemistry's role for the future, George Whitesides, professor of chemistry at Harvard University wrote an essay entitled "Reinventing Chemistry: The End of One Era and the Beginning of Another" and suggested a new role and new directions of chemistry to solve the problems of society [2]. Ronald Breslow, ACS president in 1996 emphasized that chemistry must continue to broaden fields and be a creative and useful science through his article in C&EN [3].

Before we discuss our own views on the future, let us see the views of two major intergovernmental organizations: a business point of view from the World Economic Forum (WEF) and a global point of view from the United Nations (UN).

The WEF published in 2015 the top 10 emerging technologies that will shape the agenda of future industries: (1) fuel cell vehicles, (2) next-generation robotics, (3) recyclable thermoset plastics, (4) precise genetic-engineering techniques, (5) additive manufacturing, (6) emergent artificial intelligence, (7) distributed manufacturing, (8) "sense and avoid" drones, (9) neuromorphic technology, and (10) digital genome.

It is noteworthy to see the works of the United Nations for the global future because the United Nations is an organization where experts from many countries discuss and cooperate on the global issues intersecting with local and national interests. The UN General Assembly's 69th session in 2014 adopted 17.

Sustainable Development Goals (SDGs) to ensure happiness and well-being for all human beings and to deal with poverty, education, gender equality, health and environmental sustainability (see Table 13.1) [4, 5]. These goals are based on human needs, and many of them have material aspects where chemistry is very much involved. The UN Millennium Project Team published a "2015–16 State of the Future" report [6], which described and predicted the progress of current technological trends up to 2050.

Table 13.1   Sustainable development goals

Goal 1	End poverty in all its forms everywhere
Goal 2	End hunger, achieve food security and improved nutrition and promote sustainable agriculture
Goal 3	Ensure healthy lives and promote well-being for all at all ages
Goal 4	Ensure inclusive and equitable quality education and promote lifelong learning opportunities for all
Goal 5	Achieve gender equality and empower all women and girls
Goal 6	Ensure availability and sustainable management of water and sanitation for all
Goal 7	Ensure access to affordable, reliable, sustainable and modern energy for all
Goal 8	Promote sustained, inclusive and sustainable economic growth, full and productive employment and decent work for all
Goal 9	Build resilient infrastructure, promote inclusive and sustainable industrialization and foster innovation
Goal 10	Reduce inequality within and among countries
Goal 11	Make cities and human settlements inclusive, safe, resilient and sustainable
Goal 12	Ensure sustainable consumption and production patterns
Goal 13	Take urgent action to combat climate change and its impacts*
Goal 14	Conserve and sustainably use the oceans, seas and marine resources for sustainable development
Goal 15	Protect, restore and promote sustainable use of terrestrial ecosystems, sustainably manage forests, combat desertification, and halt and reverse land degradation and halt biodiversity loss
Goal 16	Promote peaceful and inclusive societies for sustainable development, provide access to justice for all and build effective, accountable and inclusive institutions at all levels
Goal 17	Strengthen the means of implementation and revitalize the global partnership for sustainable development

*Acknowledging that the United Nations Framework Convention on Climate Change is the primary international, intergovernmental forum for negotiating the global response to climate change.

Out of the 17 goals, more than a half of the goals, 2, 3, 6, 7, 9, and 11–15 are related to environmental issues and are directly related to chemistry. The rest of the goals are also directly and indirectly related to chemistry. It means that chemistry is really necessary tool to achieve the UN's goals for all human beings.

From the viewpoints of these two important organizations, we can grasp the images and issues of our future. Let's move forward.

13.2   The Future May Come Differently

Although we usually project our future in linear progress, the future may come quite differently than we expect. For instance, if we look at the development of lighting equipment in our houses, we will realize that the lighting equipment and methods are not developed in a linear pattern and their lighting principles are quite different.

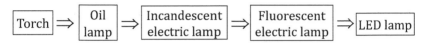

In the development of photographs, we can see a similar pattern. The most recent CMOS (complementary metal-oxide semiconductor) image sensor in digital cameras is completely different from a photographic glass plate in terms of imaging principles.

If we look at a television, we notice that the change and progress are not linear, either. The first generation of television sets were operated by CRTs (cathode-ray tubes), but current ones are LCD (Liquid-crystal-display) and OLED (organic light-emitting diode) televisions.

Development of vehicles started from animal-drawn vehicles, followed by machine-drawn vehicles, called automobiles. In the early stages, automobiles had steam engines, but now there are internal combustion engines and in the future, electric motors might become common.

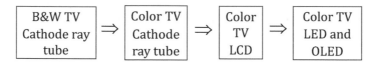

Thus, we realize that the future will not be simply an outcome of linear development and progress based on current materials and principles. This is one of the reasons why the prediction of future is complex and the future cannot be extrapolated from the current situation. Therefore, we need to continue to seek diverse directions to achieve new developments for our future.

Now, readers may try to predict their own projection for the future. Put any present item or idea in box (B) and previous version in box (A). Then, put your future projection in box (C). It may be a starting point for readers to start their own future business or research to develop their projection. It may need your insights and inspiration.

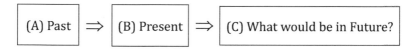

13.3 Future Issues

Foods, energy, environment, water, space exploration, health and disease, resources, sustainability and future technology are common key areas for our future as envisaged by many concerned experts in the introduction section. We will try to add new aspects and observations on the following subjects from the point of chemistry:

- Artificial Photosynthesis
- New energy sources
- Materials
- Climate control
- Human life and health
- Humans and nature
- Changes in deserts

- Use of oceans and seas
- Space travel

13.3.1 Artificial Photosynthesis

All living creatures depend on sunlight, which not only enables us to see but also activates everything in the world. All living things on Earth depend on the photosynthesis of plants and algae for respiration and nutrients for their survival [7]. Photosynthesis produces oxygen gas and nutrients by chemical reactions of carbon dioxide with water using solar energy (Fig. 13.1).

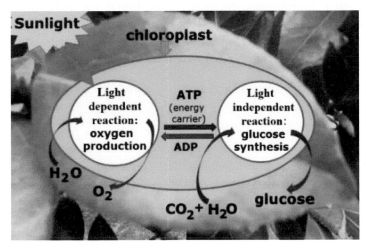

Figure 13.1 Photosynthetic steps in a chloroplast.

In the chloroplast of the cells of green plants, glucose and oxygen are synthesized (Eq. 13.1).

$$6CO_2 + 6H_2O \xrightarrow{\text{Sunlight}} C_6H_{12}O_6 + 6O_2 \qquad (13.1)$$

(six carbon dioxide molecules) (six water molecules)   (one glucose molecule) (six oxygen molecule)

This is an amazing chemical reaction that produces food (in the form of glucose) synthesized from carbon dioxide and water. In this reaction, carbon dioxide (264 kg) and water (108 kg) are converted to glucose (180 kg) and oxygen gas (192 kg) by using solar energy. Photosynthesis is a circulating system between animals and plants that exchanges byproducts of their life, carbon dioxide and oxygen, with the help of sunlight. If we can produce glucose and starch in laboratories and chemical plants using carbon dioxide and water, we will solve the problem of food shortage and reduce carbon dioxide gas, which causes global warming. Carbon dioxide will be a useful resource and not a source of problems. However, scientists are not yet successful in reproducing this reaction in laboratories.

In order to artificially duplicate photosynthesis in a laboratory, we need to know what has been found until now and what has to be further investigated. Scientists have found the mechanisms of photosynthesis in plant leaves. Photosynthesis occurs in the chloroplast of plants. The reactions are divided into two steps: light-dependent and light-independent reactions.

(1) Light-dependent reaction

When chlorophylls in the chloroplast of plant cells absorb sunlight, they are excited and transfer electrons through several sophisticated steps, eventually splitting the water molecule and producing ATP (adenosine triphosphate) and oxygen gas using $NADP^+$ (nicotineamide adenosine diphosphate) and ADP (adenosine diphosphate). We call it as a light-dependent reaction. ATP is an energy carrier, which is used to produce glucose in the next step, light independent reaction.

(2) Light-independent reaction

Carbon dioxide and water are transformed into glucose using the hydrogen ions and electrons and the energy of ATP formed in the light-dependent step in the light-independent reaction.

Artificial Leaf: Although scientists still are not able to reproduce the exact photosynthetic reaction because they are not able to duplicate the molecular structures of complex enzymes and chloroplast, scientists can now produce sources of fuel by mimicking the photosynthetic steps using an "artificial leaf " or a "bionic leaf " made of a thin device coated with various catalytic materials (Fig. 13.2).

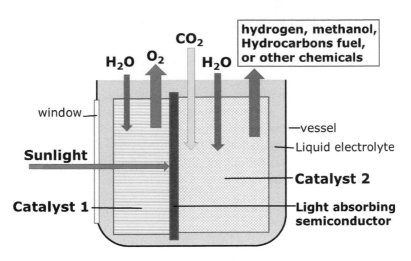

Figure 13.2 A schematic diagram of an artificial leaf.

Figure 13.2 shows a schematic diagram of an artificial leaf mimicking the processes of the photosynthesis in chloroplast of the plant cells shown in Fig. 13.1. Sunlight excites electrons in the semiconductor and the produced electrons are transferred to catalysts to split water molecules producing to produce oxygen gases and hydrogen (Eq. 13.2), hydrocarbon (Eq. 13.3), and methanol (Eq. 13.4) [8, 9]:

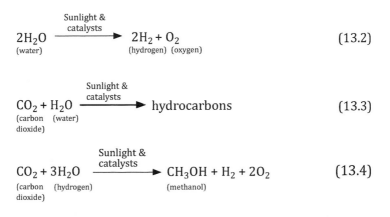

$$2H_2O \xrightarrow{\text{Sunlight \& catalysts}} 2H_2 + O_2 \qquad (13.2)$$
(water)   (hydrogen) (oxygen)

$$CO_2 + H_2O \xrightarrow{\text{Sunlight \& catalysts}} \text{hydrocarbons} \qquad (13.3)$$
(carbon   (water)
dioxide)

$$CO_2 + 3H_2O \xrightarrow{\text{Sunlight \& catalysts}} CH_3OH + H_2 + 2O_2 \qquad (13.4)$$
(carbon   (hydrogen)   (methanol)
dioxide)

The development of the light-absorbing semiconductors, specific and efficient catalysts and their matrix structures equivalent to that of the chloroplast in plant cells are essential for the economically plausible development of an artificial leaf.

Conversion of cellulose to starch: Plants produce both starch and cellulose by photosynthesis. The starting material for both is the same, glucose produced by photosynthesis. Later, somehow, plants make cellulose and starch in different ratios according to their growing periods via participation of specific enzymes under certain temperature and other conditions. If the conditions, the enzymes, and the mechanisms are identified, we can have plants produce more starch than cellulose during photosynthesis. Furthermore, we will be able to convert cellulose to starch if we understand the mechanism how they were weaved. Actually, scientists tried and succeeded. However, the processes were too expensive because the currently enzyme preparation is very costly. We are looking for better processes. We could probably solve the food problem forever if we could transform cellulose into starch. Since this transforming reaction is also a biochemical one, the chemist will eventually determine the process and this vision will turn into reality.

On December 17, 1953, the first flight of the Wright Brothers was only for 59 s and the flight distance was 256 m. Airplanes can fly a thousand miles an hour around world nowadays. Thus, the progress of scientific development is unbelievably fast and most of scientific goals are achieved. Scientists, including chemists, will hopefully discover the exact mechanism of photosynthesis at the molecular level in chemical terms and will be able to make starch for food and methanol and hydrocarbon for fuel economically from carbon dioxide and water using more advanced artificial leaf in the future.

13.3.2   New Energy Sources: Generation and Saving

Energy is one of the most important issues because every one of our activities requires energy: lighting, cooking, transportation, industries, agriculture, heating, cooling, etc. Biomass was the main fuel before the Industrial Revolution. After the Industrial Revolution, coal became an important fuel. After the invention of automobiles, oil also became a major fuel. Natural gas and hydropower were developed at the beginning of the 20th century. Nuclear power came on the scene in the 1950s. The global energy consumption is growing exponentially since the beginning of the 20th century. This is due to the increased use of automobiles, global population growth, and improvement in living conditions.

Energy demand will increase continuously in the 21st century due to the, improvement of living standards and the population expansion in the non-OECD developing countries. The past world energy consumption from 1990 and the projection to 2040 reported by the U.S. Energy Information Administration in 2016 is shown in Fig. 13.3 [10]. The report projects 48% increase of energy consumption from 549 quadrillion British thermal units (Btu) in 2012 to 815 quadrillion Btu in 2040 (note: One QBTU is equal to 172 million barrels of crude oil). During the same period, the energy demand in OECD countries will increase only by 18%, while non-OECD countries energy demand will increase by 71%.

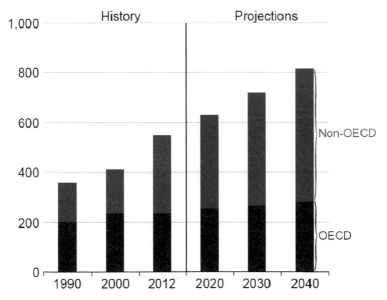

Figure 13.3 World energy consumption, 1990–2040 (quadrillion Btu) [10]. *Source*: https://www.eia.gov/outlooks/ieo/pdf/0484(2016). pdf.

Although fossil fuels, coal, and petroleum are major energy sources, we also realize that these fuels are limited and cause severe environmental problems. Burning of fossil fuels produces carbon dioxide, which is a major global greenhouse gas and one of the sources of acid rain. Consequently, we seek to develop renewable, clean, and carbon-free energy sources such as solar, geothermal, wind, and hydroelectric energy, which do not produce carbon dioxide.

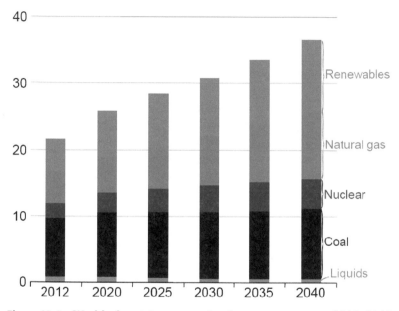

Figure 13.4 World electricity generation by energy source, 2012–2040 (trillion KWh) [10]. *Source*: https://www.eia.gov/outlooks/ieo/ pdf/0484(2016).pdf.

Electricity energy demand is growing fast because of home appliances, electronic devices, and commercial and public services, including transportation, schools, hospitals, offices, and shopping malls. Figure 13.4 shows the comparison of fuels used for electricity generation. Electricity consumption of the world increase 1.9% per year from 2012 to 2040 according to the projection. Although coal provides 40% of energy for electricity generation in 2012, its portion will be reduced to 29% in 2040. Because liquid fuel (petroleum) is expensive, the contribution of petroleum fuel to the generation of electricity will be reduced to

2% in 2040 from 5% in 2012. Renewable generation of electricity by hydropower, sun and wind increases from 22% in 2012 to 29% in 2040. Electricity generation by nuclear power will increase from 2.3 trillion kWh in 2012 to 4.5 trillion kWh in 2040. Nuclear power is becoming a major energy source in many countries because it does not emit greenhouse gas, carbon dioxide gas. Currently, nuclear power plants supply energy in more than 30 countries.

Efficient use of energy and saving energy are also very important in terms of saving resources and reduction of carbon dioxide. Future energy research will focus on (1) new energy sources replacing fossil fuels and (2) energy saving.

(1) New Energy Sources

Solar energy: The sun has been an object of human worship since ancient times. Solar energy is one of the most favorable energy sources. During cold winters, bringing solar heat into the house is very desirable. This is why the people in the northern hemisphere build houses in such a way that a large section of the house faces south. Earlier, solar collectors were actively utilized for harvesting solar energy, which collect heat by absorbing sunlight using mirrors. Solar collectors are employed to heat water and make steam for generating electricity. Currently, we focus more on solar cells than solar collectors.

Recently solar cells have become the most favored device for generating electricity. Solar cells are already applied in many areas: watches, spaceship, electricity for the home, etc. The basic principle is simple as shown in Fig. 13.5. A solar cell device is created if electrical current flows when a material exhibiting the photovoltaic effect receives sunlight.

The main issue is to find sunlight-absorbing materials exhibit the photovoltaic effect because ordinary materials such as plastic, glass, metal and wood do not do so. Although most solar cells currently used are based on silicon, other new types of solar cells have been developed: dye-sensitized, organic, perovskite, polymer, quantum dot, and thin-film flexible solar cells [11]. An example of a flexible solar cell is shown in Fig. 13.6. New types of solar cells with higher than 50% efficiency, and are also stable, have longer life times and come with lower cost will be commercially available in the near future. The US Energy

Information administration (EIA) predicted in 2015 that more stable and cheaper new types of solar cells will be commercially available in the near future with an efficiency of 50% and a longer lifetime [12].

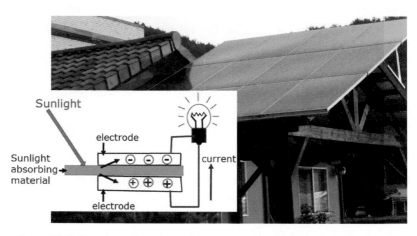

Figure 13.5 Principle of a solar cell.

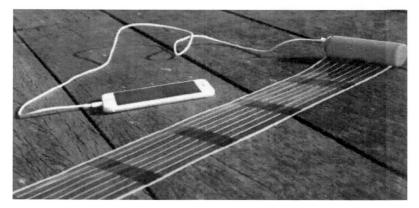

Figure 13.6 Unfurled panel of a solar cell. It can be rolled back for storage. *Source*: *C&EN*, 2016, May 2, pp. 30–35; http://richardsonroofing.us/2016/05/02/the-future-of-low-cost-solar-cells-the-biological-scene/.

Wind Power: Wind power is plentiful and available around the world (Fig. 13.7). Wind power capacity has expanded rapidly, and wind energy production was around 4% of the total worldwide electricity usage in 2014, and growing rapidly [13, 14].

If this pace of growth continues, it is predicted that one-third of the world's electricity will be generated by wind power by 2050 [15].

Other renewable energy resources: Hydroelectric power, geothermal energy, tidal power and wind power have a long history. Although these energy sources are desirable in view of environmental concerns, as of 2013, these renewable energy resources supply only 1.2% of total global energy need (Fig. 13.4). We expect further development of these renewable energy sources will be increased by many times. Hydroelectric power needs the construction of dams to confine water. In spite of environmental issues occasioned by dams, the construction of dams will help to conserve water. Water shortage is another big issue now and for the future. Geothermal energy and tidal power will also be developed and their usage will increase.

Figure 13.7 Wind power turbines.

Nuclear Fission energy: Nuclear fission energy using uranium is already a main energy sources for some developed countries such as France. The development of new nuclear power plants will follow two trends:

(1) Small nuclear power reactors, commonly referred to as small modular reactors (SMRs), of less than 300 MW capacity

based on uranium fuel will be developed, replacing many coal-burning power plants. SMR is an assembly of all components for nuclear power generation such as nuclear cores, cooling, steam turbine, generator and safety system in a steel cylinder reactor. SMRs have the advantage of low capital cost, short construction time, reduced siting cost, higher safety, lower requirement of cooling water, suitability for remote regions, and applicability for specific purposes and in situ decommissioning [16].

(2) Thorium-based nuclear reactors will be the main future nuclear energy source, replacing uranium-based nuclear reactors in the future. Thorium is a radioactive actinide metal element with atomic number 90 and atomic weight (mass) 233. It is a fertile element (Note: A fertile element is an element that is not itself *fissionable* can be converted into a fissile element by neutron absorptions). Thorium-based nuclear reactors are currently being developed in the United States, India, and several other countries [17].

Thermonuclear fusion energy: We would like to produce energy in the same way as the sun. The sun produces energy by the fusion of hydrogen nuclei: First, hydrogen and hydrogen fuse to form deuterium (D) and release fusion energy. Then deuterium and deuterium fuse either to form tritium (T) or helium (He). The reaction between deuterium and tritium is shown in Eq. 13.5. All these reactions release enormous fusion energy according to Einstein's equation, $E = mc^2$.

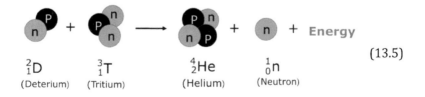

$$^2_1D \quad\quad ^3_1T \quad\quad\quad ^4_2He \quad\quad ^1_0n$$
$$\text{(Deterium)} \quad \text{(Tritium)} \quad\quad \text{(Helium)} \quad \text{(Neutron)}$$

$$(13.5)$$

Although we have succeeded in making a hydrogen bomb, we are far from harnessing nuclear fusion energy peacefully because of the difficulty of obtaining and confining an extremely high temperature for the fusion reaction. The fusion reaction occurs at a very high temperature of 15 million degrees Celsius in the core of the sun. The first International Thermonuclear Experimental Reactor (ITER) sponsored by the United States, Russia, the European Union, Japan, China, Korea and India will be built by 2025 at a southern France location [18]. Although a commercial thermonuclear fusion reactor may not be plausible in near future, we will continue to investigate and pursue this goal.

(2) Saving energy

Energy saving can be achieved by increasing fuel efficiency and reducing loss of energy from every energy consumption sector. Figure 13.8 shows projected four total energy consumption sectors, buildings and agriculture, transport, industry, and non-energy use in 2040 according to the 450 scenario based on policies needed to stabilize the long-term concentration of global greenhouse gases at 450 ppm CO_2-equivalent and to limit global average temperature increase to 2°C [10]. "Non-energy use" in the figure indicates those fuels that are used as raw materials in different sectors and petrochemical feedstocks. Various new smart energy-saving technologies applying to the energy consumption sectors will reduce the construction of many power plants.

LED lamps: Light-emitting diode (LED) lamps will consume only about 10% of the energy that an incandescent lamp and less than fluorescent lamps. Replacing old lamps with LED lamps in lighting will save huge amounts of energy because lighting consumes a lot of energy. The US Department of Energy estimates that adoption of LED lighting over the next 20 years in the United States could save about $265 billion in energy costs and that the savings in electricity is equivalent to the construction of 40 new power plants [19, 20].

Smart production processes: Factories can reduce their energy consumption dramatically by improving the production processes and smart control. Green chemistry can play a significant role. Because the portion of energy consumed by industry is 31.4% of the total energy consumption, as shown in Fig. 13.8, the savings achieved by smart production processes will be significant.

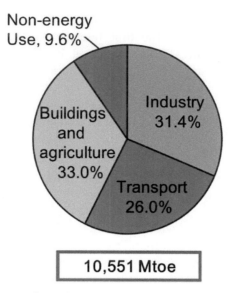

Figure 13.8 Projected total energy consumption by sector in 2040. 450 scenario [10]: https://www.iea.org/publications/freepublications/publication/KeyWorld_Statistics_2016.pdf.

Advanced insulating materials: Developing insulation materials for housing and electrical appliances will reduce consumption of heating and cooling energy. Even now, passive houses consuming almost zero energy are being built as examples. In the future, much less energy will be required for heating houses than at present.

Lighter materials: Lighter automobiles and improving fuel efficiency will reduce energy consumption greatly. The current consumption of energy for transportation is about one-fourth of the global energy consumption (26% in 2040 as shown in Fig. 13.8). Therefore, reduction of fuel in automobiles will contribute greatly to saving energy.

Reclaiming materials: Recycling scrap aluminum consumes only 5% of the energy to produce aluminum from ores. Likewise, recycling of paper products such as newspapers, and cardboards will saves energy, the environment, and trees. In the future, most products will be designed and produced in a way that they will be easily to be recyclable in other to save energy and the environment.

Materials requiring less energy: New materials demanding less energy during their service life, such as wash-and-wear fabrics, lightweight automobiles, permanently lubricated bearings, will also save energy.

13.3.3 Materials

Progress and societal developments always go together with discoveries of new materials based on the advancement of sciences and technologies. After the Stone Age, the advent of new materials such as steel, copper, aluminum and other lighter and stronger alloys, concrete, glass, plastics, rubbers, and synthetic fibers have made better forms of dwelling (houses and buildings), transportation (roads and highways, cars, airplanes and ships), and garments possible. New materials for information technology, entertainment, art, music, farming, medicine, and cosmetics have changed our life, culture and minds.

Advanced materials such as nanotubes and graphene will be assiduously pursued in the future. The carbon nanotube is a special form of carbon fiber, and it is stronger than ordinary carbon fiber, which is stronger than steel fibers. Chemistry and engineering play major roles in the development of new advanced materials for general and for specific functional purposes. Adoption of atomic and molecular level concepts of chemistry in the development of new materials will add novel properties.

Biomaterials and biodegradable plastics: Biomaterials without foreign-body rejection for artificial organs will be persistently developed and improved. Biomimetic materials using various amino acids will be developed for a variety of applications.

Plastics waste has two big problems: It does not degrade quickly, and it is difficult to recycle economically.

Examples are plastics in landfills and fishing nets and plastic debris in the oceans. Plastic waste problems can be solved in three different ways: (i) recycling, (ii) replacement with biodegradables, and (iii) use as fuel. A new environmentally friendly plastic whose biodegradation rate can be tailored so that it degrades quickly after its service life ends is under development. Application of natural polymers for plastics will be enlarged greatly to replace synthetic plastics to solve environmental problems.

Figure 13.9   Carbon nanotube model.

Batteries: Batteries are used everywhere. The advent of a variety of electric gadgets has prompted the development of new types of batteries. Currently, lead, cadmium, and lithium ion batteries have been developed and used. Emerging new electric cars have accelerated the development of batteries. Economic and safe batteries with long life, fast rechargeability, high energy, high capacity, small size, high capacity and small size will be introduced in the market. New batteries will reduce carbon dioxide emission too. Chemistry and engineering will be the main forces for the development of new types of batteries because batteries operate by chemical principles and are made of chemical materials. Advancement of new battery technology will revolutionize transportation. Most passenger cars will be run by batteries in the near future.

Catalysts and synthetic enzymes: Most industrial chemical reactions employ various catalysts to improve yields, shorten reactions and reduce byproducts and obtain products economically and safely. Improvement of catalysts will be continued. Application of enzymes and enzyme-mimetic catalysts in chemical reactions will be increased and promoted. This initiative will open new synthetic routes for the discoveries of new drugs and materials.

Reusable materials: Instead of petroleum-derived materials, household items can be made using renewable raw materials with lower energy requirement and longer life. Recycling technology will be developed to recycle most of the materials used. Materials will be bar-coded according to composition and separation for recycling. Mining for ores will be greatly reduced.

Fibers and clothing: Clothing can be made from various functional novel fibers such as nanotubes (Fig. 13.9). Functional clothing capable of sensing, receiving, and transmitting information, generating electricity, and yet protective, washable, comfortable, and lightweight will be popular for casual wear.

Sensors and detectors: Diverse chemical-sensing materials for safety and security will be developed. Sensing and detecting dangerous materials, gases, chemicals, and environmental hazards are very important. Sensors capable of detecting very small changes on the nano- and pico-scale will be developed for the safety and security of humans. Multipurpose detectors, micro and nano-scale detectors, sensors for various blood components and diseases such as Alzheimer's disease will also appear. Various biosensors for animals and plants will also be developed to detect their diseases and monitoring growth. Early warning detectors for natural disasters, earthquake activities, volcanoes, climate change and ozone holes growth will be developed.

Computer chips: Currently, silicon chips are approaching their limitation for further development. New materials for computation based on molecules similar to DNA and RNA, such as DNA-chips, will be developed. These technologies will make chips stable under severe conditions and will be available for permanent storage devices. Quantum computation materials for superfast computers will be developed for rapid calculation.

Lighter and safer transportation vehicles: New lighter, stronger, and safer materials for transportation vehicles will be continuously developed and improved further. As mentioned in the energy saving section, this will save energy and contribute to reduction in carbon dioxide emissions.

Materials for extreme conditions: Materials sustainable under extreme conditions, such as super-hot and cold temperature and cosmic rays for space travel, will be developed. Sealing and thermal insulating materials in extreme conditions will also be developed for space travel.

Advanced insulating materials: Advanced thermal insulating, **comfortable**, **economic,** and **ecological** materials for passive houses will be developed as mentioned in the energy saving section.

Electricity-conducting plastics and magnetic plastics: Conducting plastics and magnetic plastics will be developed for light-weight motors and other applications. Since there are no commercial products made of magnetic plastics and conductive plastics, research on these plastics will be continued.

13.3.4 Climate Control

Global warming and climate change due to the increase of carbon dioxide concentration in the air are the most significant challenges of the 21st century. Earth is dynamic and changes continuously. The atmospheric temperature of Earth depends on all these dynamic phenomena.

Climate change negatively affects the ecosystem of animals and plants and leads to the possible extinction of many animals and plants on Earth and under the sea [21, 22]. When the concentration of carbon dioxide in seawater increases due to the increase of carbon dioxide in the air, oceans become acidified. Ocean acidification leads to ocean calcification and eventually ocean desertification.

There were five extinction events when a widespread and rapid decrease in the number of species and living creatures happened on Earth occurred. However, if the sixth extinction takes place because of climate change, it will include us. That is why we should prevent climate change as quickly as possible. We need to remember this. Human beings have always overcome various difficulties and crises so far.

The concentration of carbon dioxide is increasing relentlessly, and the current concentration is over 400 ppm according to the data of the Mauna Loa Observatory, NOAA, Hawaii, United States (Fig. 13.10). If the current rate of carbon dioxide rise is not checked, the atmospheric temperature will rise by 4.5°C above late-19th century levels of the pre-Industrial Revolution by the end of the 21st century and will result in disastrous effects. In 2015, more than 150 nations agreed to reduce carbon dioxide the emission and to maintain the temperature increase within 2°C until the end of 21st century at the United Nations Climate Change Conference held in Paris.

We can reduce the concentration of carbon dioxide in the air in four ways: (i) reduction in fossil fuel usage as an energy source, (ii) tree planting, (iii) utilization of carbon dioxide as a resource, and (iv) capturing carbon dioxide and storing it underground.

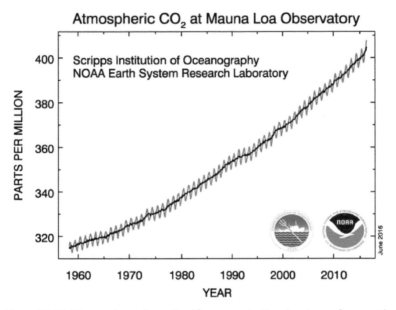

Figure 13.10 Change in carbon dioxide concentration in atmosphere with time. *Source*: https://commons.wikimedia.org/wiki/File:The_Keeling_Curve.png.

Reduction of fossil fuel usage: More than 80% of the energy we currently use comes from fossil fuels. Thus, the best and most important way of reducing carbon dioxide emission is reduction in burning fossil fuels. In addition, replacement with carbon-free clean renewable energy sources, as described in previous sections, and saving energy are essential activities.

Planting trees: Planting trees on Earth will be important and necessary. It seems to be an easy and economical way to reduce carbon dioxide emission and also assists in cleaning the air and preventing water shortage. However, we have heard regular news about the devastation of tropical rain forests and urban development and the destruction of forests. This is a paradoxical human behavior; humans want to reduce carbon dioxide but at the same time they destroy a carbon dioxide absorbing ecosystem.

Usage of carbon dioxide as a resource: Carbon dioxide can be a feedstock for producing chemicals, fuels, and materials [8, 9]. Chemists have learned about the "chemistry of carbon dioxide" and have employed carbon dioxide as a reactant in many reactions in small quantities. Carbon capture and usage (CCU) is a new theme to use carbon dioxide, and chemistry and engineering will play major roles in converting carbon dioxide into useful materials or other resources (Fig. 13.11). Many novel methods will be developed. To name but a few of the current efforts:

(1) Photosynthesis to make carbohydrates (Eq. 13.1)
(2) Methanol synthesis from carbon dioxide: carbon dioxide can be hydrogenated to methanol (Eq. 13.4)
(3) Hydrocarbons from carbon dioxide and water and sunlight (Eq. 13.3)

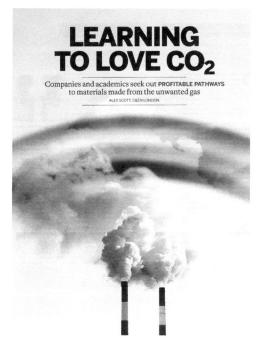

Figure 13.11 Learning to love CO_2 (*C&EN*, November 16, 2015, p. 10.)

Capturing CO_2 and storing it underground: Although it may have limited application, carbon capture, and sequestration (CCS) is another attempt to reduce atmospheric carbon dioxide.

After capturing carbon dioxide gas from smokestacks, it is pumped deep underground in a gaseous or liquid form and kept there to prevent it from escaping into the atmosphere. It is found that basalts, volcanic rocks, react with carbon dioxide and form calcite when carbon dioxide is pumped with water [23]. Because calcite is a mineral rock, carbon dioxide can be sequestered permanently. This new technology can be applied where basalts and water are abundant.

13.3.5   Human Life and Health

The contribution of chemistry in the future will be most pronounced in the medical sector. Aging and getting ill are inevitable but important fundamental issues for us. We do not know exactly why we age, although there are several theories. Typical aging phenomena are wrinkles, osteoporosis, presbyopia, diabetes, arthritis, high blood pressure, grey hair, baldness, senile dementia, and many other age-related phenomena.

Humans have many diseases. Diseases are either caused by internally dysfunctional organs or by external infection. Infectious diseases come from pathogenic agents such as viruses, bacteria, fungi, protozoa, multicellular organisms and prions. Even though some diseases have been known for a long time, such as Hansen' disease, syphilis, typhus, smallpox, tuberculosis, Spanish influenza, cholera, and malaria, they are yet to be eradicated completely. New epidemic diseases such as HIV/AIDS, ebola, avian influenza, and microcephaly caused by Zika virus carried by mosquitoes are becoming serious health issues.

We have been trying to find cures for diseases since ancient times. Instead of curing, some treatments simply relieve the symptoms. Although medical science and new medicines have made great strides, progress is slow because the problems are very complex. Currently, medical and medicinal scientists have started to understand the causes of diseases at the molecular level. Fundamental cures require knowledge of the principles of chemistry embedded in the diseases.

Recent studies have revealed that gene mismatch occurs more than thousand times in our body every day and most of the mismatches are repaired. Cancers may develop randomly in a body when the gene mismatches are not repaired. However, by comprehensive understanding of new principles and discoveries, cures of these diseases can be discovered sooner.

In our body, the brain and heart are very important organs, and consequently they have been extensively studied. As shown in Fig. 13.12, like a MRI instrument, medical diagnostic tools have been developed so that diagnosis of brain diseases becomes more accurate than the diagnosis based on symptoms (see also Fig. 7.18 in Chapter 7). Molecular- and cellular-level studies on brain and the nervous system will shed more light on the core mechanism of the diseases related to them. Consequently, the findings from these studies will be able to cure diseases of the five senses: sight, hearing, taste, smell, and touch. In the future, brain diseases such as stroke, Alzheimer's disease, dementia, Parkinson's disease, and Huntington's disease, will be cured just like the hardware of a personal computer is fixed and reprogrammed.

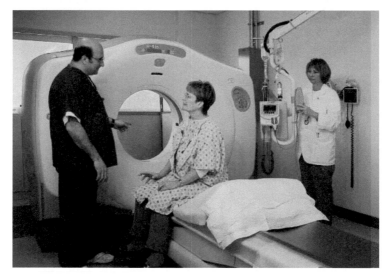

Figure 13.12 MRI instrument and medical diagnosis.

Heart attack and cardiac arrest patients will be fully recovered just like a car is jumped when a battery is drained. However, not like the mechanical parts described earlier, heart and brain malfunctions could cause problems in other parts of the body when the heart stops pumping, blood begins to coagulate, and oxygen cannot be supplied to cells, leading to death. To prevent these disastrous consequences, an emergency kit similar to the wrist pulse-oximeter (Fig. 13.13) will be developed. When the kit's sensor detects irregular or no pulses, the kit immediately injects a first-aid medication into a vein automatically. At the same time, the kit will make a call to the hospital reporting the medical conditions. These actions will extend the critical time for a patient to receive full medical attentions.

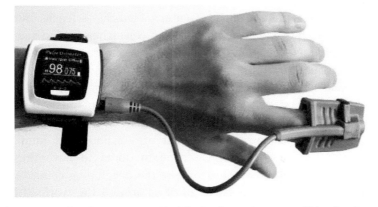

Figure 13.13 A pulse-oximeter. Modified pulse-oximeters will be developed to sense pulse and to inject first-aid medication into a vein and to call the hospital automatically. *Source*: https://commons.wikimedia.org/wiki/File:Wrist-oximeter.jpg.

Identification of the causes of aging and delaying of aging will make significant progress through the development of cell chemistry and cell biology. Since an organism starts from a single stem cell, through the development of cell biochemistry, chemistry of stem cells will reveal fundamental solutions for aging problems.

The aging of the human body can be delayed by revitalizing organs in the body with the help of advancement of stem cell study. New antibiotics can be developed by modifying of the mutation of viruses and bacilli. The development of diagnostic technology at the cellular or DNA level will enable early detection of dread diseases from a drop of blood sample, such as cancers, hereditary diseases, epidemic diseases, diabetes, arthritis, and high blood pressure. Early detection will even make prevention of the diseases possible.

Automatic synthesizer of new drugs: New drug development will be much easier and faster. If one wishes to make a new compound from certain starting materials, a computer program assisted by artificial intelligence will show the route to synthesize it. Then, if you just add the starting materials in an automated synthesizer, the synthesizer will produce the target drug after reactions such as heating, cooling, mixing, and separation and purification.

Sanitation and drinking water: Medical sciences and technologies have made great progresses. Nevertheless, and sadly enough, millions of people in developing countries still suffer from poor sanitation and shortages of safe drinking water. As a result, their health and lives are at risk. These problems should be solved in the 21st century, and this will help to improve the health, safety and happiness of people globally. Water purification for tap water and sewage treatment are technologies known for a very long time. What we need to do is to have these technologies available for the people in the developing countries. The results will be beneficial to all of us around the world. The hope for the future is to not only to explore unknown territory, but also to fully apply the known, which is good not yet disseminated or executed for the benefit of every human being.

13.3.6 Humans and Nature

Human beings have recognized the importance of better understanding of plants and animals to improve their own lives because humans, plants and animals are interconnected and share a common origin, DNA, and RNA. We have the tendency to regard plants and animals as just foods and resources for our living without recognizing that our lives are entwined with theirs and that we are actually part of nature.

If plants get diseases and die, animals cannot get food and will lose their shelters and we cannot acquire grains and timbers. Needless to say, plants produce oxygen and absorb carbon dioxide. Plants are essential for the life of animals, including us. Diseases of rice or wheat plants will create shortage of food grains. For instance, the Irish famine was caused by potato late blight disease. Prevention and treatment of plant diseases are accomplished by the elimination of pathogens such as fungi, bacteria, and nematodes using fungicides and antibiotics.

If bees disappear from Earth, we will not get bee honey (Fig. 13.14). Also it will be difficult to harvest fruits and grains from plants due to lack of pollination. If cattle get diseases such as mad cow disease, or hoof and mouth disease, meats will become scarce. If migratory birds, chickens, and ducks get influenza, we might be infected with avian influenza too. If fishes get diseases, we cannot get protein from fishes and many other animals cannot get their food either. We know mosquitoes spread many diseases, including malaria, yellow fever, dengue, and Zika virus. Can we eradicate mosquitoes? Do we understand the consequence of eradication of mosquitoes? Is it possible to eradicate rats and mice? Rats and mice also spread many diseases, including hantavirus pulmonary syndrome.

Figure 13.14 A South Korean stamp of honey bee.

There are veterinary drugs for animal diseases: livestock, birds, insects, and fishes. But they treat diseases separately. It is necessary to understand animal diseases, pathogens and why animals get sick. Thus, our prosperity and security depend on the prosperity of animals. It is necessary not only due to compassion toward plants and animals, but also from the self-preservational point of view. Research on our relationship with animals is a must for the better future of humans.

Biologists estimate that more than 8.7 million species live on Earth but only 2.3 million species are identified and named. About 15,000 are discovered every year, and they expect that most of the new species to be discovered would be bacteria and archaea [24]. This means that we do not expect to find cures for diseases caused by these new species very quickly simply because we do not know anything about these new pathogens.

13.3.7   Changes in the Desert

A desert lacks rain and is barren and is a difficult environment for humans to live in. Development of deserts for human habitation and protection from climate change will be a new frontier for human beings. The Sahara Desert in northern Africa, the Arabian Desert in the Arabian Peninsula, the Gobi Desert in China and Mongolia, the Great Basin Desert in the United States, the Patagonian Desert in South America, and the Great Victoria Desert in Australia are typical deserts. These deserts are vast and constitute about one-third of the land on Earth. The area of the Sahara Desert alone is almost equivalent to that of the United States. If we can make the deserts green and make water available there, people can set up new cities and live in these vast areas. How do we make a desert green and how do we deliver water there?

Planting plants: One method of greening a desert is to plant plants such as typical desert plants, short grasses and cacti. Water can be retained by using water-holding polymeric materials. We also can use nitrogen fixation bacteria and plants for their nutrition. Plants adaptable to dry climate will be developed by genetic modification.

Currently, we are actively trying to reduce the area of deserts and developing deserts. One example is the activities of the United Nations Convention to Combat Desertification (UNCCD). In the Gobi Desert, many countries including China, Mongolia, Korea and Japan, participate in planting plants. The Sahara Forest Project signed by the Royal Norwegian Embassy in Amman is another example. This type of project may grow further in many other countries to reduce the area of deserts. It will conserve environment, increase agricultural products and improve economic activities.

Supplying fresh water: The second method is directly supplying fresh water to deserts. Delivering fresh water costs a lot, and fresh water is difficult to obtain. Digging and pumping underground water is also very limited. In the case of Libya, groundwater was connected by pipelines; however, the source is dwindling and will not be sustainable for delivery. Groundwater is already being depleted in Arabia, China and the United States.

Desalination: Because of groundwater shortage, desalination of seawater is an option. Various desalination techniques, including reverse osmosis technology, have been established and are already employed in various areas. Actually, desalination of seawater at Ras Al Khair Desalination Plant, Saudi Arabia, has started to produce 728 million liters per day and is capable of serving approximately 3.5 million people in the city of Riyadh. Construction of the desalination plant started in early 2011 and the plant was commissioned in April 2014.

Building canals: We can build canals to flood deserts with seawater. Then, we can desalinate the seawater into fresh water at a distant place. Canals will cause a dramatic change for those living there and improve societies there. The use of the path of a dry valley ("Wadi" in Africa) will be economical in many ways. Although it requires big financial planning, the benefit of the canals will be enormous: fresh water, mild climate, increase in the number of oases, lower atmospheric temperature in the region and consequently global reduction of carbon dioxide, agriculture and fisheries, tourism, saving groundwater, and increase in employment.

Take a look at the map of Sahara Desert and Arabian Desert in Fig. 13.15. Figure 13.15 shows the examples of the several canals: SEC for the Egypt canal, SLC for the Libya canal, SAC for the Algerian canal, SGC for the canal passing many countries through the African continent and connecting the Atlantic Ocean and the Red Sea. AC-1 to AC-3 are canals for Arabian canals. Canals are dug deep from the seashore 10 km, 100 km, and 500 km continuously and made in the form of like saw teeth. The length of SGC will be approximately 7,000–7,500 km, a length that can be compared to that of the Great Wall of China built between the 8th and 5th centuries BC. If seawater arrives in remote desert areas, it will be converted into fresh water by desalination. Especially, canals AC-1 and SGC will greatly change the climate, because they will cool down the temperatures of adjacent lands and change the direction of winds. These grandiose channels are not a simple dream, but they will be worthwhile with respect to their economic, social, and direct impact on humans. Construction will go together with the building of highways, railroads, towns and cities and manufacturing plants. Towns will be parts of oasis belts along the canals and a new desert culture will rise from barren deserts.

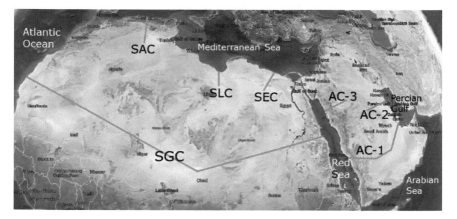

Figure 13.15 Sahara canals and Arabian canals.

13.3.8 Use of Oceans and Seas

Oceans constitute 71% (almost of three-fourths) of the surface area of Earth. The average depth of the oceans is about 3.7 km, and the deepest one is almost 11 km. There are 230,000 known living species in the ocean, and it is estimated that probably 2 million marine species are living there. Oceans will be the last and the most difficult frontier on Earth to explore and develop for human beings. We will use oceans and seas interchangeably here, although the terms ocean and sea are different in a sense.

Oceans are heat reservoirs and a very important factor to the atmospheric temperature. They are the largest carbon dioxide absorbers, and the absorption depends on temperature and carbon dioxide concentration in the atmosphere. There are many dissolved mineral resources in seawater, and on and under the sea bed. Oceans produce, hold, and carry nutrients for plants and sea animals. Oceans are the main routes for ships and boats to transport people and commodities. Most of the oceans are high seas and not well understood and controlled by us.

We are very concerned about the rise in sea level due to global warming and climate change. Earth's climate is greatly affected by ocean currents and their temperature. Climate change, including temperature increase of the oceans, will affect the population of animals and plants. Rising water temperature will reduce the concentration of dissolved oxygen, and consequently, it will suffocate fish and animals in the sea. The 2015 World Wildlife Fund's Living Blue Planet Report reveals that populations of marine vertebrates have declined by nearly half between 1970 and 2012 [24].

Absorption of carbon dioxide and photosynthesis of plankton: Oceans absorb carbon dioxide from the atmosphere, and the amount is enormous because of the volume of seawater in the oceans. In addition to the carbon dioxide absorption by seawater, phytoplankton absorb carbon dioxide via photosynthesis. Cultivating phytoplankton in oceans will reduce carbon dioxide concentration in the atmosphere and consequently reduce global warming. Furthermore, it will be biomass resources.

Production of protein through fisheries: Oceans are farming fields of proteins. If phytoplanktons are plentiful, fishes and sea animals will flourish because planktons are at the bottom of the food chain in oceans. Thus, oceans will provide proteins for humans in the long run. At present, the production of proteins in oceans exceeds the production of meat from land. Fisheries will expand further to the high seas from exclusive economic zones, and the fishing industry will provide more protein for us. According to the Food and Agriculture Organization (FAO), the current fish production is 90 million tons per year. For the purpose of conservation and sustainability of fisheries, fishing is restricted. New developments need to increase the productivity of oceans due to population increase. We currently have overfishing and ghost fishing problems, which lead to the destruction of the ecosystem caused by depleting fish populations. Rising temperature and increasing acidity due to carbon dioxide and contamination by toxic chemicals and waste products also negatively affect the ecosystem of marine life. These problems will be solved in the future because we understand the problems and we know how to manage them scientifically and technologically if there is a will.

Construction of floating cities: Oceans are the main route for ships to transport people and commodities. We will build standing floating cities for many purposes such as fisheries in the high seas, transportations, harbor, and airport, monitoring climate change currents, communications, collecting minerals in the seawater and under the seas, and ocean trash treatment (Fig. 13.16). Floating cities (islands), with sizes from 500 m × 500 m up to 2 km × 4 km, will be built by using blocks, 10 m (width) × 50 m (length) × 20 m (height) like Lego connection. Building blocks for the floating cities will be corrosion resistant to seawater for a long time and strong enough for high winds, tides, and typhoons.

Figure 13.16 A model of a floating city.

Currently, permanent dwellings at sea, the concept of seasteading is developing. For example, the SeaOrbitor project is building an ocean-going research vessel, a "floating oceanographic laboratory," under the leadership of French architect Jacques Rougerie, oceanographer Jacques Piccard and astronaut Jean-Loup Chretien. Another example is to build "Ocean Spiral" for deep sea research by the Japanese Shimazu Company. Floating cities much larger in size than current seasteading will be built and operated by international co-operative efforts.

Chemistry will play key roles analyzing data related to climate change, global warming gases, temperature, nutrients, and distribution of necessary nutrients for the cultivation of phytoplanktons and algae and for fisheries and collecting minerals in seas.

Deep-sea mining: Polymetallic nodules composed of almost pure manganese oxide, polymetallic sulfide muds containing copper, zinc, lead, iron, silver and gold; and ferromanganese crusts containing cobalt have been found on the seabed or in clusters in the deep sea. These metallic resources can be mined in the future. The International Seabed Authority (ISA), an international organization established under the 1982 United Nations Convention on the Law of the Sea, will regulate and control mining in the international seabed area.

Problems of the Oceans and Our Future: Oceans are the final destiny of all kinds of waste delivered by rivers. Marine debris are either sunken or buried in the seabed or float on the surface of the ocean. The Great Pacific Garbage Patch located over a vast region of the North Pacific Ocean covered with wastes is a good example. Most of the floating debris are plastics, and they cause many serious ecological problems. The particulate ocean garbage can be economically cleaned up if the floating cities have capabilities of collection, incineration, or recycling. Ghost fishing due to the abandoned fishing is another problem. In addition to the debris, many toxic soluble materials, e.g., PCB leached from plastic debris, are also introduced into the oceans even though some of them escape to the air.

In addition to the polluted oceans, the rise of sea level due to global warming is a great concern. Major cities around the world, such as New York, Washington D.C., Boston, Miami, San Francisco, Rio de Janeiro, Buenos Aires, Shanghai, Hong Kong, Tokyo, Bangkok, London, Amsterdam, Brussels, Stockholm, Venice, Cairo, and many other cities, too will be under the sea if all the ice of Arctic and Antarctic melted by global warming [25].

Global warming is also the cause of the rise of ocean temperature, which in turn reduces the concentration of dissolved oxygen. Consequently it could suffocate fish and animals in the sea. The 2015 World Wildlife Fund's Living Blue Planet

Report shows that populations of marine vertebrates have declined by nearly half between 1970 and 2012 [26].

Other negative effects of global warming are acidification from higher carbon dioxide concentration [27], ocean desertification, i.e., fewer sea bottom life due to a lack of nutrients in the sea bottom, methane gas released from sea bed. Any of these together or alone could be catastrophic to human and marine life.

Academic theoretical discussion only without any meaningful actions does neither solve nor prevent grave consequences. Ocean significantly affects our life many different ways. Our future is collective results of our past and present activities. Keeping clean the oceans and sustaining the various characteristics of oceans in nonpolluting state will be very important and urgent matter for us to maintain our life on Earth. We have to watch, monitor and catch signals of the oceans and care them well, too. We need actions right now, not later, for our better future.

13.3.9 Space Travel: A Never-Ending Quest

When we look up at the sky at night and see twinkling stars, we dream of reaching the stars far away. In ancient times, space was the place only for gods and for human spirits. After Nicolaus Copernicus proposed the heliocentric theory and the invention of telescope by Galileo Galilei, the concept of space has changed dramatically. The details of the movement, shape and celestial bodies with Earth have been observed. Now the Hubble telescope (it will be replaced with James-web-Space Telescope) orbiting around Earth has shown a much clearer and farther space and has made us search for the origin and extent of the universe.

French science fiction writer, Jule Verne's 1865 novel, *From the Earth to the Moon* may be the first novel describing space travel. Since then, many science fiction books and movies have been produced and have inspired people to travel to space. Most recently, Andy Weir wrote a novel *The Martian* and it was also made into a science fiction survival film with the same title.

A human being's travel to space is no longer a dream. Sputnik 1, launched by the Soviet Union on October 4, 1957 was the first human-made object sent to space and became the first satellite orbiting Earth. Since then, exploring space has been vigorously planned and executed. Two US astronauts, Neil

Armstrong and Buzz Aldrin landed on the moon for the first time via the American Apollo 11 mission on July 20, 1969. Currently, an International Space Station (ISS) has been rotating in Earth's orbit between 330 km and 435 km from Earth since 1998. Crews in the ISS are experimenting on various scientific studies. Their experiments in space will help for future space trips to the moon, Mars and outer space.

The moon and Mars are the first choices for our space destinations because they are closer than any others (Fig. 13.17).

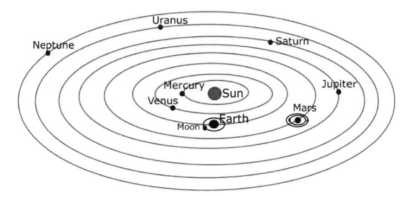

Figure 13.17 The solar system: the Sun and its planets. Earth and Mars are depicted exaggeratedly in the figure.

Preparation for a Space trip: Because it is difficult to reach there since they are a far away and very difficult to live on the moon and Mars even if we reach there. We have to prepare many things for space exploration. Oxygen, water, foods, energy, and shelter will be the minimum requirements for inhabiting them, even just for exploration. We will need to develop a Closed Ecological System (CES) where oxygen and other gases do not to leak into space. In addition, such as a CES should produce oxygen and food and must consume human wastes such as carbon dioxide through photosynthesis. Space stations will be made of mainly light polymeric film materials and adhesives and materials available there. The materials also should maintain their properties under extreme conditions such as very low and high temperatures and heavy cosmic radiation. The construction of multiple CES and energy generated by mini nuclear reactors will be a prerequisite for human travel and inhabitation of the moon and Mars. Advancement of space chemistry and engineering for material production and space medical science for human health in the space will be required for the preparation of the space travel.

Space exploration by robots: Although there are many ambitious plans to send humans for exploration to the moon and Mars, exploration by robots is preferred. Because of high risks of human life and advancements in robotics, future exploration of space will be mainly performed by robots. We will remote control them from Earth. Corps of robots, romoons (robots working on moon) and romarses (robots working on Mars) will be sent to the moon and Mars. Romoons and romarses will explore the moon and Mars. By the mid-2020s, a couple of romoons will be sent to the moon. They will survey and find mineral resources and send information back to Earth in real time. Romoons will get energy from an electrical power station operated by a micro nuclear reactor.

Each romoon and romars will have a specific working function like human workers. A dozen of robots will be sent by the mid-21st century to the moon and Mars. Romoons and romarses will be constructed like Lego modules and will be frequently upgraded. Parts will be interchanged and replaced by romoons and romars programmed to do the jobs like human medical doctors and mechanics. They will be operated by batteries and have shelters for recharging and repairs. We on Earth will receive reports online or periodically and program robots' daily working orders and send robots the orders and they will uploaded with the new orders. Next generations of romoons and romarses will be sent every 10 years. It will be economical, efficient and safer way to explore the moon and Mars, and space.

Moon Exploration: The moon orbits around Earth in an elliptical pattern. Its closest distance to Earth is 363,104 km, and the greatest distance is 406,696 km. It takes a little longer than 1 second for light to travel from Earth to the moon. The average radius of the moon is 1,737 km and it is about one-fourth the size of Earth, whose average radius is 6,371 km (Fig. 13.18). The atmosphere of the moon is almost a vacuum, and water is not found there. The gravitational pull of the moon is about 16.6% that of Earth. Because the moon rotates with Earth in a synchronous motion, we always see the same side of the moon.

Figure 13.18 DSCOVR spacecraft's Earth Polychromatic Imaging Camera's view of moon transiting Earth. *Source*: http://www. csmonitor.com/Science/2015/0806/NASA-spacecraft-snaps-spectacular-images-of-moon-transiting-Earth-video.

The moon was explored by telescopes on Earth and unmanned space crafts, moon orbiters, landers, and rovers. Two US astronauts landed on the moon on July 20, 1969, for the first time in the human history. Russia, the European Space Agency (ESA), India, China, and Japan are also currently participating in the moon exploration by sending their satellites, landers, and rovers. We will operate orbiters around the moon first. We already have the experiences. Before we inhabit the moon, robotic exploration will be performed many times. We will start to inhabit the moon by 2050 and the first moon baby will be born by the end of the 21st century.

Mars Exploration: Mars is the next planet to Earth and the fourth planet from the sun. The average radius of Mars is 3390 km, about one-half that of Earth. The gravitational pull of Mars is 38% of that of Earth, and it means that a weight of 100 kg on Earth will be 38 kg on Mars. The distance between Earth and Mars varies greatly because the orbits and orbital periods of Earth and Mars around the sun are different.

The closest distance between the two planets is about 54.6 million km and the farthest distance is 401 million km.

The travel time for a spacecraft to reach Mars from Earth will be from 6 months to a year depending on the relative positions of the planets. The desirable timing for travel is when the flight distance between the two is shortest. It comes closest only about every 26 months. A Martian day is 24 h and 37 min and similar to that of Earth. A Martian year is 686 days and, Mars has four seasons. The atmosphere of Mars consists of 96% of carbon dioxide, 1.9% of argon, 1.9% of nitrogen, 0.15% of oxygen and 0.05% of carbon monoxide. The amount of sunlight on Mars is about 43% of that on Earth because Mars orbits the sun 1.52 times farther from the sun than Earth. In addition, dust storms weaken the solar beam on Mars, which makes it difficult to use solar cells. Although there is no surface water because of rapid water evaporation due to very low atmospheric pressure, NASA's Mars Odyssey orbiter found water deposits in the polar ice cap of Mars in 2002 and Mars rover, Phoenix confirmed the presence of underground water ice in 2008.

Exploration of Mars started by *Mariner 4* in 1965 by NASA, United States. Since then, NASA has sent spacecraft to Mars on a regular basis to search for life and human habitability: Viking left Earth on August 1975 and landed on June 20, 1976 after 11 months of travel; Mars Pathfinder on July 4, 1997; Spirit and Opportunity in January 2004; Phoenix in January 2008; and Curiosity in August 2012. NASA plans to send another rover to Mars in 2018 and send spacemen by 2030. Many other international organizations have participated in the space exploration programs, including the ESA, the Russian Federal Space Agency (Roscosmos), the Indian Space Research Organization (ISRO) and the Japan Aerospace Exploration Agency (JAXA) [28–31].

Orbiters similar to ISS, landers, rovers, and multiple CES will be experimenting and operating by the mid-21st century before human's travel to the Mars. First, electrical power stations operated by a micro nuclear reactor will be established. Wind turbines will also produce electric energy using the winds and storms on Mars. Water obtained by melting ice on Mars will be supplied through pipelines to the station where humans will stay.

The presence of water and carbon dioxide on Mars will make it easy to cultivate vegetables, moss, and lichen to produce oxygen and food. The carbon monoxide in the atmosphere of Mars will be used as a reducing agent in producing steel by reducing iron ore. It will be the first example of producing materials in Mars directly. Many metallic materials will be produced on Mars, too. Organic compounds and plastics will be produced using, carbon dioxide and carbon monoxide as carbon sources, and water. Metallic materials, organic compounds and plastics and ceramics will be produced on Mars, and these works will be performed by robots. Humans will inhabit Mars when oxygen, water, food and various materials are produced self-sufficiently on Mars.

Chemical sensors, analytical instruments, diagnostic instruments, and various other equipment to explore Mars will be developed. Preparation for human habitation on Mars will start from the mid-21st century and humans will inhabit it at the end of the 21st century.

Beyond Mars and toward outer space: After Mars exploration, our curiosity will be expanded further to deep space. So far, we are only exploring a planet next to Earth, Mars. Other planets and their moons and beyond the solar system are awaiting our visit. In addition to exploration of the moon and Mars, our curiosity and quest for exploring the solar system and beyond will continue. We will never stop trying to reach outer space. The farther we explore the space, the better we will understand Earth for humans to live safely and comfortably. We should take care of Earth and make it more sustainable and healthier.

13.4   Prospects: All Is a Matter of Chemistry and Us

The quality of our living standards has been improved dramatically during the last century because foods, clothing, housing, transportation, communication, medicine, and education have been developed to an unprecedented degree and have been made available to us broadly and globally.

Fertilizers made it possible to produce more foods. Electric refrigerators using refrigerants and thermal insulators keep food safer and edible for a longer time. Synthetic fibers such as nylons, polyesters, and spandex keep us warm and fashionable, and synthetic fur and leather save wild animals. Many antibiotics and new drugs save and extend the lifespan of humans. Use of computer chips and optical fibers in diagnostic devices has advanced. Airplanes, automobiles, and ships have made travel easier and safer. Internet, TV, and phones have made communication among people worldwide easier, helped people live more freely, and made the world a global village.

At the same time, some unwanted byproducts also emerged, such as depletion of resources, perturbation of the ecosystem, and overpopulation, which may produce problems in the future. Many problems created by the development during the last century should be solved or at least modulated so as not to become more problematic in the future. We begin to understand and realize that there are risks and limitations in the progress of development. Predictions for future should be more prudent, and consequently plans for the future should be more accurate like weather forecasting.

Nanotechnologies, robotics, electric cars, 3D printers, artificial intelligence, self-driving cars, robotic surgeries, application of stem cells, curing of many diseases such as cancers, search for new materials for information storage, and challenging climate change are no longer futuristic issues and are the parts of new continents. We see the beams of the light-houses of these new continents beckoning, and all we need is to sail to these harbors and develop and explore them. Some of these are already around us and will be fully grown soon. Development of various sensors, finer imaging and imaging analyzers, lighter and long-lasting, heavy-duty batteries will be key factors for the success of these issues and goals in the future. The detailed road maps and paths for the goals may be different from what we see now. Our minds always sail to find unknown harbors and territory beyond the horizon.

Although the world is made of atoms and molecules whose sizes are on the nano- and pico-scales, until now we could only observe and handle materials and living cells very superficially. Because of recent developments and understanding of materials and cells at the molecular level and dramatic enhancement in the imaging resolution of microscopes, we can observe materials and cells in more detail. It means in the future the characteristics of materials and cells become more precise than we see now and the methods of solutions for these will be fundamentally different. We are going to see the reactions of molecules and living cells. Needless to say, these developments are largely dependent on the development of chemistry.

The 20th century was a period of development of synthetic materials, introduction of molecules, and the conquest of nature. The 21st century and the future will be a period of wisdom, maturing the understanding and application of molecules and toward nature and space for the sustainable prosperity of humans and Earth. Global and international cooperation and sharing knowledge and wisdom and the value of human dignity will become more important than ever because many problems are interwound and not restricted to local places. Solutions to problems can be obtained, which will benefit citizens globally.

We are born with blank brains but are capable of learning the unlimited knowledge of science and technology. We, humans, may have a fate just like Sisyphus who was forced to roll a boulder up a hill and then to watch it roll back to him, repeating this action for eternity. There is a remark, too, "History repeats itself." Progress is not granted to humans. There are lessons to learn from the fact that the human civilizations have been destroyed many times by humans and nature in the past. Therefore, we should be able to sustain and improve our civilization and culture and to solve unexpected problems from generation to generation.

We have a nature of unstoppable and unending quest to seek the fundamental origins of human nature and reach stars beyond space just like our ancestors journeyed from Africa to other continents many hundred thousands of years ago. We always dream of reaching unexplored frontiers for a brighter and better future. We also may face unexpected problems but will be able to find solutions for those problems.

References

1. https://www.weforum.org/agenda/archive/fourth-industrial-revolution/
2. G. M. Whitesides, *Angew. Chem. Int. Ed.,* 2015, **54**, 3196–3209.
3. R. Breslow, *C&EN*, 2016, **94**(18), 28–29.

4. https://sustainabledevelopment.un.org/topics/sustainable-developmentgoals.

5. N. Tarasova, *Chem. Int.*, 2015, **37**(1), 4–8.

6. http://www.millennium-project.org/millennium/2015-SOF-ExecutiveSummary-English.pdf.

7. https://en.wikipedia.org/wiki/Artificial_photosynthesis.

8. A. Scott, *C&EN*, 2015, Nov. 16, pp. 10–16.

9. K. Bourzac, *C&EN*, 2016, Nov. 21, pp. 32–38.

10. U.S. Energy Information Administration, International Energy Outlook 2016, https://www.eia.gov/outlooks/ieo/pdf/0484(2016).pdf.

11. M. Jacoby, *C&EN*, 2016, **94**(18), 30–35.

12. http://www.eia.gov/forecasts/aeo/pdf/0383(2015).pdf

13. The World Wind Energy Association. (2014). *2014 Half-year Report.* WWEA, pp. 1–8.

14. http://eis.anl.gov/guide/basics/

15. http://environment.nationalgeographic.com/environment/global-warming/wind-power-profile/

16. http://www.world-nuclear.org/information-library/nuclear-fuel-cycle/nuclear-power-reactors/small-nuclear-power-reactors.aspx.

17. https://en.wikipedia.org/wiki/Thorium-based_nuclear_power.

18. https://www.iter.org/.

19. http://www.consumerenergycenter.org/lighting/bulbs.html.

20. J. D. Bergesen, L. Tähkämö, T. Gibon, and S. Suh, *J. Ind. Ecol.*, 2016, **20**(2), 263.

21. https://www.climateinteractive.org/programs/scoreboard/

22. https://www3.epa.gov/climatechange/basics/

23. J. M. Matter, M. Stute, S. Ó. Snæbjörnsdottir, E. H. Oelkers, S. R. Gislason, E. S. Aradottir, B. Sigfusson, I. Gunnarsson, H. Sigurdardottir, E. Gunnlaugsson, G. Axelsson, H. A. Alfredsson, D. Wolff-Boenisch, K. Mesfin, D. Fernandez de la Reguera Taya, J. Hall, K. Dideriksen, and W. S. Broecker, *Science*, 2016, **352**(6291), 1312–1314.

24. *Scientific American*, 2016, March, p. 68.

25. *National Geographic*, 2013, Sept., p. 30.

26. http://assets.worldwildlife.org/publications/817/files/original/Living_Blue_Planet_Report_2015_Final_LR.pdf.

27. D. R. Bishop, *C&EN*, **2014**, July 21, pp. 26-27.

28. http://mars.nasa.gov/allaboutmars/facts/#

29. A. Extance, *Chem. World*, 2015, **12**(8), 42–45.

30. K. Sanderson, *Chem. World*, 2015, **12**(8), 54–57.

31. E. Stoye, *Chem. World*, 2015, **12**(8), 58–61.

Chapter 14

Chemistry in Africa: Progress and Application

Temechegn Engida

President of FASC (2006-2013) and Editor-in-Chief of AJCE

14.1   Introduction

The progress of industrialization in African countries is slow due to many reasons. These countries usually have shortage of various crucial materials necessary to improve their living standards and provide information. The activities and development of chemical industries and public perception on chemicals and chemistry are different from those of developed countries. Activities, plans, and philosophy of these countries related to chemistry and chemical information for their future will be described.

This chapter will include the methods and problems in their chemistry education and how they benefit from chemical products. It would also include how various natural materials were used for health and agricultural purposes even before the role of chemical components in those natural products was actually known. The chapter also describes what type of research is being done to investigate needed materials for the special circumstances in those countries. It is, however, important that

Chemistry: Our Past, Present, and Future
Edited by Choon Ho Do and Attila E. Pavlath
Copyright © 2017 Pan Stanford Publishing Pte. Ltd.
ISBN 978-981-4774-08-6 (Hardcover), 978-1-315-22932-4 (eBook)
www.panstanford.com

if certain failures and negligence, such as environmental contamination and safety, caused problems in the developed countries during the development, the chapter describes consecutive actions and research that remedied the situation.

14.2   Traditional Medicine, Indigenous Practices, and Chemistry in Africa

Traditional medicine is conceptualized as health practices, approaches, knowledge, and beliefs incorporating plant-, animal-, and mineral-based medicines, spiritual therapies, manual techniques and exercises, applied singularly or in combination to treat, diagnose, and prevent illnesses and maintain well-being (WHO, 2001). Traditional medicine has been playing an important role in Africa. The philosophical clinical care embedded in these traditions, culture, and taboos has contributed to making traditional medicine practices acceptable and hence highly demanded by the African population.

About 80% of people in Africa use traditional medical systems for much or all of their health care. There are diversities in the African traditional medicines, but many of them share characteristics that distinguish them from biomedicine (Science Museum 1, nd). As long as health is concerned, equal importance is given to both spiritual and physical aspects of the body. With regard to the causes of illness, the tradition recognizes spirit-world intervention, or family or community conflict.

The traditional "doctors" in Africa are well known for treating patient holistically. They attempt to reconnect the social and emotional equilibrium of patients based on community rules and relationships. They also act as an intermediary between the visible and invisible worlds, between the living and the dead or ancestors, sometimes to determine which spirits are at work and how to bring the sick person back into harmony with the ancestors (Abdullahi, 2011). The whole community is involved in treating illness. Treatment is through medicines, physical treatments, or divination by the traditional healer. In divination, ancestors are asked to reveal the meaning of an illness and ways to resolve it. Diviners may suggest a ceremony in which the community sings, dances, and performs rituals together.

The majority of medicinal plants are herbs, and leaf is the most preferred plant part in remedy preparations. There are many botanical remedies that have survived scientific scrutiny, and there are many more still awaiting systematic investigation. Let us see some examples.

Ocimum lamifolium (Fig. 14.1) has been used to treat coughs and colds; the fresh leaves are squeezed, and the juice is sniffed. The juice can also be used as an eye rinse for eye infection. In Ethiopia, it is used for mich, an infection of fever with headache and mouth blisters. As such it is called *Michi Medhanit* (or *Damakesse*) in Amharic (the Ethiopian official language).

Figure 14.1   *Ocimum lamifolium. Source*: Author's home garden.

As reviewed by Mann (2012), several ethnobotanical surveys show that *Ocimum gratissimum* is among the plants reported in Nigerian communities that are used traditionally to treat bacterial infections such as enteric diseases, diarrhea, dysentery, and other gastrointestinal infections; upper respiratory tract infections associated with coughing, pneumonia, asthma, and bronchitis; urogenital infections, including sexually transmitted diseases, skin infections (dermatitis, eczema, scabies), wounds and ulcers; headache, ophthalmic, insect bites, nasal bleeding, stroke, measles, paludism; bacterial fevers such as typhoid fever; and diabetes and veterinary problems. It is also used in the treatment of epilepsy, shigellosis, trypanosomiasis, convulsion, pile, and anemia in Nigeria.

Rosemary is one of the most commonly used herbs for roasting meat. It is an excellent flavoring agent for various sauces. Mostly people grow it in their home gardens. For instance, the following picture is taken from my home garden and is in frequent use (Fig. 14.2).

Figure 14.2 Rosemary. *Source*: Author's home garden.

The essential oil of *Artemisia afra* (Fig. 14.3) has antimicrobial properties. In South Africa, it is one of the most popular and commonly used herbal medicines for treating various ailments, from coughs and colds to malaria and diabetes. In Ethiopia, it is called *Ariti*. The juice of the crushed leaves of this plant is mixed with water or honey and administered orally to address stomach pain in traditional medicine practices. It has also been commonly used in rituals, especially during festivities on the Ethiopian new year (September 11) and *Meskel* (September 27, the finding of the "true cross").

Black seed's medicinal use has a very rich traditional history that goes beyond ancient Egyptians and Biblical times. Traditionally, black seed has been used for a variety of conditions and treatments related to respiratory health, stomach and intestinal complaints, kidney and liver support, circulatory and immune system support, improvement in general health, different skin conditions, dryness, and joint and scalp massage. In Ethiopia, it is referred to as *Tikur Azmud*. When people catch cough, they put the seeds in a piece of cloth (usually cotton) and tie a knot, as in Fig. 14.4. They then sniff it as frequently as possible, opening their respiratory tubes well so that they can breathe normally.

Figure 14.3 *Artemisia afra. Source*: https://www.google.com.et/sear ch?q=ariti+plant+image&biw=1366&bih=635&tbm=isc h&imgil=4t_MjyoE9P7xqM%253A%253BVGi_upJqig7yt M%253Bhttp%25253A%25252F%25252Fwww.hilishi. com%25252F%27%27YIGERMAL&source=iu&pf=m&fir=4t _MjyoE9P7xqM%253A%252CVGi_upJqig7ytM%252C_&usg=_ _XzeAkXulF11hfX-iUZdR945FCj0%3D&ved=0CDAQyjdqFQoT CL6hparA9McCFQnWFAod25oBNQ&ei=7631Vf7tFImsU9u1h qgD#imgrc=4t_MjyoE9P7xqM%3A&usg=_XzeAkXulF11hfX- iUZdR945FCj0%3D.

Figure 14.4 Black seeds prepared for individual sniffing. *Source*: Author's own collection.

Another medicinal plant is *Taverniera abyssinica*. The species is known to occur in Northeast Africa. In Ethiopia, it is commonly known under the Amharic name as *Dingetegna*, literally meaning remedy against sudden illness. *T. abyssinica* has been traditionally used for the treatment of various diseases. Dagne et al. and Balcha et al. (2010) state that a small bundle of roots is chewed and the juice is swallowed for immediate relief of fever, discomfort, and pain. The root extracts are used locally as antipyretic and analgesic.

Rue is used in North and Northeast Africa. Especially, Ethiopia uses rue extensively in food preparation like red pepper and in hot drinks. Locally, it is known as *Tena Adam*, literally meaning the Health of Adam. Rue is used to flavor coffee. It gives the coffee a very nice refreshing flavor. One usually finds the plant in home gardens, as in Fig. 14.5.

Figure 14.5 Rue. *Source*: Author's home garden.

The scientific community in Africa has long realized the potential of natural products in modern health care. As such African chemists have been exploring the traditionally used medicine by way of validating their use for health care. In fact substantial components of conventional medicine are actually based on therapeutic agents derived from plants and other natural sources. However, the efforts toward discovery of drugs from natural products are still based on the traditional approach and hence surrounded by many challenges, inherent to all natural product drug discovery and those challenges emanating from the

African setting (Chibale et al., 2012). Problems such as possible seasonal or environmental variations in the content of the bioactive principle, problems of guaranteed access and supply of the source material, loss of source through extinction or legislation, complexity of extracts for fractionation, frequent loss of biological activity following fractionation or purification, isolation of very small quantities of bioactive substance, and challenging physicochemical properties such as solubility and stability are inherent to all natural products. Problems related to the African context include inadequate financial resources, poor infrastructure, and a lack of skills and competency in several key areas.

However, knowledge about the extent and characteristics of traditional healing practices and practitioners in Africa is limited and has frequently been ignored in the national health systems of those countries. Researchers, policymakers, pharmaceutical companies, and traditional healers are joining forces to bring traditional medicine into the 21st century (Shetty, 2010). Nearly a quarter of all modern medicines is derived from natural products, many of which have been first used in traditional remedies. The abundant natural products present on the continent (many of which possess known medicinal properties) should be a powerful tool for African scientists as they could leverage on to improve the scope and quality of their research, thereby enhancing the potential for meaningful findings and eventual success. The HIV/AIDS epidemic has been a great challenge to the African countries. Sub-Saharan Africa has been hit hardest by the AIDS epidemic. It is hoped that a better understanding oflocal African medical traditions could help design prevention programs and treatments that make sense to ordinary people.

Chemistry in Africa is also exploring the potential of certain local herbs for the development of medicines for the treatment of neglected diseases. Neglected diseases are a group of infectious diseases that are endemic in poor communities of developing countries (Midiwo, 2010). Such diseases kill about 500,000 to 1,000,000 people every year. Chemistry's role in this regard is commendable and needs to be strengthened. It is also worth mentioning that biologically active natural products are finding their way for the invention of new herbicides, fungicides, and insecticides (Clough, 2010). The development of such crop protection chemicals is useful for the African population whose main livelihood is based on agriculture.

If indigenous knowledge is examined using the powerful tool of science, we will be in a better position to validate indigenous knowledge (Dagne, 2011). This validation will in turn help us derive innovative products from indigenous knowledge. This argument was elaborated using six important natural products, namely, coffee, khat, civet, kebericho (*Echinops kebericho*), dingetegna (*Taverniera abyssinica*), and black seed, some of which were discussed earlier. Such traditional African medicines have sometimes been turned into pharmaceutical products. For instance, bioprospecting, the development of traditional medicines as commercial products, looks for chemically active ingredients in traditional remedies that can be developed into commercial pharmaceutical products (Science Museum 2, nd). Of course, this practice is controversial in the sense that critics consider it "bio-piracy" since companies may attempt to take out a patent on a medicine derived from a traditional cure without recognizing the indigenous people.

14.3 Chemistry Research in Africa

In terms of global chemistry research, the contribution of developing countries is limited. Lack of infrastructure and funding for research and graduate studies is one of the factors for this trend. However, it is worth mentioning that there have been some progresses in areas such as research on natural products, agriculture, food science, water quality, environment, etc., which have immediate impact on the lives of people in developing countries. Green chemistry research is also coming up as an emerging field of investigation. Most of these research works are done in collaboration with universities and professional societies in developed nations. For instance, since its inception in 2006 with UNESCO and the Royal Society of Chemistry (RSC) support, the Federation of African Societies of Chemistry (FASC) has conducted four Pan African conferences on green chemistry—in Ethiopia (2007), Egypt (2009), South Africa (2011), and Morocco (2013)—as a way of sharing recent research works and best practices on green chemistry among African chemists and beyond. Individual societies of chemistry in Africa have also conducted meetings of their own. While these efforts are highly appreciated, they will only have a sustainable impact on African societies if green chemistry permeates all echelons of the education system from the primary to tertiary levels. This is a challenge ahead of us.

It is true that universities and other stakeholders in Africa are pushing legislation to reduce toxicity and to ensure that safety guidelines are respected in the development and use of chemical products. For instance, the Chemical Society of Ethiopia was working with other NGOs to raise awareness on hazards related to pesticides used in agriculture. The Kenyan Chemical Society was studying the extent to which DDT permeates the Kenyan environment. It has uncovered evidence suggesting that DDT remains a huge problem in Kenya (Engida, 2011a). At the University of Johannesburg in South Africa, scientists are hoping to use nanosponges to purify water, SciDev.net reported in May 2011. If the technology can be perfected, the idea would be to clamp the sponges onto the end of a tap or pipe to filter impurities such as pesticides or pollutants while allowing the purified water to continue on its way. Nanosponges were invented more than a decade ago at the Los Alamos National Laboratory in the United States.

Chemistry research in Africa often faces problems of funding and infrastructure, but by concentrating on fields relevant to local problems or resources, the research can have significant impacts. Now it is a common understanding among African chemists that chemistry has a key role to play in maintaining and improving quality of life and should, therefore, help in the development of African countries. As a consequence, they are engaged in exploring local resources. A few examples are as follows (Chemistry World, 2011):

- In Ethiopia, some chemistry research projects focus on synthesizing biodegradable materials from the oil of a plant called *Vernonia galamensis* and on identifying and characterizing zeolites–microporous, aluminosilicate minerals commonly used as commercial adsorbents, catalysts, or for water purification. The latter research is involved in characterizing the mineral and chemical properties of zeolites to assess the potential for their exploitation and commercialization.

- In Ghana, there are chemistry research projects focused on developing biofuels, as well as materials for solar cells and treatment of drinking and waste water. Biofuels are good candidates as alternatives to fossil fuels as energy sources. The work on biofuels involves characterizing different oils (including edible, non-edible, and used vegetable oil) for biodiesel production. The work has been searching for local raw materials such as cocoa husks and plantain peels for catalyst production and trying to use corn cobs, corn husks, rice husks, and other agricultural wastes to produce bio-ethanol. The research work on the treatment of drinking and waste water is based on the need for water quality since one of the biggest problems in Ghana is contaminated water from rivers, streams, and shallow wells. It is recognized that application of simple water-treatment techniques using local materials can help reduce incidences of water-borne diseases to a large extent.

- In Kenya, since water is contaminated with sediments, biological pathogens, and chemicals from agriculture and industry, chemistry researchers believe that water quality is an area where chemistry can have a real impact. The researchers are, therefore, developing analytical techniques that can be applied at the household and community level to identify and remove chemical pollutants from water. The work involves developing catalysts that can oxidize chlorinated organic compounds in water. The researchers are also involved in the development of a climate change adaptation institute that will encourage research into climate adaptation technologies and provide a framework for national and regional policy assessment and advice to governments regarding climate change issues, in addition to providing education and training.

- In Mauritius, the big problem is waste since wastes are disposed of in open dumps around the island and have created a high risk to the environment and public health. Researchers look at this waste as a source of materials or energy. Their work primarily involves hydrolyzing ligno-cellulosic biomass–plant material composed of cellulose, hemicellulose, and lignin through acid, alkali, and enzymatic hydrolysis with the aim of finding environmentally friendly waste disposal alternatives and explore readily available feedstock in Mauritius for ethanol production. There is also a growing work on optimizing the enrichment of compost and the composting process, with the aim of using compost as an alternative to chemical fertilizers.

As argued in the previous paragraphs, research on improving water quality has been one of the priorities of chemists in Africa. To this effect, the Pan Africa Chemistry Network (PACN) held the *Sustainable Water Conference*, hosted by the University of Nairobi, Kenya, and sponsored by the RSC and Syngenta in 2009. The various research works in this conference highlighted the severity of lack of water quality in the continent. For instance, it was recognized that around half of all patients occupying African hospital beds suffer from water-borne illnesses due to lack of access to clean water and sanitation (PACN, 2010). Africa's water resources are being degraded due to high demand and untreated waste water entering the environment from industrial and domestic sources. The PACN report further highlights that adaptation and planning of water resources are difficult, since many African countries have no established water quality monitoring programs. It is, therefore, mandatory that African chemists engage themselves in research works that alleviate these problems in the continent. A strong focus on developing and improving technologies to conserve and reuse water for agriculture is required. It is also necessary to engage in research that optimizes water use, treatment of contaminated water, recycling water, desalinating water, and harvesting water for irrigation.

African chemists are also engaged in computational chemistry research. In fact, not only researchers but also chemistry students are getting access to computational chemistry labs that enable them turn molecules on the screen and see the excitement (Wang, 2013). Young chemistry researchers in Africa (like in Kenya) believe that the computational chemistry facility has opened up alternative ways of doing research and with that they can now write and ask for research grants.

Some African countries have also established research centers related to chemical sciences. For instance, in South Africa, the National Centre for Nano-structured Materials (NCNSM) was created in 2007 as part of the implementation of Government's National Nanotechnology Strategy. The objectives of the center are to:

- Conduct leading research into the design, modeling, and synthesis of nanomaterials with targeted properties and various possible applications
- Effectively disseminate the outcomes of its research activities
- Facilitate the application of its research outputs and outcomes in support of national priorities and needs
- Make a meaningful contribution to strengthening the national science base and developing strategic human capital and human resources

A similar case is the National Research Institute for Chemical Technology (NARICT) of Nigeria. The Institute aims to carry out research of international standard to solve problems both locally and internationally. Renewable energy, minero-organic fertilizer, bio-fertilizer, production of specialty chemicals, and chemical catalysis are some of the current areas of focus. The trend of organic farming is finding its way in Africa. The use of inorganic fertilizers, which is injurious to the soil and human health as a result of consumption of foods produced using the fertilizer, is being gradually discouraged in Nigeria, allAfrica (2014) reports. Awareness of the benefits of organic fertilizer is said to be growing in the country, and it is now considered to be the best alternative to inorganic fertilizer by farmers, consumers, experts, and other stakeholders in Nigerian agriculture.

The more African universities engage in advanced chemistry research, the more they will need safe, secure facilities and related skills. Development partners who share Africa's goal of sustainable development need to collaborate with African chemical societies and FASC to institutionalize the principles and practices of green chemistry in African universities and industries. For instance, development partners can introduce "advance market commitments," by which they ensure a market for the green products and processes developed by African scientists.

There is already an existing platform on which we can build our future regional initiatives. It is the African Ministerial Council on Science and Technology (AMCOST), established in November 2003 under the auspices of the New Partnership for Africa's Development (NEPAD) and the African Union (AU). It is a platform not just for chemistry but for the whole sciences and technology, and hence it meets the sustainable development agenda as seen from the broader perspective of science technology. AMCOST is a high-level platform for developing policies and setting priorities on science, technology, and innovation for African development. AMCOST provides both political and policy leadership for the implementation of Africa's Science and Technology Consolidated Plan of Action (CPA). The vision of the CPA is well integrated into the global economy and an Africa free of poverty. CPA aims at enabling Africa to harness and apply science, technology, and related innovations to eradicate poverty and achieve sustainable development and at ensuring that Africa contributes to the global pool of scientific knowledge and technological innovations.

The AU's approach to the post-2015 development agenda and the forthcoming Agenda 2063 highlight the promotion of science, technology, and innovation (STI) as a key driver of change and recognize that Africa's sustained growth, competitiveness, and economic transformation will require investments in new technologies and innovations. Such innovations could not be envisaged without the key role of chemistry since it provides the basis for understanding the atomic and molecular aspects of the science and technology disciplines, AU reports.

14.4 Chemistry for Sustainable Development in Africa

"Chemistry for Sustainable Development in Africa" has been the motto of the FASC since its inception in February 2006. Sustainable development has been conceptualized in different ways, but the most widely used definition, as articulated by the World Commission on Environment and Development, is "development that meets the needs of the present without compromising the ability of future generations to meet their own needs." Meeting the needs of the future depends on how well we balance social, economic, and environmental objectives when making decisions today (this can be considered a tri-archaic model of sustainable development). In other words, sustainable development refers to some form of modern technological society, with business taking responsibility for its impact on society and the environment (Engida, 2011b).

Most developing countries possess abundant natural resources, and the majority of their population lives in agriculture-based economies. Yet they are not self-sufficient in food production. Worse, most of the people who die from hunger are in these countries. Perhaps in response, many African universities have opened units or departments devoted to food science, in which chemistry is an essential ingredient. Like food chemistry, environmental chemistry makes major contributions to sustainable development, in this case primarily through understanding and monitoring our impact on the environment. Unfortunately, I suspect that many standard chemistry textbooks used in Africa do not directly deal with the science of environmental issues, including climate change, water pollution, and renewable energy.

In this regard, chemistry can play a very positive role. For instance, chemists are well placed to appreciate the scientific issues underlying sustainable development. Chemistry also contributes to sustainable development via economic growth and improved social well-being. For instance, not long ago, many Africans (mostly females) were (perhaps still) leading a very labor-intensive, economically less supportive activities to feed the family, get reasonable housing, and have health leaving. A typical activity is like the case in Fig. 14.6, in which women go to a bush, collect some bundles of twigs and tree branches, carry them over their shoulders, sell them to households who need a source of energy, and try to fulfill the daily subsistence of their family.

Figure 14.6 A typical economic activity of African Women. *Source*: http://whatsoutaddis.com/index.php/around-addis/good-deeds/200-women-fuelwood-carriers-project.

Similarly, the majority of African men were using traditional tools and animals for plowing their farms, as in Fig. 14.7.

Figure 14.7 Traditional farming practices in many parts of Africa. *Source*: https://en.wikipedia.org/wiki/Ox#/media/File:Traditional_ Farming_Methods_and_Equipments.jpg.

Obviously, it is hard to lead sustainable development if the majority of the population is living in such conditions. It is thus clear that chemistry, and science and technology in general, in Africa should do more to help people in the continent lead better life. Some of the obvious contributions chemistry/science could make include better pharmaceuticals, high-purity materials for use in the electronics industry, better housing with clean water, and jobs for a large number of people. In fact, some signs for sustainability is emerging. For instance, in Addis Ababa/ Ethiopia (the political capital of Africa, where the African Union Commission is housed), it was very common to see the majority of residences being century old cottage like areas. Now it is not uncommon to see public residence areas like the one in Fig. 14.8 (apart from the privileged few luxurious residences) in northeastern part of Addis Ababa.

It is also important to consider the contribution of chemistry to sustainable development in a broader context of the natural sciences since current advanced researches are multidisciplinary in their nature and as chemistry provides the basis for understanding the atomic and molecular aspects of these disciplines (Matlin and Abegaz, 2011). Like the contributions of chemistry to such areas of human concern in Africa like food/agriculture and health/medicine, the industrial sector also relies on chemistry since industrial chemistry serves as the backbone of economic growth and improved social well-being as well as a major source of employment. It is true that the industrial sector in Africa is now in its infancy, but more and more African governments plan to evolve their economies so that they rely increasingly on the industrial sector.

Figure 14.8 Public apartments in Addis Ababa. *Source*: Author's own collection.

However, it should be kept in mind that the beneficial economic and social results of industrial growth have been accompanied by the reversal effect of global environmental crisis. The technology-based industry has a critical environmental impact since it is the major consumer of natural resources and the major contributor to the overall pollution load (Miertus and Clerici, 2002). Given the way in which chemical industry developed/is developing, many processes were not designed taking environmental requirements into account. They are still operational in industries, because they are the only ones currently available. Such a paradox between economic development and environmental global crisis can be addressed by sustainable development programs, notably through green chemistry. In

Africa and other developing countries, however, the perception of environmental issues can be weakened by the relative priority assigned to economic growth. The lack of technical and financial resources identification, assessment, and development of clean technologies could also be a big problem.

In spite of such challenges, recognizing the potential green chemistry can achieve the triple-bottom-line benefits of economic, environmental, and social improvement. African chemical societies have been conducting a number of workshops and conferences on green chemistry in the past decade under the auspices of FASC and in collaboration with the University of Nottingham, the Pan African Chemistry Network, UNESCO, and IUPAC.

Since green chemistry is concerned with issues related to industry, technologies, and production processes, it requires an adequate number of researchers in Africa selecting this area for their activity. This, in turn, requires that young people include green chemistry in the range of areas they take into consideration for their future professions (Mammino, 2002).

In order to enhance the role of chemistry for sustainable development in Africa, I believe that various stakeholders in the continent need to contribute their share as in the following selected suggestions:

- The publication rate of research articles on chemistry by African chemists should be enhanced if Africa is to meet its aspirations of sustainable national and regional development. It is generally believed that there is a direct relationship between the production of science in a given country and its economic growth. This will be possible only if African scientists are engaged in high-level, original research in the basic and applied sciences.
- African chemists should also engage themselves extensively in educating the young to be scientifically literate and to fall in love with science. This requires us to work more on understanding our immediate local environment from the scientific point of view and systematically contextualizing the resulting scientific knowledge and skills to the African youth. This also helps to promote the public understanding of science.

- The professional scientific societies in Africa should be able to work upstream or policy level in order to influence science and technology policy planners and decision makers. FASC, as a regional professional association, should try its best to work closely with those regional initiatives such as the AU-NEPAD initiative on science and technology, the United Nations Economic Commission for Africa (UNECA) programs on science and technology for Africa, and many others that are linked with bilateral and multilateral donors, and UN agencies working in Africa. In this regard, we strongly acknowledge the link we have with the RSC-PACN, UNESCO, IUPAC, CSP, OPCW, and others. But at the same time, we feel that a lot has to be done in the near future. It is to be noted that the International Year of Chemistry (IYC, 2011) served as our platform to expand the horizon of our link with organizations dedicated in promoting science and technology in general and chemical sciences in Africa, in particular.

References

Abdullahi, A. A. (2011). Trends and challenges of traditional medicine in Africa. *Afr J Tradit Complement Altern Med.*, **8**(5 Suppl): 115–123.

allAfrica (2014). Nigeria: Organic farming—Is Nigeria lagging behind? http://allafrica.com/stories/201405090987.html, accessed on 12 February 2015.

AU. Science and technology: Key for Africa's socio-economic advancement, accessed 3 February 2015, http://www.africa-eu-partnership.org/newsroom/all-news/science-and-technology-key-africas-socio-economic-advancement

Abera, B., Negash, L., Kumlehn, J., and Feyissa, T. (2010). In vitro regeneration of *Taverniera abyssinica* A. Rich: A threatened medicinal plant. *Ethiop. J. Educ. Sc.*, **6**(1).

Chemistry World. (2011). Wealth of opportunities. www.chemistryworld.org, pp. 40–44.

Chibale, K., Davies-Coleman, M., and Masimirembwa, C. (2012). *Drug Discovery in Africa*. Springer-Verlag Berlin Heidelberg.

Clough, J. M. (2010). Natural products as leads for new and innovative crop protection chemicals. In J. O. Midiwo and J. M. Clough (eds.), *Aspects of African Biodiversity: Proceedings of the Pan African Chemistry Network Biodiversity Conference*, Nairobi, 10–12 September 2008.

Dagne, E. (2011). Validating traditional medicines: Adding value to indigenous resources. African Laboratory for Natural Products (ALANP), http://knowledge.cta.int/Dossiers/S-T-Issues/Indigenous-knowledge-systems/Relevant-publications/IKS-General/Validating-traditional-medicines-Adding-value-to-indigenous-resources.

Engida, T. (2011a). Greening Chemistry in Africa. *A World of SCIENCE*, **9**(3): 8.

Engida, T. (2011b). Chemistry boosts global sustainable development. *C&EN*, **89**(26): 41–45.

Mammino, L. (2002). The challenges of green chemistry education. In P. Tundo and L. Mammino (eds), *Green Chemistry in Africa, Green Chemistry Series N. 5*, pp. 188–208.

Mann, A. (2012). Phytochemical constituents and antimicrobial and grain protectant activities of clove basil (*Ocimum gratissimum* L.) grown in Nigeria. *Int. J. Plant Res.*, **2**(1): 51–58.

Matlin, S. A., and Abegaz, B. M. (2011). Chemistry for development. In J. Garcia-Martinez and E. Serrano-Torregrosa (eds.), *The Chemical Element: Chemistry's Contribution to Our Global Future*, 1st edn. Wiley-VCH Verlag GmbH & Co. KGaA.

Midiwo, J. O. (2010). Natural products from plant biodiversity and their use in the treatment of neglected diseases. In J. O. Midiwo and J. M. Clough (eds.), *Aspects of African Biodiversity: Proceedings of the Pan African Chemistry Network Biodiversity Conference*, Nairobi, 10–12 September 2008.

Miertus, S., and Clerici, M. G. (2002). Green chemistry program relevant to sustainable industrial development. In P. Tundo and L. Mammino (eds), *Green Chemistry in Africa, Green Chemistry Series N. 5*, pp. 174–186.

NEPAD. Advancing science and technology in Africa, accessed 3 February 2015. http://www.nepad.org/humancapitaldevelopment/news/1581/advancing-science-and-technology-africa.

PACN (2010). *Africa's Water Quality: A Chemical Science Perspective*.

Science Museum 1 (nd.) African Medical Traditions, accessed on 2 February 2015. http://www.sciencemuseum.org.uk/broughttolife/techniques/africanmedtrad.aspx.

Science Museum 2 (nd.) Bioprospecting, accessed on 10 February 2015. http://www.sciencemuseum.org.uk/broughttolife/techniques/~/link.aspx?_id=3C7C7261C5FA4D3582B2AD7F73E0ECDB&_z=z,

Shetty, P. (2010). Integrating modern and traditional medicine: Facts and figures. *SciDevNet*, http://www.scidev.net/global/indigenous/feature/integrating-modern-and-traditional-medicine-facts-and-figures.html.

UNESCO (2011). Small is beautiful. *A World of SCIENCE*, **9**(3): 17–19.

Vandenbrink, D., Martin, B., and Mahaffy, P. (2013). Weaving together climate science and chemistry education in an African context. *AJCE*, **3**(2): 3–27.

Wang, L. (2013). Computational chemistry takes root in Kenya. *C&EN*, **91**(38): 33.

WHO (2001). *Legal Status of Traditional Medicine and Complementary/Alternative Medicine: A World Wide Review*. WHO, Geneva.

Index